W0235088

Frontiers in Mathematics

Advisory Editorial Board

Vladimir Dragović
Milena Radnović

Poncelet Porisms
and Beyond

Integrable Billiards,
Hyperelliptic Jacobians and
Pencils of Quadrics

 Birkhäuser

Vladimir Dragović
Milena Radnović
Mathematical Institute SANU
Kneza Mihaila 36
11001 Belgrade, p.p. 367
Serbia
vladad@mi.sanu.ac.rs
milena@mi.sanu.ac.rs

2010 Mathematics Subject Classification: 37J35, 58E07, 65T20, 70H06, 94A08

ISBN 978-3-0348-0014-3   e-ISBN 978-3-0348-0015-0
DOI 10.1007/978-3-0348-0015-0

Library of Congress Control Number: 2011926874

Cover design: deblik, Berlin

Printed on acid-free paper

Springer Basel AG is part of Springer Science+Business Media

www.birkhauser-science.com

Contents

Chapter 1

Introduction to Poncelet Porisms

Figure 1.1: Jean Victor Poncelet

"One of the most important and also most beautiful theorems in classical geometry is that of Poncelet (...) His proof was synthetic and somewhat elaborate in what was to become the predominant style in projective geometry of last century. Slightly thereafter, Jacobi gave another argument based on the addition theorem for elliptic functions. In fact, as will be seen below, the Poncelet theorem and addition theorem are essentially equivalent, so that at least in principle Poncelet gave a synthetic derivation of the group law on an elliptic curve. Because of the appeal of the Poncelet theorem it seems reasonable to look for higher-dimensional analogues... Although this has not yet turned out to be the case in the Poncelet-type problems..."

These introductory words from [GH1977], written by Griffiths and Harris exactly 30 years ago, serve as a motto of the present book.

In a few years, we are going to reach a significant anniversary, the bicentennial of Jean Victor Poncelet's proof of one of the most beautiful and most important theorems of projective geometry. As is well known, he proved it during his captivity in Russia, in Saratov in 1813, after Napoleon's wars against Russia. The first proof was in a sense an analytic one. In 1822, Poncelet published another, purely geometric, synthetic proof in his *Traité des propriétés projectives des figures* [Pon1822]. Suppose that two ellipses are given in the plane, together with a closed polygonal line inscribed in one of them and circumscribed about the other one. Then, Poncelet's theorem states that infinitely many such closed polygonal lines exist – every point of the first ellipse is a vertex of such a polygon. Besides, all these polygons have the same number of sides. Later, using the addition theorem for elliptic functions, Jacobi gave another proof of the theorem in 1828 (see [Jac1884a]). Essentially, Poncelet's theorem is equivalent to the addition theorems for elliptic curves and his proof represents a synthetic way of deriving the group structure on an elliptic curve. Another proof of Poncelet's theorem, in a modern, algebro-geometrical manner, was done quite recently by Griffiths and Harris (see [GH1977]). There, they also presented an interesting generalization of the Poncelet theorem to the three-dimensional case, considering polyhedral surfaces both inscribed and circumscribed about two quadrics.

If we have in mind the geometric interpretation of the group structure on a cubic (see Figure 1.2), then the question of finding an analogous construction of the group structure in higher genera arises.

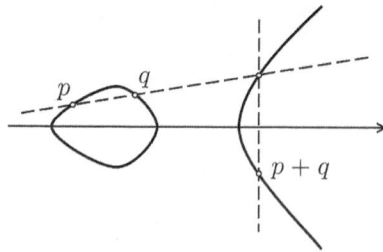

Figure 1.2: The group law on the cubic curve

Thus, thirty years ago, Griffiths and Harris announced a program of understanding higher-dimensional analogues of Poncelet-type problems and a synthetic approach to higher genera addition theorems.

The main aim of the present book is to report on progress made in settling and completing of this program. We will also present in a quite systematic way the most important results and ideas around Poncelet's theorem, both classical and modern, together with their historical origins and natural generalizations.

A natural question connected with Poncelet's theorem is to find an analytical condition determining, for two given conics, if an n-polygon inscribed in one and circumscribed about the second conic exists. In a short paper [Cay1854], Cayley derived such a condition in 1853, using the theory of Abelian integrals. He had dealt with Poncelet's porism in a number of other papers [Cay1853, Cay1855, Cay1857, Cay1858, Cay1861]. Inspired by [Cay1854], Lebesgue translated Cayley's proof to the language of geometry. Lebesgue's proof of Cayley's condition, derived by methods of projective geometry and algebra, can be found in his book *Les coniques* [Leb1942]. In modern settings, Griffiths and Harris derived Cayley's theorem by finding an analytical condition for points of finite order on an elliptic curve [GH1978a].

It is worth emphasizing that Poncelet, in fact, proved a statement that is much more general than the famous Poncelet theorem [Ber1987, Pon1822], then deriving the latter as a corollary. Namely, he considered $n + 1$ conics of a pencil in the projective plane. If there exists an n-polygon with vertices lying on the first of these conics and each side touching one of the other n conics, then infinitely many such polygons exist. We shall refer to this statement as the *Full Poncelet theorem* and call such polygons *Poncelet polygons*.

A nice historical overview of the Poncelet theorem, together with modern proofs and remarks is given in [BKOR1987]. Various classical theorems of Poncelet type with short modern proofs are reviewed in [BB1996], while the algebro-geometrical approach to families of Poncelet polygons via modular curves is given in [BM1993, Jak1993].

Figure 1.3: Elliptical billiard table

The Poncelet theorem has an important mechanical interpretation. *An Elliptical billiard* [KT1991, Koz2003] is a dynamical system where a material point of the unit mass is moving under inertia, or in other words, with a constant velocity inside an ellipse and obeying the reflection law at the boundary, i.e., having congruent impact and reflection angles with the tangent line to the ellipse at any bouncing point. It is also assumed that the reflection is absolutely elastic. It is well known that any segment of a given elliptical billiard trajectory is tangent to the same conic, confocal with the boundary [CCS1993]. If a trajectory becomes closed after n reflections, then the Poncelet theorem implies that any trajectory of the billiard system, which shares the same caustic curve, is also periodic with the period n.

The Full Poncelet theorem also has a mechanical meaning. The configuration dual to a pencil of conics in the plane is a family of confocal second-order curves [Arn1978]. Let us consider the following, a little bit unusual billiard. Suppose n confocal conics are given. A particle is bouncing on each of these n conics respectively. Any segment of such a trajectory is tangent to the same conic confocal with the given n curves. If the motion becomes closed after n reflections, then, by the Full Poncelet theorem, any such trajectory with the same caustic is also closed.

The statement dual to the Full Poncelet theorem can be generalized to the d-dimensional space [CCS1993] (see also [Pre1999, Pre]). Suppose vertices of the polygon $x_1 x_2 \ldots x_n$ are respectively placed on confocal quadric hypersurfaces $\mathcal{Q}_1$, $\mathcal{Q}_2$, ..., $\mathcal{Q}_n$ in the d-dimensional Euclidean space, with consecutive sides obeying the reflection law at the corresponding hypersurface. Then all sides are tangent to some quadrics $\mathcal{Q}^1$, ..., $\mathcal{Q}^{d-1}$ confocal with $\{\mathcal{Q}_i\}$; for the hypersurfaces $\{\mathcal{Q}_i, \mathcal{Q}^j\}$, an infinite family of polygons with the same properties exist.

But, more than one century before these quite recent results, in 1870, Darboux proved the generalization of Poncelet's theorem for a billiard within an ellipsoid in the three-dimensional space [Dar1870]. It seems that his work on this topic is completely forgotten nowadays.

Darboux was occupied by Poncelet's theorem for almost 50 years, and many of his results and ideas, in one way or another, are going to be incorporated throughout the book.

Let us mention that in the same year, 1870, appeared another very important work: [Wey1870] of Weyr. It can be treated as the historic origin of the modern Griffits–Harris Space Poncelet Theorem. A few years later, Hurwitz used Weyr's results to get a new proof of the standard Poncelet theorem (see [Hur1879]).

It is natural to search for a Cayley-type condition related to generalizations of the Poncelet theorem. Such conditions for the billiard system inside an ellipsoid in the Eucledean space of arbitrary finite dimension were derived in [DR1998a, DR1998b]. In recent papers [DR2004, DR2005, DR2006b, DR2006a], algebro-geometric conditions for existence of periodical billiard trajectories within k quadrics in d-dimensional Euclidean space were derived. The second important goal of these papers, actually for the present book as well, was to offer a thorough historical overview of the subject with a special attention on the detailed analysis of ideas and contributions of Darboux and Lebesgue. While Lebesgue's work on this subject has been, although rarely, mentioned by experts, on the other hand, it seems to us that relevant Darboux's ideas are practically unknown in contemporary mathematics. We give natural higher-dimensional generalizations of the ideas and results of Darboux and materials presented by Lebesgue, providing the proofs also in the low-dimensional cases if they were omitted in the original works. Besides other results, interesting new properties of pencils of quadrics are established – see Theorems 5.30 and 5.33. The latter gives a nontrivial generalization of the Basic Lemma from Lebesgue's book.

In our presentation of the development connected with the Griffiths–Harris program, we follow the recent paper [DR2008]. We present a geometric construction generalizing a summation procedure on the elliptic curve for the case of hyperelliptic Jacobians. These ideas are continuations of those of Reid, Donagi and Knörrer, see [Rei1972], [Knö1980], [Don1980]. Further development, realization, simplification and visualization of their constructions is obtained by using the ideas of billiard dynamics on pencils of quadrics developed in [DR2004].

The projective geometry nucleus of that billiard dynamics is the Double Reflection Theorem, see Theorem 5.27 below. There are four lines belonging to a certain linear space and forming *the Double reflection configuration*: these four lines reflect to each other according to **the billiard law at some confocal quadrics**.

In higher genera, we construct the corresponding, more general, billiard configuration, again by using the Double Reflection Theorem. This configuration, which we call *s-brush*, is in one of the equivalent formulations, a certain billiard trajectory of length $s \leq g$ and the sum of s elements in the brush is, roughly speaking, the final segment of that billiard trajectory.

The milestones of this presentation are [Knö1980] and [DR2004] and the key observation, from [DR2008], giving a link between them is *that the correspondence $g \mapsto g'$ in Lemma 4.1 and Corollary 4.2 from [Knö1980] is* **the billiard map at the quadric $\mathcal{Q}_\lambda$**.

Thus, after observing and understanding the billiard nature behind the constructions of [Rei1972], [Knö1980], [Don1980], we become able to use the billiard tools to construct and study hyperelliptic Jacobians, and particularly their real part. It may be realized as a set T of lines in $\mathbf{R}^d$ simultaneously tangent to given $d-1$ quadrics $\mathcal{Q}_1, \ldots, \mathcal{Q}_{d-1}$ of some confocal family. It is well known that such a set T is invariant under the billiard dynamics determined by quadrics from the confocal family. By using the Double Reflection Theorem and some other billiard constructions we construct a group structure on T, a *billiard algebra*. The usage of billiard dynamics in algebro-geometric considerations appears to be, as usual in such a situation, of a two-way benefit. We derive a fundamental property of T: *any two lines in T can be obtained from each other by at most $d-1$ billiard reflections at some quadrics from the confocal family*. The last fact opens a possibility to introduce new hierarchies of notions: of *s-skew lines in T*, $s = -1, 0, \ldots, d-2$ and of *s-weak Poncelet trajectories of length n*. The last are natural quasi-periodic generalizations of Poncelet polygons. By using billiard algebra, we obtain complete analytical descriptions of them. These results are further generalizations of our recent description of Cayley's type of Poncelet polygons in arbitrary dimension, see [DR2006b]. Let us emphasize that the method used in [DR2008], based on billiard algebra, differs from the methods exposed in [DR2006b], see also [DR2010]. Both of the methods will be presented in the sequel.

The interrelations between billiard dynamics, subspaces of intersections of quadrics and hyperelliptic Jacobians developed in [DR2008], enable us to obtain higher-dimensional generalizations of several classical results. To demonstrate the power of the methods, generalizations of Weyr's Poncelet theorem (see [Wey1870]) and also the Griffiths–Harris Space Poncelet theorem (see [GH1977]) in arbitrary dimension are derived and presented here. We also give an arbitrary-dimensional generalization of the Darboux theorem [Dar1914].

Let us mention at the end of a brief outline of main results which are going to be presented here, that the line we are going to establish and follow, is to demonstrate the deep intimate relationship between on one hand general hyperelliptic Jacobians and integrable billiard systems generated by pencils of quadrics on the other hand. This can be seen as a very simple and specialized level of general ideology of integrable systems which culminated with the so-called Novikov's conjecture, solved by Shiota in 1985.

Let us recall that Novikov's conjecture demonstrates the deepest relationship between the theory of integrable dynamical systems and theory of algebraic curves. It solved a century old, general and important Riemann–Schottky problem of description of period matrices of Jacobians among Riemannian matrices through the solutions of the Kadomtsev–Petviashvili integrable hierarchy.

There is another, very important connection of our subject with some of the most prominent parts of contemporary mathematics.

The Euler–Chasles correspondences, or symmetric (2-2)-correspondences play one of the main roles in our exposition. They were used by Jacobi, then by Trudi [Tru1853, Tru1863] and finally, Darboux extended their use in the theory of Poncelet porisms essentially.

One of the central objects in mathematical physics in the last 25 years is the R-matrix, or the solution $R(t,h)$ of the quantum Yang–Baxter equation

$$R^{12}(t_1 - t_2, h)R^{13}(t_1, h)R^{'23}(t_2, h) = R^{23}(t_2, h)R^{13}(t_1, h)R^{12}(t_1 - t_2, h),$$

as a paradigm of modern understanding of the addition relation. Here t is a so-called *spectral parameter* and h is the *Planck constant*. If the h dependence satisfies the quasi-classical property $R = I + hr + O(h^2)$, the classical r-matrix r satisfies the classical Yang–Baxter equation. Classification of the solutions of the classical Yang–Baxter equation was done by Belavin and Drinfeld in 1982 [BD1982]. The problem of classification of the quantum R-matrices is still open. However, some important results of classification have been obtained in the basic 4×4 case by Krichever in [Kri1981], and following his ideas in [Dra1992a, Dra1993].

Krichever in [Kri1981] applied the idea of "finite-gap" integration to the theory of the Yang equation:

$$R^{12}L^{13}L^{'23} = L^{'23}L^{13}R^{12}.$$

The principal objects that are considered are $2n \times 2n$ matrices L, understood as 2×2 matrices whose elements are $n \times n$ matrices; $L = L_{j\beta}^{i\alpha}$ is considered as a linear operator in the tensor product $\mathbf{C}^n \otimes \mathbf{C}^2$. The theorem from [Kri1981] uniquely characterizes them by the following spectral data:

1. the vacuum vectors, i.e., vectors of the form $X \otimes U$, which L maps to vectors of the same form $Y \otimes V$, where $X, Y \in \mathbf{C}^n$ and $U, V \in \mathbf{C}^2$;

2. the vacuum curve Γ : $P(u,v) = \det L = 0$, where $L_j^i = V^\beta L_{j\beta}^{i\alpha} U_\alpha$, $(V^\beta) = (1, -v)$, $X_n = Y_n = U_2 = V_2 = 1$; $U_1 = u$, $V_1 = v$;

3. the divisors of the vector-valued functions $X(u,v)$, $Y(u,v)$, $U(u,v)$, $V(u,v)$, which are meromorphic on the curve Γ.

It appeared that vacuum curves in 4×4 case are exactly Euler–Chasles correspondences. The Yang–Baxter equation itself provides the condition of commutation of the two Euler–Chasles correspondences. The classification follows by application of the Euler theorem in the general case, and by studying possible degenerations.

This is practically the same picture we meet in the study of the Poncelet theorem. The hope is that our study of higher-dimensional analogues of the Poncelet theorem could provide us the intuition that will help us in classification of higher-dimensional solutions of the Yang–Baxter equation.

Thus, we include the story about Krichever's algebro-geometric approach to 4×4 solutions of the Quantum Yang–Baxter equation in the last chapter. We explained there the relationship between the Poncelet theorem for a triangle and the Darboux theorem from one side and Krichever's commuting relation of vacuum curves from another side (see Theorem 10.12). We underline connection of classification results for 4×4 R-matrices to the classification of pencils of conics, see Theorem 10.12 and Proposition 10.13. Pencils of conics and their classification played a crucial role in previous chapters. Finally, we point out a sort of billiard construction within the Algebraic Bethe Ansatz associated to four-dimensional R-matrices, see Lemma 10.14 and Theorem 10.15.

The Poncelet theorem is usually called the Poncelet *porism*. Let us give some explanation of the meaning of the word **porism**. It has roots in ancient Greek mathematics, and it is usually translated in two ways. The first one is *lemma* or *corollary*. The second one goes deeper into the philosophy of ancient Greek mathematics. Scientists of that time used to divide mathematical statements into two categories:

- *Theorems* – where something has to be proven, and
- *Problems* – where something needs to be constructed.

Nevertheless, they recognized the third, intermediate, class as well, called *Porisms*, directed to finding what is proposed. The most famous collection of porisms of ancient times was the book *The Porisms* of Euclid. Unfortunately, this work is lost, and the trace which survived leads through *The Collection* of Pappus of

Alexandria. Even then, there was much discussion about the definition of the notion of porism as well as about Euclid's porisms. These discussions continue today. In the XVII century, important contributions were made by Albert Girard and Pierre Fermat. In the XVIII century, we can mention Robert Simson and John Playfair. Here is Simson's definition of a porism.

> *"Porisma est propositio in qua proponitur demonstrate rem aliquam vel plures Batas ease, cui vel quibus, ut et cuilibet ex rebus innumeris non quidem datis, sed quae ad ea quae data sunt eandem habent relationem, convenire ostendendum est affectionem quandam communem in propositione descriptam. Porisma etiam in forma problematis enuntiari potest, si nimirum ex quibus data demonstranda aunt, invenienda proponantur."*

Playfair, continuing the work of Simson, tried to understand the probable origin of porisms, to find out what led the ancient geometers to the discovery of them. He remarked that the careful investigation of all possible particular cases of a proposition would show that:

(1) under certain conditions a problem becomes impossible;

(2) under certain other conditions, indeterminate or capable of an infinite number of solutions.

For more details see [1911, E.B.].

This is exactly the situation we recognize in the Poncelet theorem. For two given conics, there are two possibilities. Either, a polygon inscribed in one of them and circumscribed about the other has an infinite number of sides, or the number of sides is finite. If it is finite, then the number of sides does not depend on an initial point. We want to stress here that the idea of porism of Poncelet type, in a very special case, existed almost 70 years before Poncelet. This case of Poncelet's theorem is the one with two circles, inscribed and circumscribed about the same triangle. We come to such a situation starting from an arbitrary triangle, and considering its inscribed and circumscribed circle. Denote by r and R their radii respectively, and by d the distance between the centers of the circles. The formula connecting these three values, sometimes referred as "Euler's formula" is well known:

$$d^2 = R^2 - 2rR.$$

However, this relation was discovered by English mathematician Chapple in 1746, and he caught sight of the *poristic* nature of the problem: if there are two circles satisfying the last *Chapple formula*, **then there are infinitely many triangles inscribed in one and circumscribed about the other circle**. Probably, this is the first known appearance of porisms of Poncelet type.

The Euler school was also interested in that subject. Nicolas Fuss, one of Euler's personal secretaries, and after Euler's death the secretary of St. Petersburg

Academy of Sciences, published several works on study of bicentric polygons. In 1797 he published the formula for bicentric quadrilaterals:

$$(R^2 - d^2)^2 = 2r^2(R^2 + d^2).$$

But, although it was 50 years after Chapple, Fuss did not understand the poristic nature of the problem.

It was Jacobi in 1828 who understood the relationship between Poncelet porism in general and study of bicentric polygons of Fuss, Steiner and others.

Some parts of the material presented here were used by the authors for graduate courses they taught: V. D. in 2002/2003 in the International School of Advanced Studies in Trieste [Dra2003], and M. R. in 2006 in the Weizmann Institute of Science in Rehovot. Both authors read mini-courses on the subject, M. R. in the Weizmann Institute of Sciences in 2005 and V. D. at the University of Lisbon in 2007. Also, both authors gave several lectures on seminars and conferences in Italy, France, Germany, Serbia, Spain, Portugal, Montenegro, Israel, Czechia, Poland, Hungary, Great Britain, Austria, Russia, Brazil, USA, Canada, and Bulgaria. One of our observations was that there was a visible division between the communities of Algebraic and Projective Geometry, although some 50 years ago these fields were quite a unified subject. Having this experience in mind, we decided to include introductions to both subjects in order to make the book self-contained as much as possible and usable for both communities and for the mathematical community at large.

Acknowledgement

For many years we felt support and constant care about our work from Professor Boris Dubrovin and he was the one who suggested us to write this book. Enthusiastic discussions about the subject and presentations by some of the leading world experts in the fields such as Philip Griffiths, Marcel Berger, and Valery Kozlov were very encouraging and stimulating for us. We learned a lot from numerous discussions with our distinguished colleagues: Alexander Veselov, Alexey Bolsinov, Victor Buchstaber, Igor Krichever, Yuri Fedorov, Emma Previato, Borislav Gajić, Božidar Jovanović, Rade Živaljević, Gojko Kalajdžić, Vered Rom-Kedar, Jean-Claude Zambrini, Simonetta Abenda, Alexey Borisov, Armando Treibich, Nikola Burić... It is our great pleasure to thank them all.

The research was partially supported by the Serbian Ministry of Science and Technology, Project *Geometry and Topology of Manifolds and Integrable Dynamical Systems* and by Mathematical Physics Group of the University of Lisbon, Project *Probabilistic approach to finite- and infinite-dimensional dynamical systems, PTDC/MAT/104173/2008*. The last part of the book was written during the visit of one of the authors (V. D.) to the IHES and he uses the opportunity to thank the IHES for hospitality and outstanding working conditions.

Chapter 2

Billiards – First Examples

2.1 Introduction to billiards

Let us start from the following well-known problem.

Suppose that a railway is passing near two neighbouring villages, and that a new railway station, common for both of them, is about to be built. Where to place the station, in order to minimize the length of the road connecting the villages with it? (See Figure 2.1.)

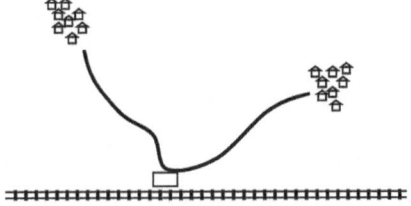

Figure 2.1.

In other words, on a given line (i.e., the railway), we need to find a point such that the sum of its distances from two fixed points is smallest possible.

Denote the given points (villages) by A and B, and by r the line (railway). Let B' be the point symmetric to B with respect to r. The intersection point S of r with AB' has the requested properties. Indeed, notice that $AS + SB = AS + SB' = AB'$, while for any other point $S' \in r$, $AS' + S'B = AS' + S'B' > AB'$. (See Figure 2.2.)

It is easy to see that segments AS and BS form the same angles with the line r, i.e., the segment BS is the billiard reflection of AS on the line r. In other words, *the minimal trajectory from A to B that meets the line r is exactly the billiard trajectory, with the reflection point on r.*

Exercise 2.1. Let M be a point inside a convex angle α. Find the points K, L on the sides of α such that the triangle KLM has the minimal perimeter. Prove that

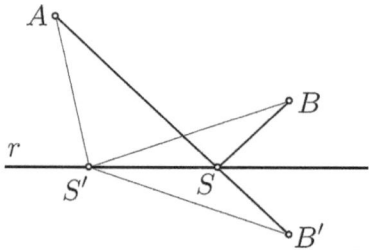

Figure 2.2.

segments MK,KL and KL,LM satisfy the billiard reflection law on the sides of the angle.

2.2 Triangular billiards

Now, we are going to investigate the billiards within a triangle in the Euclidean plane. A trajectory of such a billiard is a polygonal line, finite or infinite, with vertices on the sides of the triangle, such that consecutive edges of the trajectory satisfy the billiard law: i.e., they form the same angle with the side of the triangle on which their common vertex lies. The reflection is not defined only at the vertices of the triangle – thus we omit from our consideration trajectories falling at a vertex of a triangle.

Let us try to find out if closed trajectories of a billiard within a triangle exist. Denote by A, B, C the vertices of the triangle. It is clear, from Section 2.1, that the edges of the triangle with minimal perimeter, whose vertices are inner points of the sides $\triangle ABC$, will represent a billiard trajectory.

Theorem 2.2. *Let $\triangle ABC$ be an acute angled triangle. If $\triangle KLM$ is the triangle with minimal perimeter inscribed in $\triangle ABC$, then its vertices are the feet of the altitudes of $\triangle ABC$. Moreover, inside $\triangle ABC$, KLM is a unique closed billiard trajectory with 3 bounces.*

Proof. Let M be a fixed point on the edge AB. We want to find points $K \in BC$, $L \in AC$ such that the triangle KLM has the minimal perimeter. Denote by M', M'' points symmetric to M with respect to the sides BC, AC. It is easy to see that K,L are intersection points of $M'M''$ with BC,AC respectively (see Figure 2.3).

The perimeter of $\triangle KLM$ is equal to the segment $M'M''$. Notice that $M'M''$ is a side of the isosceles triangle $CM'M''$, with $CM' \cong CM'' \cong CM$ and $\angle M'CM'' = 2\angle BCA$. It follows that $M'M''$ will be the shortest for CM being an altitude of the triangle ABC, i.e., M being its foot. Similarly, we prove K, L are also feet of the corresponding altitudes (see Figure 2.4).   □

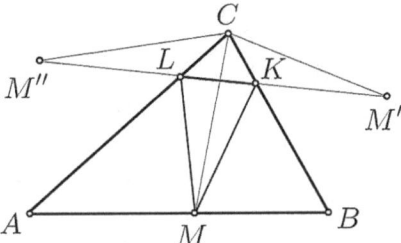

Figure 2.3.

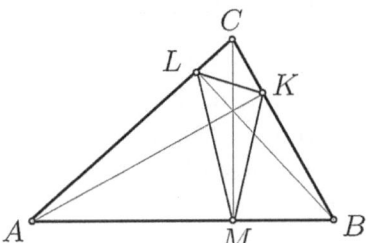

Figure 2.4.

After having this periodic trajectory inside an acute triangle, it is easy to see that there is an infinity of other closed billiard trajectories (see Figure 2.5).

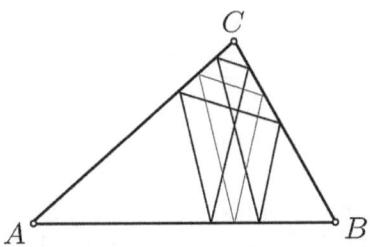

Figure 2.5.

There are also closed billiard trajectories inside a right triangle. One of them, the polygonal line $KLMNMLK$, is shown on Figure 2.6.

For obtuse triangles in general, the existence of periodic trajectories is not proved. There are only examples for some special cases.

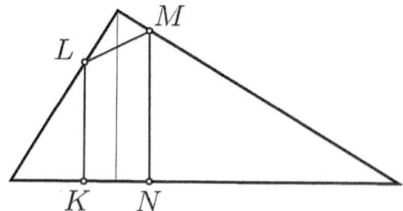

Figure 2.6.

2.3   Billiards within an ellipse

In this section, we are going to discuss in an elementary way the most important aspects of billiards within an ellipse in the plane.

A billiard trajectory within an ellipse is a polygonal line with the vertices lying on the ellipse and with consecutive edges satisfying *the billiard law*, i.e., forming the same angles with the tangent line to the ellipse at the joint vertex of the edges (see Figure 2.7).

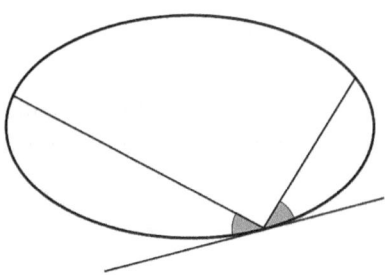

Figure 2.7.

Proposition 2.3 (Focal property of the ellipse). *Let $\mathcal{E}$ be an ellipse with foci F_1, F_2 and $A \in \mathcal{E}$ an arbitrary point. Then segments AF_1, AF_2 satisfy the billiard law on $\mathcal{E}$. (See Figure 2.8.)*

Proof. It is enough to prove that for any point C on the tangent, the sum $CF_1 + CF_2$ is greater than $AF_1 + AF_2$. Let B be the intersection of the segment CF_1 with the ellipse. Then $AF_1 + AF_2 = BF_1 + BF_2 < BF_1 + BC + CF_2 = CF_1 + CF_2$. (See Figure 2.9.) □

As an immediate consequence of this proposition, we have: if one segment of a billiard trajectory within ellipse $\mathcal{E}$ contains a focus of $\mathcal{E}$, then all segments of the trajectory contain one or the other focus, alternately.

Now, we are going to prove the following important property of the billiard within an ellipse.

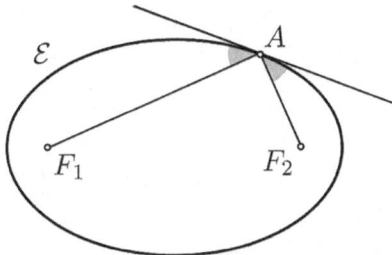

Figure 2.8: Focal property of the ellipse

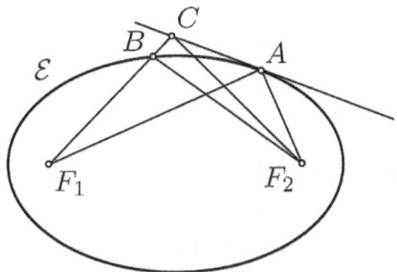

Figure 2.9.

Proposition 2.4. *Let two lines satisfy the billiard law on the ellipse $\mathcal{E}$. If one of the lines is tangent to the ellipse $\mathcal{E}'$ that is confocal with $\mathcal{E}$, then the other one is also tangent to $\mathcal{E}'$. (See Figure 2.10.)*

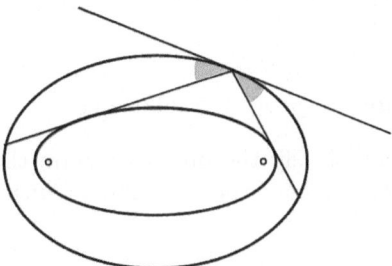

Figure 2.10.

Proof. Let A, B, C be points on $\mathcal{E}$ such that segments AB and BC satisfy the reflection law at B. Suppose that AB is tangent to $\mathcal{E}'$. Denote by F_1, F_2 the foci of the two ellipses and by T_1, T_2 points on the tangent to $\mathcal{E}$ in B such that angles $\angle F_1 B T_1$ and $\angle F_2 B T_2$ are acute. By Proposition 2.3, these angles are congruent. Notice that then AB is placed inside one of these angles, say $\angle F_1 B T_1$. Since, by the billiard law, $\angle ABT_1 \cong \angle CBT_1$, the segment BC is placed inside angle $\angle F_2BT_2$, as shown on Figure 2.11. Let D_1, D_2 be points symmetric to F_1, F_2 with

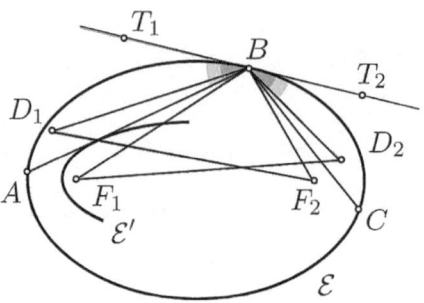

Figure 2.11.

respect to lines AB, BC respectively.

We are going to show that $\triangle D_1BF_2 \cong \triangle F_1BD_2$. We have that $D_1B \cong F_1B$, $F_2B \cong D_2F$, since the corresponding segments are symmetric with respect to AB, BC. Also, $\angle D_1BF_2 = \angle D_1BF_1 + \angle F_1BF_2$ and $\angle F_1BD_2 = \angle F_2BD_2 + \angle F_1BF_2$. Since $\angle D_1BF_1 = 2\angle F_1BA = 2(\angle F_1BT_1 - \angle ABT_1) = 2(\angle F_2BT_2 - \angle CBT_2) = 2\angle F_2BC = \angle D_2BT_2)$, we have that $\angle D_1BF_2 \cong \angle F_1BD_2$, which proves the congruence of triangles $\triangle D_1BF_2$ and $\triangle F_1BD_2$. Hence, $D_1F_2 \cong F_1D_2$.

The segment D_1F_2 is equal to the minimal sum of distances from F_1 and F_2 of a point on line AB, i.e., to the sum of distances of an arbitrary point on ellipse $\mathcal{E}'$ from its foci, since this line is touching $\mathcal{E}'$. Similarly, F_1D_2 is equal to the minimal sum of distances from F_1 and F_2 of a point on line BC, thus BC is also tangent to $\mathcal{E}'$.   □

We leave to the reader to prove the following

Exercise 2.5. Let two lines satisfy the billiard law on the ellipse $\mathcal{E}$. If one of the lines is tangent to the hyperbola $\mathcal{H}$ that is confocal with $\mathcal{E}$, then the other one is also tangent to $\mathcal{H}$.

From Propositions 2.4 and Exercise 2.5, immediately we derive

Corollary 2.6. *Let $\mathcal{T}$ be a billiard trajectory within ellipse $\mathcal{T}$. If $\mathcal{C}$ is a conic confocal to $\mathcal{E}$ such that one segment of $\mathcal{T}$ is tangent to $\mathcal{C}$, then all segments of $\mathcal{T}$ are tangent to $\mathcal{C}$.*

Corollary 2.7. *Let B be a point outside the ellipse $\mathcal{E}'$ with focal points F_1 and F_2. Denote tangents to the ellipse from the point B as BB_1 and BB_2, where B_i are points of contact with the ellipse. Then the angles B_1BF_1 and B_2BF_2 are equal.*

2.4   Periodic orbits of billiards and Birkhoff's theorem

We saw in Section 2.3 that billiards within an ellipse have remarkable properties. It would be interesting to examine closer periodical trajectories of these billiards. As a first step, let us present a classical result due to Birkhoff on periodic trajectories of a more general class of billiards.

Consider a billiard table bounded by a closed convex curve in the plane. Assume that the length of the curve is equal to 1 and introduce a natural parameter φ. Suppose that $\varphi_1, \ldots, \varphi_n$ represents the sequence of bouncing points of an n-periodic trajectory. Additionally, we may choose the values $\varphi_1, \ldots, \varphi_n, \varphi_{n+1}$ such that the differences $\varphi_2 - \varphi_1, \varphi_3 - \varphi_2, \ldots, \varphi_n - \varphi_{n-1}, \varphi_{n+1} - \varphi_n$ are between 0 and 1, and $\varphi_{n+1} \equiv \varphi_1 \mod 1$.

The integer $k = \varphi_{n+1} - \varphi_1$ is called the *rotation number* of the periodic trajectory.

Theorem 2.8 (Birkhoff). *Suppose there is given a smooth, closed, convex plane curve having nonzero curvature at any point. Then for any numbers $n, k \in \mathbf{N}$, $n > k$, there exist at least two geometrically different periodic trajectories, with the rotational number k and n bounces, of the billiard within the given curve. For one of these trajectories, the corresponding polygonal line has maximal length among all nearby closed polygonal lines inscribed in the curve. If this maximum is an isolated critical point of the length function on the set of inscribed polygonal lines with n vertices, then the polygonal line corresponding to the other trajectory is not an isolated maximum of the length function.*

2.5   Bicentric polygons

We have shown in Section 2.3 that billiard trajectories within an ellipse have a caustic, which is a conic confocal to the boundary. Notice that any pair of conics in a plane can be, by a projective mapping, transformed into a confocal pair. Thus, it is natural to consider two general conics and polygonal lines inscribed in one and circumscribed about the other one.

The case when these two conics are circles can be analyzed in an elementary way.

Triangles

It is easy to prove, even to students of elementary schools, that it is possible to inscribe a circle in a triangle and also to circumscribe another one about it. Harder is to find out, for two given circles, if they are inscribed and circumscribed about some triangle. In the next proposition, a sufficient and necessary condition on two circles is given.

Theorem 2.9 (Chapple–Euler formula). *Let k and K be two given circles with radii r and R. Denote by d the distance between their centers. Then there exists a triangle inscribed in K and circumscribed about k if and only if*

$$d^2 = R^2 - 2Rr. \tag{2.1}$$

Moreover, if condition (2.1) is satisfied, then every point of K is a vertex of such a triangle.

Proof. Let ABC be a triangle inscribed in K and circumscribed about k and denote by O, S the centers of these circles. Let N, M be the second intersection points of lines AS, NO with K, as shown on Figure 2.12. It is easy to see that

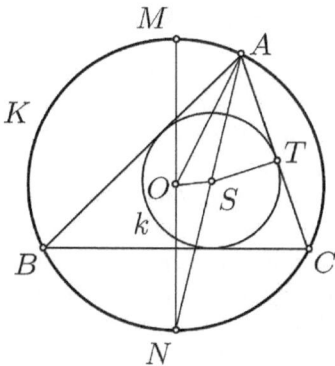

Figure 2.12.

ON is a bisector of BC.

The power of S with respect to circle K equals $d^2 - R^2 = -SA \cdot SN$.

From triangle SNC, we see that $SN = NC$. Namely, $\angle SCN = \angle SCB + \angle BCN = \dfrac{\angle BCA}{2} + \angle BAN = \dfrac{\angle BCA + \angle CAB}{2}$ and $\angle CSN = \angle SCA + \angle SAC = \dfrac{\angle BCA + \angle CAB}{2}$.

Thus, $SO^2 - R^2 = -SA \cdot SN = -SA \cdot NC$.

Notice that $\triangle AST \sim \triangle MNC$, where T is a common point of k and AC. Therefore, $AS \cdot NC = ST \cdot MN = r \cdot 2R$.

Thus, the relation (2.1) follows.

Now, suppose the equality (2.1) holds. Let A be an arbitrary point on K.

Denote by N the intersection of AS with K, and with B, C points on K such that $NB = NC = NS$. $\triangle ABC$ is inscribed in K and its inscribed circle k' is concentric with k. Then, $SO^2 = R^2 - 2Rr'$, where r' is the radius of k'. Since we also have $SO^2 = R^2 - 2Rr$, it is $r = r'$, i.e., $k = k'$. $\qquad\square$

Quadrilaterals

In this section, we are going to prove the following

Theorem 2.10. *Let k and K be two given circles with radii r and R. Denote by d the distance between their centers. Then there exists a quadrilateral inscribed in K and circumscribed about k if and only if*

$$d^2 = r^2 + R^2 - r\sqrt{r^2 + 4R^2}. \tag{2.2}$$

Moreover, if condition (2.2) is satisfied, then every point of K is a vertex of such a quadrilateral.

First, let us prove several lemmata.

Lemma 2.11. *Let $ABCD$ be a cyclic quadrilateral. If K, L, M, N are normal projections of the intersection of the diagonals to the sides of $ABCD$, then quadrilateral $KLMN$ is circumscribed about a circle.*

Proof. Let $P = AC \cap BD$. Since $ABCD$, $AKPN$ and $KBLP$ are cyclic quadrilaterals, we have

$$\angle PKN \cong \angle PAN = \angle CAD \cong \angle CBD = \angle LBP \cong \angle LKP,$$

i.e., $\angle PKN \cong \angle LKP$. Thus KP is a bisector of $\angle NKL$. Similarly, LP, MP, NP

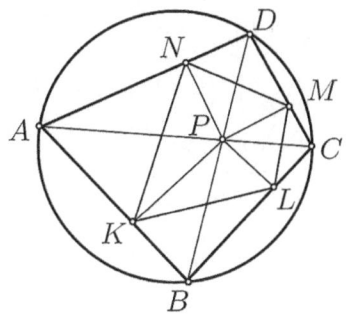

Figure 2.13.

are bisectors of corresponding angles, hence P is the center of the circle inscribed in $KLMN$. □

Lemma 2.12. *Let $ABCD$ be a cyclic quadrilateral and K, L, M, N as in Lemma 2.11. Additionally, suppose that AC is perpendicular to BD, $P = AC \cap BD$. Let O be the circumcenter and R_1 the circumradius of $ABCD$, r the inradius of $KLMN$, and $d_1 = OP$. Then*

$$r = \frac{R_1^2 - d_1^2}{2R_1}. \tag{2.3}$$

Proof. Let $\alpha = \angle DAC$, $\beta = \angle CAB$. Then we have

$$
\begin{aligned}
r &= PK \cdot \sin\alpha \\
&= PB \cdot \sin(90° - \beta) \cdot \sin\alpha \\
&= PB \cdot PD \cdot \frac{\sin(90° - \beta) \cdot \sin\alpha}{PD} \\
&= |p_{P,k}| \cdot \frac{\sin(90° - \beta) \cdot \sin\alpha}{PD} \\
&= (R_1^2 - d_1^2) \cdot \frac{\sin(90° - \beta) \cdot \sin\alpha}{AD \cdot \sin\alpha} \\
&= (R_1^2 - d_1^2) \cdot \frac{\sin(90° - \beta)}{2R_1 \cdot \sin(90° - \beta)} \\
&= \frac{R_1^2 - d_1^2}{2R_1}. \qquad\qquad\qquad\qquad \square
\end{aligned}
$$

Lemma 2.13. *Let $ABCD$ be a cyclic quadrilateral with $AC \perp BD$, P, K, L, M, N, R_1, O, d_1 as in Lemmata 2.11 and 2.12, and S_1, S_2, S_3, S_4 midpoints of AB, BC, CD, AD (see Figure 2.14). Then $K, L, M, N, S_1, S_2, S_3, S_4$ belong to the*

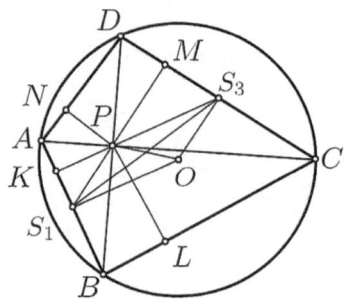

Figure 2.14.

same circle. If R is its radius, then

$$
R = \frac{1}{2}\sqrt{2R_1^2 - d_1^2}. \tag{2.4}
$$

Proof. First, let us prove that $PS_3 \perp AB$, i.e., S_3, P, K are collinear. Namely,

$$
\begin{aligned}
\angle S_3 PK &= \angle KPA + \angle APD + \angle DPS_3 \\
&= 90° - \angle PAK + 90° + \angle DPS_3 \\
&= 180° - \angle PDC + \angle DPS_3.
\end{aligned}
$$

Since S_3 is the midpoint of CD, we have $\angle DPS_3 = \angle PDS_3$. Thus $\angle S_3 PK = 180°$.

It follows that $S_3P \parallel OS_1$, because both lines OS_1, S_3P are perpendicular to AB. Similarly, $OS_3 \parallel PS_1$. It follows that OS_3PS_1 is a parallelogram. Thus

$$S_1S_3^2 + OP^2 = 2 \cdot \left(PS_3^2 + PS_1^2\right)$$

$$= 2 \cdot \left(\left(\frac{CD}{2}\right)^2 + \left(\frac{AB}{2}\right)^2\right)$$

$$= \frac{1}{2}\left((2R_1 \sin \alpha)^2 + (2R_1 \sin(90° - \alpha))^2\right)$$

$$= 2R_1^2,$$

with $\alpha = \angle DAC$. Hence, $S_1S_3^2 = 2R_1^2 - OP^2 = 2R_1^2 - d_1^2$.

Now, we are going to prove that all points K, L, M, N, S_1, S_2, S_3, S_4 belong to the circle with radius $S_1S_3/2$ and center at the midpoint of OP. Clearly, S_1S_3 is a diameter of this circle, and K, M belong to it because $\angle S_3MS_1 = \angle S_3KS_1 = 90°$. We get the same for S_2, S_4, L, N.

Finally, we have $R = \frac{1}{2}S_1S_3 = \frac{1}{2}\sqrt{2R_1^2 - d_1^2}$.   □

Now, let us return to the bicentric quadrilateral from Theorem 2.10 – denote it by $KLMN$.

Construct lines perpendicular to the bisectors of angles of $KLMN$ at the vertices. These lines determine a quadrilateral $ABCD$. Let P be the incenter of $KLMN$.

It is easy to prove that $ABCD$ is inscribed in a circle, that P is the intersection of its diagonals and that the diagonals are perpendicular to each other.

Denote as in previous lemmata: O – the center of the circle circumscribed about $ABCD$; R_1 – its radius; r, R – the radii of inscribed and circumscribed circle of $KLMN$; $d_1 = PO$; S – the midpoint of PO; $d = SO = d_1/2$.

By Lemmata 2.12 and 2.13,

$$r = \frac{R_1^2 - 4d^2}{2R_1} \quad \text{and} \quad R = \frac{1}{2}\sqrt{2R_1^2 - 4d^2}. \tag{2.5}$$

Eliminating R_1 from (2.5), we get

$$\frac{1}{r^2} = \frac{1}{(R+d)^2} + \frac{1}{(R-d)^2}. \tag{2.6}$$

From (2.6) it is possible to express d and obtain equation (2.2).

Now, let us prove the opposite part of Theorem 2.10. Suppose that $k(P, r)$ and $K(S, R)$ are given circles, such that equation (2.2) holds, $d = PS$.

Construct the point O such that S is the midpoint of OP and the circle $K_1(O, R_1)$, where R_1 satisfies (2.3) with $d_1 = 2d$. Notice that equation (2.3) is quadratic with respect to R_1, but only one of its solutions is positive.

Take A to be an arbitrary point on K_1 and construct $B, C, D \in K_1$ such that $AC \cap BD = P$, $AC \perp BD$. Denote by K, L, M, N the normal projections of P to the sides of $ABCD$. By Lemmata 2.11 and 2.12, $KLMN$ is circumscribed about k. By Lemma 2.13, $KLMN$ is also inscribed in a circle with center O and radius equal to $\frac{1}{2}\sqrt{2R_1^2 - d_1^2}$. Eliminating r from (2.2) and (2.3), we obtain that this is equal to R.

Thus, we constructed a quadrilateral $KLMN$ inscribed in K and circumscribed about k. In the construction, choosing arbitrarily the initial point A on K_1, we can get that any point on K can be a vertex of such a quadrilateral.

2.6   Poncelet theorem

In this section, we are going to state the Poncelet theorem and give its mechanical interpretation. The proof of the theorem will be given at the end of Chapter 4 and in Chapter 5.

Poncelet theorem

As we already mentioned, the Poncelet theorem is one of the most beautiful and deepest theorems of geometry, with numerous consequences and interrelations in a wide range of areas of mathematics. It was proved by Jean Victor Poncelet, while he was imprisoned in Russia, in 1813. He published another proof in 1822 in [Pon1822].

Theorem 2.14 (Poncelet Theorem). *Let C and D be two conics in the plane. Suppose that there is a polygon inscribed in C and circumscribed about D. Then there are infinitely many such polygons and all of them have the same number of sides. Moreover, each point of C is a vertex of such a polygonal line.*

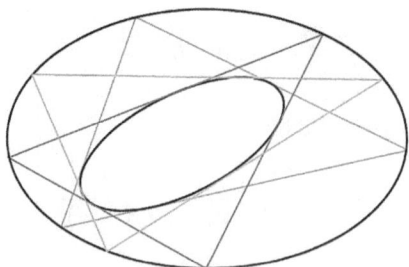

Figure 2.15: Three triangles inscribed in an ellipse and circumscribed about the other one

Mechanical interpretation of the Poncelet theorem

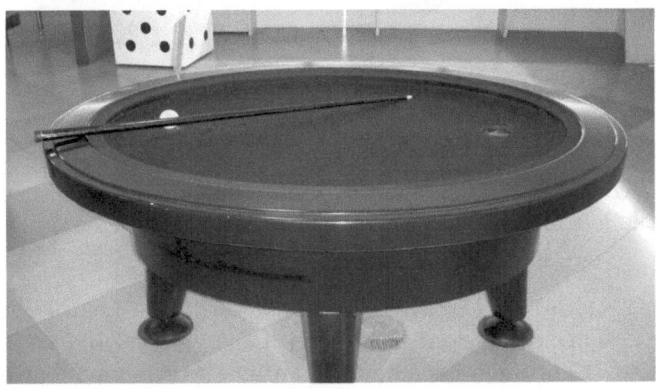

Figure 2.16: Elliptical billiard table

The Poncelet theorem obtains a natural and beautiful mechanical interpretation, if we take that $\mathcal{C}$ is an ellipse and $\mathcal{D}$ a conic confocal to $\mathcal{C}$. Then, as was shown in Section 2.3, the polygonal lines inscribed in $\mathcal{C}$ and circumscribed about $\mathcal{D}$ are trajectories of the billiard motion within $\mathcal{C}$.

In other words, consider a billiard trajectory within ellipse $\mathcal{C}$. Suppose that a line containing one segment of the trajectory is tangent to a conic $\mathcal{D}$, confocal with $\mathcal{C}$. Then, all segments of the trajectory are also tangent to $\mathcal{D}$ (see Figure 2.17).

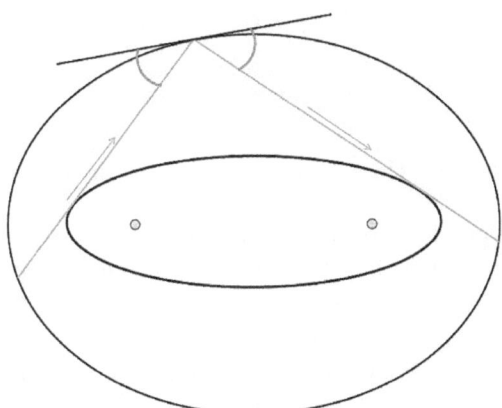

Figure 2.17: Billiard system and confocal conics

If a billiard trajectory is not periodic, then it will densely wind in the region bounded with the caustic and the billiard boundary as is shown on Figure 2.18.

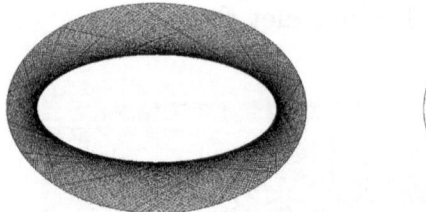 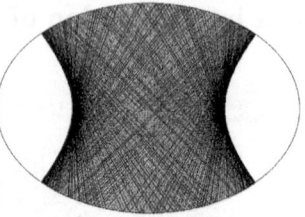

Figure 2.18: Two billiard trajectories within an ellipse: with another ellipse on the left and with a hyperbola on the right as caustics

Consider a closed trajectory of a billiard within $\mathcal{C}$. Then each billiard trajectory inside $\mathcal{C}$ sharing the same caustic with the given closed trajectory, is also closed. Moreover, all of them become closed after the same number of bounces on the boundary $\mathcal{C}$.

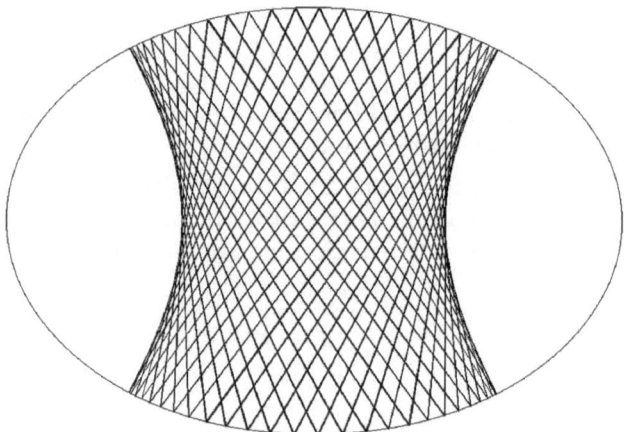

Figure 2.19: Example of a closed billiard trajectory with a hyperbola as caustic

In a mathematical description of billiard dynamical systems, we impose several assumptions: the billiard particle is a mass point of unit mass, friction is neglected and reflection at the boundary is absolutely elastic.

For a study of a more realistic situation, we recommend that the reader tries out real billiard tables with the conic shape, of an ellipse, a hyperbola or a parabola, as shown on Figures 2.16, 2.21, 2.20.

Back to Poncelet's theorem; in order to prove it and study its generalizations, we need certain mathematical tools. Chapters 3 and 4 are devoted to these necessary prerequisites.

Figure 2.20: Parabolic billiard table

Figure 2.21: Hyperbolic billiard table

Chapter 3

Hyperelliptic Curves and Their Jacobians

The theory of algebraic curves and their Jacobians is a vast subject with a long history and huge applications. In this chapter, we are going to present only some basic notions and facts, as necessary for further use. Balancing a reasonable length of an introductory chapter and the wish to provide a self-contained exposition, we refer the reader to [GH1978b, Mum1983, Dub1981, Gun1966] and references therein. Since elliptic and hyperelliptic curves and their Jacobians play a predominant role in what follows, our exposition will be focused on these classes, but not reduced only to them.

3.1 Riemann surfaces

Definition of Riemann surfaces

Study of algebraic curves over $\mathbf{C}$ is closely related to Riemann surfaces. These surfaces appeared in the study of multi-valued complex functions. As a good starting example one may consider the square root function

$$f(x) = \sqrt{x}.$$

Although Riemann surfaces historically and even methodologically used to come together with a marked function, we will start in a more invariant manner and introduce Riemann surfaces as complex manifolds of complex dimension 1. More precisely, we have

Definition 3.1. *A Riemann surface* is a connected Hausdorff topological space with an open covering $\{U_\alpha\}$ and a family of mappings $z_\alpha : U_\alpha \to \mathbf{C}$, such that:

(a) each z_α is a homeomorphism between U_α and the open set $z_\alpha(U_\alpha)$;

(b) each function $z_\beta \circ z_\alpha^{-1}$ is holomorphic on $z_\alpha(U_\alpha \cap U_\beta)$, whenever $U_\alpha \cap U_\beta \neq \emptyset$.

Mappings z_α are called *local coordinates*. We set

$$\varphi_{\alpha\beta} = z_\beta \circ z_\alpha^{-1} : \; z_\alpha(U_\alpha \cap U_\beta) \to z_\beta(U_\alpha \cap U_\beta)$$

and call them *transition functions*.

In other words, Riemann surfaces are connected one-dimensional complex manifolds.

Example 3.2. The set of complex numbers $\mathbf{C}$ is a Riemann surface.

Example 3.3 (Riemann sphere). Consider the unit sphere $\mathbf{S}^2$ in the real three-dimensional space $\mathbf{R}^3$:

$$\mathbf{S}^2 = \{(\xi, \eta, \zeta) \mid \xi^2 + \eta^2 + \zeta^2 = 1\},$$

and its covering $\{U_0, U_1\}$, $U_0 = \mathbf{S}^2 \setminus \{(0,0,-1)\}$, $U_1 = \mathbf{S}^2 \setminus \{(0,0,1)\}$. Define the local coordinates by

$$z_0(\xi, \eta, \zeta) = \frac{\xi - i\eta}{1 + \zeta}, \quad z_1(\xi, \eta, \zeta) = \frac{\xi + i\eta}{1 - \zeta}.$$

Transition function φ_{01} is of the form $\varphi_{01}(z_0) = 1/z_0$.

Exercise 3.4. Prove that the Riemann sphere can be identified with the extended set of complex numbers $\mathbf{C} \cup \{\infty\}$.

Definition 3.5. Suppose that M is a Riemann surface, and $\{(U_\alpha, z_\alpha)\}$ is its coordinate covering. We say that the function $f : M \to \mathbf{C} \cup \{\infty\}$ is *meromorphic* if each function $f \circ z_\alpha^{-1}$ is meromorphic on $z_\alpha(U_\alpha)$. Additionally, if all of them are holomorphic, we also say that f is *holomorphic*.

It is straightforward to check that the set of all meromorphic functions on a Riemann surface is a linear space of $\mathbf{C}$, while the holomorphic functions constitute a subspace.

Theorem 3.6. *Each holomorphic function on a compact Riemann surface is constant.*

Proof. Suppose f is a holomorphic function on M. $|f|$ is continuous thus there is $a \in M$ such that it is a maximum point of $|f|$ on compact set M. Suppose $a \in U_\alpha$. Then, by the maximum module principle, function $f \circ z_\alpha^{-1}$ is constant in a neighbourhood of $z_\alpha(a)$. By the analytic continuation principle then we get that f is constant. $\qquad\qquad\square$

Exercise 3.7. Let f be a meromorphic non-constant function on a compact Riemann surface. Prove that, for each $z \in \mathbf{C} \cup \{\infty\}$, the set $f^{-1}(z)$ is not empty.

Example 3.8. Each rational function $r(z)$ on $\mathbf{C}$ determines a meromorphic function on the Riemann sphere. Indeed, we may define

$$\hat{r}(p) = \begin{cases} r(z_0(p)), & \text{if } p \in U_0, \\ r\left(\frac{1}{z_1(p)}\right), & \text{if } p \in U_1. \end{cases}$$

Exercise 3.9. Check that function $\hat{r}$ is well defined on $U_0 \cap U_1$.

We are going to prove that all meromorphic functions on the Riemann sphere are rational. More precisely, we formulate:

Proposition 3.10. *Each meromorphic function on the Riemann sphere is equal to $\hat{r}$, for some rational function r, as defined in Example* 3.8.

Proof. Let f be a meromorphic function of the Riemann sphere, $f \not\equiv 0$. Define a rational function r in the following way: r has the same zeroes and poles, counting their multiplicities, as $f \circ z_0^{-1}$. Then the function $f/\hat{r}$ has neither zeroes nor poles in U_0 and from Exercise 3.7, it follows that it is equal to a nonzero constant c on the Riemann sphere. Thus, $f = c \cdot \hat{r}$. □

Definition 3.11. Let M be a Riemann surface. *A meromorphic differential* on M is a family $\{U_i, z_i, \omega_i\}$ such that:

(a) $\{U_i, z_i\}$ is a holomorphic covering of M, and $\omega_i = f_i(z_i)dz_i$, where f_i are holomorphic functions;

(b) if φ_{ij} is the transition function on $U_i \cap U_j$, then

$$f_i(\varphi_{ij}(z_j))\frac{d\varphi_{ij}(z_j)}{dz_j} = f_j(z_j).$$

Additionally, if all the functions $f_i(z_i)$ are holomorphic, we also say that the differential is *holomorphic*.

Both sets of meromorphic and holomorphic differentials on a given surface are linear spaces over $\mathbf{C}$. Moreover, we have:

Theorem 3.12. *Let Γ be a compact Riemann surface of genus g. Then the dimension of the space of all holomorphic differentials on Γ is equal to g.*

This theorem will be proved for a special class of Riemann surfaces, so-called *hyperelliptic surfaces* later on in this chapter. However, at the moment, we are ready to demonstrate it for the Riemann sphere:

Proposition 3.13. *If ω_0, ω_1 are meromorphic differentials on a Riemann surface, such that $\omega_0 \not\equiv 0$, then ω_1/ω_0 is a holomorphic function on the surface.*

Proof. If $\omega_{0\alpha} = f_\alpha(z_\alpha)dz_\alpha$, $\omega_{1\alpha} = g_\alpha(z_\alpha)dz_\alpha$ are local representations on U_α, then the function $\psi = \omega_1/\omega_0$ is defined as

$$\psi(p) = \frac{g_\alpha(z_\alpha(p))}{f_\alpha(z_\alpha(p))},$$

for each $p \in U_\alpha$. We need only to check that the function is well defined for $p \in U_\alpha \cap U_\beta$:

$$\frac{g_\alpha(z_\alpha(p))}{f_\alpha(z_\alpha(p))} = \frac{g_\beta(z_\beta(p)) \cdot d\varphi_{\alpha\beta}(z_\alpha)/dz_\alpha}{f_\beta(z_\beta(p)) \cdot d\varphi_{\alpha\beta}(z_\alpha)/dz_\alpha}$$
$$= \frac{g_\beta(z_\beta(p))}{f_\beta(z_\beta(p))}. \qquad \square$$

Example 3.14. If $r(z)$ is a rational function on $\mathbf{C}$, then $r(z)dz$ defines a meromorphic differential on the Riemann sphere.

Exercise 3.15. Prove that all meromorphic differentials on the Riemann sphere are of the form $r(z)dz$, where $r(z)$ is a rational function.

Proposition 3.16. $\omega_0 \equiv 0$ *is the only holomorphic differential on the Riemann sphere.*

Proof. Let $\omega = r(z)dz$ be a holomorphic differential on the Riemann sphere. By Exercise 3.15, $r(z)$ is a polynomial. On the other chart of the Riemann sphere, ω is represented as

$$\omega = -\frac{r\left(\frac{1}{z_1}\right)}{z_1^2}dz_1.$$

It follows that $r(1/z_1)/z_1^2$ is also a polynomial, which is possible only for $r \equiv 0$. $\quad\square$

However, using the Stokes theorem, we may show Theorem 3.12 for all Riemann surfaces of genus 0:

Exercise 3.17. Let Γ be a compact Riemann surface of genus 0, and ω its holomorphic differential.

(a) If $P_0 \in \Gamma$ is a fixed point, then

$$f(P) = \int_{P_0}^{P} \omega,$$

is a well-defined holomorphic function on Γ.

(b) Show that $\omega = 0$.

Definition 3.18. Mapping $f : M \to N$ between Riemann surfaces $(M, \{U_i, z_i\})$ and $(N, \{V_j, w_j\})$ is *holomorphic* if each composition $w_j \circ f \circ z_i^{-1}$ is holomorphic on $z_i(U_i \cap f^{-1}(V_j))$.

Exercise 3.19. Prove that each meromorphic function on a Riemann surface can be identified with a holomorphic mapping between the surface and the Riemann sphere.

3.2 Algebraic curves

First examples of algebraic curves

Riemann surfaces are essentially connected with *algebraic curves*, namely sets in the plane given by equations of the form

$$p(x, y) = 0,$$

with p being a polynomial in x, y, and we summarize their basic properties. We start with the examples of curves given by equations of degree less than three.

Lines. The simplest algebraic curves are those given by equations of the form

$$ax + by + c = 0,$$

where a, b, c are given constants, $(a, b) \neq (0, 0)$.

Conics. First examples of algebraic curves, after lines, are those of second order, which are known from antique times as *conic sections*. They are simple enough so that much of their properties may be studied by elementary means, and yet they give rise to such interesting structures, challenging even for modern mathematical tools.

Consider a curve in $\mathbf{R}^2$ given by the equation of the form

$$ax^2 + by^2 + 2cxy + dx + ey + f = 0, \tag{3.1}$$

where a, b, c, d, e, f are real constants, $(a, b, c) \neq (0, 0, 0)$.

It is well known that such a curve may be:

- Empty set, for example: $x^2 + y^2 + 1 = 0$;
- A point, for example: $x^2 + y^2 = 0$;
- Union of two lines. This happens when the polynomial on the left-hand side of equation (3.1) can be reduced to a product of two linear terms:

$$(a_1 x + b_1 y + c_1)(a_2 x + b_2 y + c_2) = 0. \tag{3.2}$$

  Here, we may distinguish the three subcases: two lines may coincide, they may be disjoint or intersect each other at a point;

- An ellipse: in appropriately chosen coordinates, equation (3.1) may be rewritten as

$$\frac{x^2}{\alpha^2} + \frac{y^2}{\beta^2} = 1;$$

- A hyperbola. The equation in the canonical form is

$$\frac{x^2}{\alpha^2} - \frac{y^2}{\beta^2} = 1;$$

- A parabola. The equation may be reduced to

$$y^2 = 2px, \quad p \neq 0.$$

Exercise 3.20. What types of second-order curves exist over the field of complex numbers **C**?

Newton's diverging parabolae. The curves given by equation of the form

$$y^2 = x^3 + ax^2 + bx, \quad a, b \in \mathbf{R}, \tag{3.3}$$

form one class in Newton's classification of cubic curves. This class is divided into five subclasses, depending on roots of the right-hand side of equation (3.3):

- All roots are real and distinct. The curve is composed of two components, an oval and an unbounded part, see Figure 3.1.
- All roots are real, but two smaller ones are equal to each other. The curve is composed of a point and an unbounded part, see Figure 3.2.

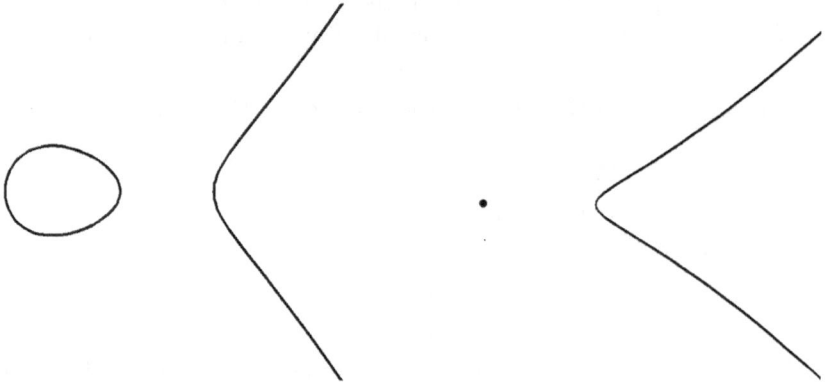

Figure 3.1: Curve $y^2 = x(x-1)(x-2)$ Figure 3.2: Curve $y^2 = x^2(x-1)$

- All roots are real, but two larger ones are equal to each other. The curve has a point of self-intersection, see Figure 3.3.
- All roots are equal: the equation is $y^2 = x^3$. This curve, known as a *semicubical parabola* is composed of two smooth parts, symmetric with respect to the x-axis, and has a *cusp* at the origin, where the two parts meet, see Figure 3.4.

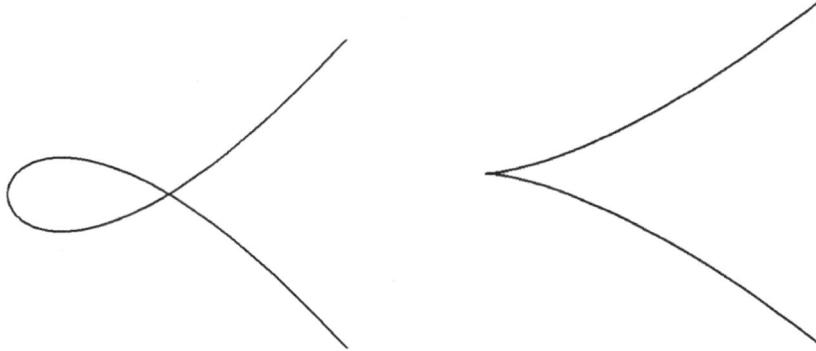

Figure 3.3: Curve $y^2 = x(x-1)^2$ Figure 3.4: Semicubical
                                                   parabola: $y^2 = x^3$

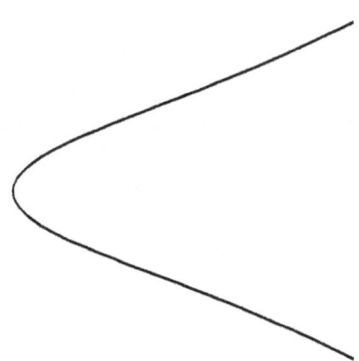

Figure 3.5: Curve $y^2 = x(x^2 + 1)$

- Two roots are imaginary. The curve has only one component, which is smooth and bell-like, see Figure 3.5.

Exercise 3.21. Show that each of the two smooth parts of the semicubical parabola has its x-axis as a tangent at its endpoint.

Exercise 3.22. Consider the Newton parabola over complex numbers, i.e., the set in $\mathbf{C}^2$ given by the equation: $y^2 = P_3(x)$, where P_3 is a third degree polynomial. Show that:

- If the roots of P_3 are distinct, then the Newton parabola is homeomorphic to a torus without one point.
- If P_3 has one double zero, then the Newton parabola is homeomorphic to a pinched sphere without one point.
- If P_3 has a triple zero, then the Newton parabola is homeomorphic to a plane.

Bifolium.   A *bifolium* is a curve of the form

$$(x^2 + y^2)^2 = 4ax^2y, \quad a > 0.$$

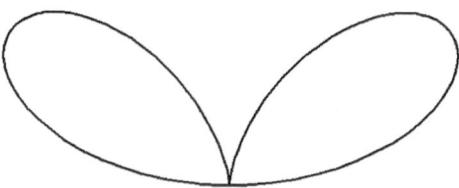

Figure 3.6: Bifolium

Exercise 3.23. Show that, in a neighborhood of the origin, the bifolium is the union of three smooth parts: two of them forming a cusp, and the third one passing through the cusp point and being orthogonal to them. (See Figure 3.6.)

Trifolium.   A *trifolium* is a curve of the form

$$(x^2 + y^2)^2 = a(x^3 - xy^2), \quad a > 0.$$

Figure 3.7: Trifolium

Exercise 3.24. Show that, in a neighborhood of the origin, the trifolium is the union of three smooth parts that intersect each other transversally. (See Figure 3.7.)

Algebraic curves in the complex plane

Definition 3.25. Let $p(x, y)$ be a polynomial with complex coefficients. *The algebraic curve* corresponding to this polynomial is the set of all $(x, y) \in \mathbf{C}^2$ such that $p(x, y) = 0$. The degree of polynomial $p(x, y)$ is then also called *the degree of the algebraic curve.*

Example 3.26. The algebraic curve given by the polynomial $p(x, y) = y^2 - x$ is a Riemann surface: consider the whole curve as the unique chart of a covering and the mapping $(x, y) \mapsto y$ as a local coordinate. The mapping $(x, y) \mapsto x$ is a holomorphic mapping on the curve.

However, on most algebraic curves it is not possible to define a structure of Riemann surface. The reducibility of the polynomial may appear as the first obstacle to the existence of such a structure on a curve. Thus we define:

Definition 3.27. The algebraic curve is *irreducible* if the polynomial $p(x, y)$ is irreducible.

Note that the curves defined by reducible polynomials split into the union of several irreducible ones. Thus, in many cases, it possible to limit oneself to study of irreducible curves only.

Definition 3.28. Point (x_0, y_0) on the algebraic curve given by the polynomial $p(x, y)$ is called *singular* if

$$
\frac{\partial p}{\partial x}(x_0, y_0) = \frac{\partial p}{\partial y}(x_0, y_0) = 0.
$$

Otherwise, we say that this point is *regular* or *nonsingular*.

Exercise 3.29. Prove that an irreducible curve has a finite number of singular points.

Proposition 3.30. *Let $\mathcal{C}$ be an irreducible algebraic curve and Σ the set of its singular points. Then the set $\mathcal{C} \setminus \Sigma$ has a natural structure of a Riemann surface.*

Proof. Let (x_0, y_0) be a regular point of the algebraic curve $p(x, y) = 0$. Suppose that $\frac{\partial p}{\partial y}(x_0, y_0) \neq 0$. By the implicit function theorem, there exists a unique function $y = y(x)$ such that $p(x, y(x)) = 0$, $y(x_0) = y_0$ and $y(x)$ is analytic in some neighbourhood U_{x_0} of x_0. The set $\{(x, y(x)) \mid x \in U_{x_0}\}$, together with the projection $(x, y) \mapsto x$ is a chart on $\mathcal{C}$.

Since the whole $\mathcal{C}$ can be covered with such charts, and it is easy to see that the transition functions are holomorphic, the proof is complete. $\square$

Singular points of algebraic curves

In this section, we are going to study more closely singular points on algebraic curves.

Let P be a point on the algebraic curve $\mathcal{C}$ given by a polynomial $p(x, y)$. Since we may translate P to $(0,0)$, we may suppose that $P = (0,0)$ without losing generality.

Let

$$p(x, y) = p_1(x, y) + p_2(x + y) + \cdots + p_d(x, y),$$

where $p_j(x, y)$ is a homogeneous polynomial of degree j, and $d \geq 1$. It is easy to see that p is singular if and only if $p_1 \equiv 0$.

If p is a regular point, then the tangent to the curve at this point is given by the equation $p_1(x, y) = 0$. Otherwise, let $k > 1$ be the smallest number such that $p_k(x, y) \not\equiv 0$. Then we will say that P is *a k-tuple point* on $\mathcal{C}$. $p_k(x, y)$ is a product of k linear factors, each one determining a tangent line to the curve at p. If all these tangents are distinct, then we say that p is *an ordinary k-tuple point*.

Exercise 3.31. Determine all singular points of the curves listed at the beginning of this section and find the tangents at these points.

Algebraic curves in the complex projective plane

An algebraic curve of degree d and a line in $\mathbf{C}^2$, generally, have d intersecting points, counting multiplicities.

Consider a point $P = (x_0, y_0) \in \mathbf{C}$. All lines containing P can be represented by parametric equations of the form

$$x = x_0 + \alpha t, \quad y = y_0 + \beta t \quad (t \in \mathbf{C}).$$

A line is determined by the pair $(\alpha : \beta) \in \mathbf{CP}^1$, and the intersecting points with the curve correspond to the zeroes of the following polynomial of t:

$$p(x_0 + \alpha t, y_0 + \beta t), \tag{3.4}$$

taking into account the multiplicities of the zeroes.

Notice that, on the right-hand side of the equation, we have a polynomial of t of a degree not greater than d. If the polynomial is of a degree smaller than d, then the line and the curve will also have less than d common points.

Exercise 3.32. Prove that, for each point (x_0, y_0), the following equality holds:

$$\sum (d - \deg p(x_0 + \alpha t, y_0 + \beta t)) = d,$$

where the sum is taken over the set of all lines containing (x_0, y_0). Moreover, each member of the sum, i.e., the degree of the polynomial $p(x_0 + \alpha t, y_0 + \beta t)$, does not depend on point (x_0, y_0), but only on pair (α, β).

The previous exercise shows that it is natural to complete an algebraic curve in the projective plane, where it will have exactly d intersecting points with every line. The d "missing" points from Exercise 3.32 will appear on the infinite line.

Definition 3.33. Let $P(x, y, z)$ be a homogeneous polynomial with complex coefficients. *The algebraic curve* corresponding to this polynomial is the set of all $(x : y : z) \in \mathbf{CP}^2$ such that $P(x, y, z) = 0$. The degree of polynomial $P(x, y, z)$ is then also called *the degree of the algebraic curve.*

Exercise 3.34. Prove that algebraic curves in the projective plane are compact.

The affine part of the curve represented by the equation $P(x, y, z) = 0$ is given by $P(x, y, 1) = 0$. On the other hand, to a curve in $\mathbf{C}^2$ that is given by polynomial $p(x, y)$, we can naturally join the following curve in $\mathbf{CP}^2$:

$$P(x, y, z) = z^d p \left(\frac{x}{z}, \frac{y}{z} \right), \quad d = \deg p(x, y).$$

Thus, projective and affine representations of an irreducible curve are completely equivalent and it is easy to switch between them.

Exercise 3.35. Let $P(x, y, z)$ be a homogeneous polynomial and

$$P(x_0, y_0, z_0) = 0, \quad z_0 \neq 0.$$

If $p(x, y) = P(x, y, 1)$, prove that

$$\frac{\partial P}{\partial x}(x_0, y_0, z_0) = \frac{\partial P}{\partial y}(x_0, y_0, z_0) = \frac{\partial P}{\partial z}(x_0, y_0, z_0) = 0$$

is equivalent to

$$\frac{\partial p}{\partial x} \left(\frac{x_0}{z_0}, \frac{y_0}{z_0} \right) = \frac{\partial p}{\partial y} \left(\frac{x_0}{z_0}, \frac{y_0}{z_0} \right) = 0.$$

Exercise 3.36. Prove that there is a bijection between complex torus $T(\omega_1, \omega_2)$ and a curve of the form $y^2 z = 4x^3 + axz^2 + bz^3$. (See Exercise 3.88.)

3.3 Normalization theorem

We have already seen, in previous sections, that there is a firm connection between Riemann surfaces and algebraic curves: we have demonstrated that all regular points on an algebraic curve form a Riemann surface. On the other side, we proved in Exercise 3.88, that each complex torus can be represented as a third-order curve.

Here, we are going to show that a Riemann surface can be naturally joined to each algebraic curve. Moreover, every Riemann surface is a normalization of a certain curve. Further, we give an effective procedure for resolving singularities of curves.

Normalization Principle

Theorem 3.37 (Normalization Principle). *Let $\mathcal{C}$ be an irreducible algebraic curve and Σ the set of its singular points. Then there exists a compact Riemann surface $\widetilde{\mathcal{C}}$ and a mapping $\sigma : \widetilde{\mathcal{C}} \to \mathbf{CP}^2$ such that $\sigma(\widetilde{\mathcal{C}}) = \mathcal{C}$ and σ is holomorphic and injective on $\sigma^{-1}(\mathcal{C} \setminus \Sigma)$.*

According to the Normalization Principle, we may define:

Definition 3.38. The pair $(\widetilde{\mathcal{C}}, \sigma)$ is *the normalization* of curve $\mathcal{C}$.

It is possible to prove that two normalizations of a given curve are equivalent.

Exercise 3.39. Prove that a Riemann sphere, with a suitable projection σ, is the normalization of the curve $y^2 = x^2(x - a)$. Find the mapping σ.

Exercise 3.40. Prove that the curves $y^2 = P(x)$ and $y^2 = x^2 P(x)$ have the same normalizations.

Theorem 3.37 provides a powerful tool for studying algebraic curves.

Theorem 3.41. *Any compact Riemann surface is a normalization of a certain plane algebraic curve with at most ordinary double points.*

Having in mind Theorems 3.37 and 3.41, we are going to use all notions introduced for Riemann surfaces, also for algebraic curves. For example, if we say that function φ is meromorphic on curve $\mathcal{C}$, we mean that the corresponding function $\tilde{\varphi}$, such that $\varphi = \tilde{\varphi} \circ \sigma$, is meromorphic on the normalization $(\widetilde{\mathcal{C}}, \sigma)$.

Resolution of singularities of curves

In this section, we are going to present an effective procedure of resolution of the singularities of a plane algebraic curve.

Definition 3.42. *The blow-up of $\mathbf{C}^2$ at point $(0,0)$ is a subset on $\mathbf{C}^2 \times \mathbf{CP}^2$ given by the equation: $xz_1 = yz_2$, where (x,y) are coordinates in $\mathbf{C}^2$ and $(z_1 : z_2)$ in $\mathbf{CP}^1$.*

Denote by X the blow-up of $\mathbf{C}^2$ at $(0,0)$, and let $\varphi : X \to \mathbf{C}^2$ be the restriction of the projection $\mathbf{C}^2 \times \mathbf{CP}^1 \to \mathbf{C}^2$. It is easy to see that $\varphi^{-1}(P)$ contains exactly one point, for each $P \neq (0,0)$. On the other hand $\varphi^{-1}(0,0)$ is a line. In fact, each point in $\varphi^{-1}(0,0)$ corresponds to a line in $\mathbf{C}^2$ passing through $(0,0)$.

Definition 3.43. *Let $\mathcal{C}$ be an algebraic curve in $\mathbf{C}^2$, $\mathcal{C} \ni (0,0)$. The blow-up of $\mathcal{C}$ at $(0,0)$ is the closure $\overline{\varphi^{-1}(\mathcal{C} \setminus \{(0,0)\})}$ in X.*

Exercise 3.44. Find blow-ups at singular points of curves listed at the beginning of Section 3.2.

The resolution of a singularity of a plane curve, a so-called σ-process, is a sequence of blow-ups at singular points. At each singular point, we make as many blow-ups as needed, until we obtain a regular curve. Notice that this procedure is always finite.

3.4 Further properties of Riemann surfaces

In this section, we are going to present most important properties of Riemann surfaces. Some of them will be proved for the general case, for some we are going to postpone the proof for the next section, where we will show them only for some special classes of Riemann surfaces.

Divisors

Definition 3.45. A *divisor* on Riemann surface Γ is a finite sum of the form

$$D = m_1 P_1 + \cdots + m_k P_k,$$

where $m_1, \ldots, m_k \in \mathbf{Z}$, and $P_1, \ldots, P_k$ are points on Γ. *The degree* of divisor D is

$$\deg D = m_1 + \cdots + m_k.$$

In the set of all divisors on the surface, the operation of addition can be defined in the obvious way. In this way, all divisors form an Abelian group. This group is, in fact, the free Abelian group generated by the points of Γ.

Next, we show that there is a natural way to join a divisor to each meromorphic function of the Riemann surface.

Let f be a non-zero meromorphic function on the Riemann surface Γ, $P \in \Gamma$ an arbitrary point, and z a local coordinate around P. Then, in a neighbourhood of P, function f may be represented as

$$f = (z - z_0)^\nu h(z),$$

where h is holomorphic, $z_0 = z(P)$, $h(z_0) \neq 0$, $\nu \in \mathbf{Z}$.

Exercise 3.46. Prove that the integer ν does not depend on the choice of the local coordinate z.

Thus, we write $\nu = \nu_P(f)$.

Definition 3.47. The value $\nu_P(f)$ is called *the order* or *the multiplicity* of f at point P. If $\nu_P(f) > 0$, then P is called *a zero of f of order $\nu_P(f)$*. If $\nu_P(f) < 0$, then P is called *a pole of f* and $|\nu_P(f)|$ *the order of the pole P*.

Definition 3.48. *The divisor of a meromorphic function f on Γ is*

$$(f) = \sum_{P \in \Gamma} \nu_P(f)P.$$

Exercise 3.49. If Γ is either the Riemann sphere or a complex torus, prove that

$$\deg(f) = 0.$$

Then show that the same holds for any compact Riemann surface.

Exercise 3.50. Let f, g be two meromorphic functions on Riemann surface Γ. Prove that

$$(fg) = (f) + (g), \quad \left(\frac{1}{f}\right) = -(f).$$

In a similar way, we may define divisors of meromorphic differentials.

Let ω be a meromorphic differential on Γ, $P \in \Gamma$ a point, z a local coordinate around P, and f a function defined in a neighbourhood U of P such that $\omega = f(z)dz$ in U. We write $\nu = \nu_P(f)$.

Exercise 3.51. Prove that the integer ν does not depend on the choice of the local coordinate z.

Thus, we define

$$\nu_P(\omega) = \nu,$$

and refer to it as to *the order of the differential ω at P*. Zeroes end poles of a differential, as well as their orders now can be introduced exactly in the same way as they are defined for functions. *The divisor of a meromorphic differential ω on Γ is*

$$(\omega) = \sum_{P \in \Gamma} \nu_P(\omega)P.$$

Exercise 3.52. Show that

$$(f\omega) = (f) + (\omega), \quad \left(\frac{\omega_1}{\omega_2}\right) = (\omega_1) - (\omega_2).$$

Exercise 3.53. Prove that

- $\deg(\omega) = -2$ for each meromorphic differential on the Riemann sphere.
- $\deg(\omega) = 0$ for each meromorphic differential on a torus.

Definition 3.54. We say that divisors D, E on a Riemann surface are *equivalent*, and denote by $D \sim E$ if there is a meromorphic function f such that $(f) = D - E$.

Exercise 3.55.

- On a compact Riemann surface, equivalent divisors are of equal degree.
- If ω_1, ω_2 are meromorphic differential on a Riemann surface, prove that $(\omega_1) \sim (\omega_2)$.
- The divisor of a meromorphic function f is equivalent to 0.

Ramification index and Riemann–Hurwitz formula

The previous exercise shows that the number of zeros of a meromorphic function on a compact Riemann surface is equal to the number of poles, counting their multiplicities.

Definition 3.56. Suppose that Γ, Γ' are Riemann surfaces with coverings $\{U_i, z_i\}$, $\{(U'_\alpha, z'_\alpha)\}$. The mapping $f : \Gamma \to \Gamma'$ is *holomorphic* if $z_\alpha \circ f \circ z_i^{-1}$ is holomorphic on $f(U'_\alpha) \cap U_i$ whenever $f^{-1}(U'_\alpha) \cap U_i \neq \emptyset$.

Remark 3.57. Each meromorphic function on a Riemann surface Γ is a holomorphic mapping from Γ to the Riemann sphere.

Exercise 3.58. Find all holomorphic involutions of the Riemann sphere and a torus.

Let $f : \Gamma \to \Gamma'$ be a holomorphic mapping, such that $f(P) = Q$. There exist local coordinates z, w around P, Q respectively, such that $z(P) = 0$, $w(Q) = 0$ and f can be represented as $w = z^\mu$ in these coordinates, $\mu \in \mathbf{Z}^+$.

Exercise 3.59. Prove the coordinates z, w exist and that μ does not depend on their choice.

Definition 3.60. The constant μ is called *the ramification index of f at point P*. We denote it $\nu_f(P)$.

Exercise 3.61. Let Γ, Γ' be compact Riemann surfaces, $f : \Gamma \to \Gamma'$ a non-constant holomorphic function. Prove that the sum: $\sum_{f(P)=Q} \nu_f(P)$ does not depend on point $q \in \Gamma'$.

Definition 3.62. *The degree* of a holomorphic mapping $f : \Gamma \to \Gamma'$ between compact Riemann surfaces Γ, Γ', $f \not\equiv$ const is

$$\deg f = \sum_{f(P)=Q} \nu_f(P).$$

Exercise 3.63. Find the degree of the Weierstrass function.

Theorem 3.64 (Riemann–Hurwitz formula). *Let $f : \Gamma \to \Gamma'$ be a non-constant holomorphic mapping between compact Riemann surfaces Γ, Γ'. Denote the ramification divisor by*

$$R = \sum_{P \in \Gamma} (\nu_f(P) - 1)P.$$

Then

$$\deg R = 2(g + n - ng' - 1),$$

where $n = \deg f$ and g, g' are genera of surfaces Γ, Γ'.

Riemann–Roch theorem

Let Γ be a compact Riemann surface, and

$$D = m_1 P_1 + \cdots + m_k P_k$$

a divisor on Γ.

Definition 3.65. We say that D is *an effective divisor* if $m_1 \geq 0,\ \ldots,\ m_k \geq 0$, and denote it by $D \geq 0$. For two divisors, D and E, we write $D \geq E$ if $D - E \geq 0$.

To any divisor D on Γ, we may join two important sets of meromorphic functions and meromorphic divisors:

Definition 3.66. For a divisor D on Γ, we define

$$\mathcal{L}(D) = \{f - \text{meromorphic function on } \Gamma \mid (f) + D \geq 0\},$$
$$\Omega(D) = \{\omega - \text{meromorphic differential on } \Gamma \mid (\omega) \geq D\}.$$

It is easy to see that $\mathcal{L}(D)$ contains exactly meromorphic functions that have

- a pole of order at most m_j at point P_j if $m_j > 0$,
- a zero of order at least $|m_j|$ at point P_j if $m_j < 0$,

while $\Omega(D)$ contains the differentials having

- P_j as a zero of order at least m_j if $m_j > 0$,
- P_j as a pole of order at most $|m_j|$ if $m_j < 0$,

for each $j = 1, \ldots, k$.

Exercise 3.67.

(a) Show that $\mathcal{L}(D)$ and $\Omega(D)$ are linear spaces over $\mathbf{C}$.
(b) If $D \geq E$, then $\mathcal{L}(D) \supset \mathcal{L}(E)$ and $\Omega(D) \subset \Omega(E)$.
(c) If $\deg(D - E) = 1$ and Γ is compact, than the quotient spaces $\mathcal{L}(D)/\mathcal{L}(E)$ and $\Omega(E)/\Omega(D)$ are of dimension at most 1.

Now, we are ready to formulate the Riemann–Roch theorem.

Theorem 3.68 (Riemann–Roch). *Let Γ be a compact Riemann surface of genus g, and D its divisor. Then*

$$\dim \mathcal{L}(D) = \deg D - g + \dim \Omega(D) + 1.$$

We will prove the Riemann–Roch theorem in the next section, only for hyperelliptic surfaces. In the meantime, we demonstrate some applications of this strong and important statement.

Exercise 3.69. Using the Riemann–Roch theorem, derive Theorem 3.12.

Proposition 3.70. *Each compact Riemann surface of genus* 0 *is a Riemann sphere.*

Proof. Choose an arbitrary point P on the surface. Since, by Exercise 3.17, the zero differential is the only holomorphic differential on Γ, we obtain that $\dim \Omega(P) = 0$, thus, by the Riemann–Roch theorem $\dim \mathcal{L}(P) = 2$. As a basis of $\mathcal{L}(P)$, choose $\{1, f\}$, where f is a function having P as a unique and simple pole. According to Remark 3.57, f is an isomorphism between Γ and the Riemann sphere. $\qquad\square$

Let us finish this subsection with the important notion of *special divisors*.

Definition 3.71. A divisor D is called *special* if $\dim \Omega(D) > 0$. If $\dim \Omega(D) = 0$, the divisor is *nonspecial*.

Exercise 3.72. Describe all special divisors on the Riemann sphere and a torus.

Bézout theorem

We have already mentioned, in Section 3.2, that in the complex projective plane, an algebraic curve of degree d and a line have exactly d intersecting points, counting multiplicities. In other words, two curves of degrees d and 1 respectively, have $d \cdot 1$ intersections. The Bézout theorem is a generalization of this property to the pairs of curves of arbitrary degree in the plane.

Before we formulate the theorem, we need to define the multiplicity of intersection of two curves.

Let $\mathcal{C}$, $\mathcal{D}$ be two algebraic curves in the complex projective plane, such that they do not have common components. Suppose that P is their intersecting point. We may introduce an affine coordinate frame, such that the coordinates of P in this frame are $(0, 0)$. Denote by $p(x, y)$, $q(x, y)$ polynomials such that the corresponding equations of the curves are

$$\mathcal{C} \; : \; p(x, y) = 0, \quad \mathcal{D} \; : \; q(x, y) = 0.$$

Let us first consider the case when P is a regular point of $\mathcal{C}$. Let t be a holomorphic local coordinate in a neighborhood $U \subset \mathcal{C}$ of P, such that $t(P) = 0$. Remark that we may take $t = x$ or $t = y$, depending on whether $\partial p / \partial y \neq 0$ or $\partial p / \partial x \neq 0$ in point P.

Polynomial q, restricted to U, is a holomorphic function, having P as a zero. We define *the intersection number* of curves $\mathcal{C}$ and $\mathcal{D}$ at P as a multiplicity of this zero:

$$(\mathcal{C} \cdot \mathcal{D})_P = \nu_P\big(q(x(t), y(t))\big).$$

Exercise 3.73. Let P be a regular point of both curves $\mathcal{C}$ and $\mathcal{D}$. Prove that $(\mathcal{C} \cdot \mathcal{D})_P = (\mathcal{D} \cdot \mathcal{C})_P$.

Exercise 3.74. Let ℓ be a line containing point $P = (0, 0)$. Calculate $(\ell \cdot \mathcal{C})_P$, if $\mathcal{C}$ is

- a semicubical parabola $y^2 = x^3$,

- a bifolium $(x^2 + y^2)^2 = 4ax^2y$, $a > 0$,
- a trifolium $(x^2 + y^2)^2 = a(x^3 - xy^2)$, $a > 0$.

Now, consider the case when P is a singular point of $\mathcal{C}$. Let $(\widetilde{\mathcal{C}}, \sigma)$ be the normalization of $\mathcal{C}$. The composition $q^* = q \circ \sigma$ is a meromorphic function. Each preimage of P on $\widetilde{\mathcal{C}}$ is, obviously, a zero of q^*, and it is possible to define the intersection number as the sum of multiplicities of q^* at these points:

$$(\mathcal{C} \cdot \mathcal{D})_P = \sum_{\widetilde{P} \in \sigma^{-1}(P)} \nu_{\widetilde{P}}(q^*).$$

Exercise 3.75. Prove that $(\mathcal{C} \cdot \mathcal{D})_P = (\mathcal{D} \cdot \mathcal{C})_P$.

Exercise 3.76. Find the intersection number at $(0,0)$ of each two of the following curves: semicubical parabola, bifolium and trifolium.

Having defined the intersection number, we are ready to state

Theorem 3.77 (Bézout Theorem). *Suppose that two plane algebraic curves $\mathcal{C}$ and $\mathcal{D}$ do not have common components. Then*

$$\sum_{P \in \mathcal{C} \cap \mathcal{D}} (C \cdot D)_P = \deg \mathcal{C} \cdot \deg \mathcal{D}.$$

3.5   Complex tori and elliptic functions

In this Section, we are going to present the most important properties of an extremely significant class of Riemann surfaces, so-called *elliptic curves*.

Consider a *lattice* in the complex plane, i.e., a set of the form

$$\Lambda = \{2m\omega_1 + 2n\omega_2 \mid m, n \in \mathbf{Z}\},$$

where ω_1, ω_2 are complex numbers such that

$$\operatorname{Im} \frac{\omega_1}{\omega_2} \notin \mathbf{R}.$$

Since Λ is an additive subgroup of $\mathbf{R}^2$, the set

$$\mathbf{T}(\omega_1, \omega_2) = \mathbf{R}^2/\Lambda$$

has a natural structure of a compact two-dimensional manifold. This smooth manifold is diffeomorphic to the torus $\mathbf{S}^1 \times \mathbf{S}^1 = \mathbf{R}/\mathbf{Z} \times \mathbf{R}/\mathbf{Z}$.

Meromorphic functions on the torus $\mathbf{T}(\omega_1, \omega_2)$ are doubly periodic meromorphic functions on $\mathbf{C}$, with periods $2\omega_1$, $2\omega_2$. Such functions are called *elliptic functions*.

Exercise 3.78. Prove that $\omega = dz$ is a holomorphic differential on the torus.

Exercise 3.79 (Theorem 3.12 for tori). Prove that each holomorphic differential of the torus is of the form $c\,dz$, where $c \in \mathbf{C}$.

Example 3.80. Let f be an elliptic function. Then $f(z)dz$ is a meromorphic differential on the torus.

Definition 3.81. *The Weierstrass function $\wp$ is the elliptic function with periods $2\omega_1$, $2\omega_2$ all of whose poles are on Λ and its Laurent expansion in a neighbourhood of 0 is*

$$\wp(z) = \frac{1}{z^2} + zh(z),$$

where h is holomorphic.

The difference of two functions satisfying Definition 3.81 is a holomorphic doubly-periodic function, thus it is constant and, moreover, equal to 0, since it is zero at the lattice points. This proves the uniqueness of the Weierstrass function. The existence follows from the following

Exercise 3.82. Prove that the series

$$\frac{1}{z^2} + \sum_{\lambda \in \Lambda \setminus \{0\}} \left(\frac{1}{(z-\lambda)^2} - \frac{1}{\lambda^2} \right)$$

converges uniformly on compact sets in $\mathbf{C} \setminus \Lambda$, and that it satisfies the definition of the Weierstrass function.

Proposition 3.83. *Prove that: $\wp(-z) = \wp(z)$ and $\wp'(-z) = -\wp'(z)$ for all $z \in \mathbf{C}$.*

Proof. All poles of function $\wp(-z)$ also lie on Λ. Since its Laurent expansion around 0 is of the form $\wp(-z) = \frac{1}{z^2} - zh(-z)$, it follows that $\wp(z) - \wp(-z)$ is holomorphic and thus constant on the torus. Moreover, since

$$(\wp(z) - \wp(-z))\,|_{z=0} = 0,$$

it follows that $\wp(-z) \equiv \wp(z)$. By differentiating, we get $\wp'(-z) = -\wp'(z)$. $\square$

Exercise 3.84. Prove that all solutions of the equation $\wp'(z) = 0$ are the half-periods of the lattice: $\{\omega_1, \omega_2, \omega_1 + \omega_2\}$.

Elliptic tori and cubic curves

Proposition 3.85. *Prove that the Weierstrass function satisfies the following differential equation:*

$$(\wp')^2 = 4\wp^3 + a\wp + b, \tag{3.5}$$

for some $a, b \in \mathbf{C}$.

Proof. From the definition of Weierstrass function and Proposition 3.83, we have that functions $\wp$, $\wp'$ have Laurent expansions around $z = 0$ of the following form:

$$\wp(z) = \frac{1}{z^2} + \alpha z^2 + \beta z^4 + \cdots,$$

$$\wp'(z) = -\frac{2}{z^3} + 2\alpha z + 4\beta z^3 + \cdots.$$

From there, it is easy to calculate:

$$(\wp'(z))^2 - 4\wp^3(z) = \frac{a}{z^2} + b + \cdots,$$

for some constants a, b. It follows that

$$(\wp')^2 - 4\wp^3 - a\wp - b$$

is a doubly periodic holomorphic function and thus constant. Since it is equal to 0 for $z = 0$, equation (3.5) is satisfied. $\qquad\square$

We have just proved that the Weierstrass function $\wp$ and its derivative $\wp'$ satisfy a simple algebraic relation (3.5). This is the motive to study in more detail curves of the form $y^2 = 4x^3 + ax + b$. Introduction to algebraic curves in general is given in Section 3.2.

Exercise 3.86. Prove that polynomial $4x^3 + ax + b$ has 3 distinct zeroes, where a, b are as in Proposition 3.85.

Exercise 3.87. Prove that

$$a = -60 \sum_{\lambda \in \Lambda \setminus \{0\}} \frac{1}{\lambda^4}, \quad b = -140 \sum_{\lambda \in \Lambda \setminus \{0\}} \frac{1}{\lambda^6}.$$

Let us remark that Proposition 3.85 implies that we may map torus $\mathbf{T}(\omega_1, \omega_2)$ to cubic curve $y^2 = 4x^3 + ax + b$, by $z \mapsto (\wp(z), \wp'(z))$. Note that the curve can be completed in $\mathbf{CP}^2$ by adding one point on the infinite line, that is the point with the homogeneous coordinates $[x : y : z] = [0 : 1 : 0]$. When this is done, the mapping $z \mapsto (\wp(z), \wp'(z))$ is naturally extended to the whole torus: point 0 is mapped to the infinite point $[0 : 1 : 0]$ of the completed cubic.

Theorem 3.88. *The mapping* $z \mapsto (\wp(z), \wp'(z))$ *is a bijection between* $\mathbf{T}(\omega_1, \omega_2)$ *and the Riemann surface* $\mathcal{C}$ *of the algebraic function* $y^2 = 4x^3 + ax + b$.

Proof. Let us construct the inverse mapping. For point $P \in \mathcal{C}$, we define

$$\mathcal{A}(P) = \int_\infty^P \frac{dx}{y} = \int_\infty^P \frac{dx}{\sqrt{4x^3 + ax + b}}.$$

Although the value of $\mathcal{A}(P)$ depends on the integration path, it is determined up to the sum of the form

$$m \oint_{c_1} \frac{dx}{y} + n \oint_{c_2} \frac{dx}{y}, \quad m, n \in \mathbf{Z}.$$

Here, c_1, c_2 are basis cycles on the surface $\mathcal{C}$, as shown on Figure 3.8.

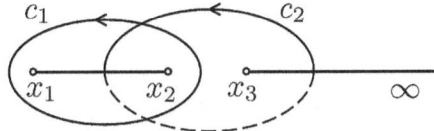

Figure 3.8: Riemann surface of function $y^2 = 4x^3 + ax + b$. x_1, x_2, x_3 are zeroes of the polynomial $4x^3 + ax + b$, and c_1, c_2 are basis cycles.

We may calculate:

$$\oint_{c_1} \frac{dx}{y} = 2 \int_{x_1}^{x_2} \frac{dx}{y}$$
$$= 2 \int_{\mathcal{A}(x_1,0)}^{\mathcal{A}(x_2,0)} \frac{d\wp(z)}{\wp'(z)}$$
$$= 2 \int_{\mathcal{A}(x_1,0)}^{\mathcal{A}(x_2,0)} dz$$
$$= 2 \left(\mathcal{A}(x_1,0) - \mathcal{A}(x_2,0)\right)$$
$$= 2 \left((\omega_1 + \omega_2) - \omega_1\right)$$
$$= 2\omega_1.$$

Similarly,

$$\oint_{c_2} \frac{dx}{y} = 2\omega_2.$$

Thus, the mapping $\mathcal{A}$ from curve $\mathcal{C}$ to torus $\mathbf{T}(\omega_1, \omega_2)$ is well defined. Moreover, it is inverse to the mapping $z \mapsto (\wp(z), \wp'(z))$, which proves the theorem. $\square$

Definition 3.89. The mapping

$$\mathcal{A} \; : \; \mathcal{C} \to \mathbf{T}(\omega_1, \omega_2)$$

is called the *Abel mapping*.

Abelian group structure on cubic curve

We may naturally introduce the following operations on the cubic curve $y^2 = 4x^3 + ax + b$. For points P, Q on the curve, consider the line PQ. This line has exactly one more intersecting point with the curve – denote it by $P \circ Q$. If $P = Q$, then the line is the tangent to the curve.

The curve also has a natural involution τ which is the symmetry with respect to the x-axis.

Now, we define addition as: $P + Q := \tau(P \circ Q)$, see Figure 3.9.

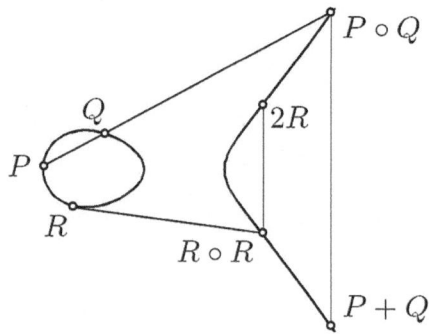

Figure 3.9: Addition on the cubic curve.

Exercise 3.90. Prove that the operation $+$ has the following properties:

- $P + Q = Q + P$;
- the infinite point of the curve is the neutral element;
- P and $\tau(P)$ are inverse to each other.

In order to prove that the cubic with the operation $+$ is an Abelian group, we need to prove that the associative law is satisfied:

$$P + (Q + R) = (P + Q) + R.$$

In order to do this we will need the following

Lemma 3.91. *If three cubic curves in the plane* $\mathbf{CP}^2$ *have eight common points, then they have one more common point.*

Proof. Denote the cubic curves by $\mathcal{C}_1, \mathcal{C}_2, \mathcal{C}_3$. Notice that the equation of a general plane cubic is the sum of 10 terms: x^3, y^3, x^2y, xy^2, x^2, y^2, xy, x, y, 1, i.e., such an equation is given by 10 coefficients.

By the Bézout theorem, which has been formulated in Section 3.4 as Theorem 3.77, two cubics in the plane have always $3 \cdot 3 = 9$ intersection points.

Since $\mathcal{C}_3$ contains eight points common to $\mathcal{C}_1$ and $\mathcal{C}_2$, and having in mind that two proportional equations determine the same curve, it follows that the equation of $\mathcal{C}_3$ is a linear combination of the equations for $\mathcal{C}_1$ and $\mathcal{C}_2$. Thus, $\mathcal{C}_3$ contains all nine intersection points.   □

Theorem 3.92. *The cubic curve $y^2 = 4x^3 + ax + b$ with the operation $P + Q = \tau(P \circ Q)$ is an Abelian group.*

Proof. Consider points P, Q, R on the curve. We need to prove that $P + (Q + R) = (P + Q) + R$.

Consider the following two degenerated cubic curves, each one being a union of three lines: $(P, Q) \cup (Q \circ R, Q + R) \cup (R, P + Q)$ and $(Q, R) \cup (P \circ Q, P + Q) \cup (P, Q + R)$.

These two curves and the cubic $y^2 = 4x^3 + ax + b$ have eight common points: P, Q, R, $P \circ Q$, $Q \circ R$, $Q + P$, $Q + R$ and the infinite point. By Lemma 3.91, they also have the nineth common point, that is $P \circ (Q + R) = (P + Q) \circ R$. $\square$

Recall that torus $\mathbf{C}/\Lambda$ has a natural structure of Abelian group inherited from complex numbers. Now, we are going to prove that the Abel map preserves the group structure.

Theorem 3.93 (Abel theorem for elliptic curves). *The Abel mapping is an isomorphisam of the Abelian groups defined on $\mathbf{C}/\Lambda$ and $\mathcal{C}$.*

Proof. Consider the mapping from $\mathbf{CP}^{2*}$ to $\mathbf{T}^2$, that maps each line ℓ of the plane $\mathbf{CP}^2$ to the sum $\mathcal{A}(P) + \mathcal{A}(Q) + \mathcal{A}(R)$ on $\mathbf{T}^2$, where P, Q, R are intersection points of ℓ with the cubic. Since $\mathbf{CP}^{2*}$ is simply connected, this mapping can be lifted up to the universal covering $\mathbf{C}$ of $\mathbf{T}^2 = \mathbf{C}/\Lambda$.

Since this mapping is holomorphic, and $\mathbf{CP}^{2*}$ is compact, it follows that it is constant. Now, consider the infinite line ℓ_∞ of $\mathbf{CP}^2$. It has a triple intersection with the cubic: the infinite point $[0 : 1 : 0]$. Since this point is the image of $0 \in \mathbf{C}$, it follows that each line of the plane is mapped to zero.

Since, on the cubic curve $P + Q + R = 0$ if and only if points P, Q, R are collinear, it follows that the Abelian groups defined on the torus and on the corresponding cubic curve are isomorphic. $\square$

Addition theorem for the Weierstrass function

Theorem 3.94 (Addition theorem). *The Weierstrass function satisfies the following addition relation:*

$$\wp(z + \zeta) + \wp(z) + \wp(\zeta) = \frac{1}{4} \left(\frac{\wp'(z) - \wp'(\zeta)}{\wp(z) - \wp(\zeta)} \right)^2.$$

Proof. By the Abel theorem (Theorem 3.93), points

$$(\wp(-z), \wp'(-z)), \quad (\wp(-\zeta), \wp'(-\zeta)), \quad (\wp(z + \zeta), \wp'(z + \zeta))$$

are intersection of a line with the cubic. Thus, they are solutions of the following system:

$$\det \begin{pmatrix} 1 & \wp(z) & -\wp'(z) \\ 1 & \wp(\zeta) & -\wp'(\zeta) \\ 1 & x & y \end{pmatrix} = 0, \quad y^2 = 4x^3 + ax + b.$$

If we eliminate y from the system, we get the polynomial

$$4x^3 - \left(\frac{\wp'(z) - \wp'(\zeta)}{\wp(z) - \wp(\zeta)} \right)^2 x^2 + a'x + b' = 0,$$

having $\wp(z)$, $\wp(\zeta)$, $\wp(z + \zeta)$ as the roots. Now, the addition formula follows from Viéte's formulae.   □

Elliptic integrals and Jacobi functions

Besides the Weierstrass function, there are other important elliptic functions. Here, we give a brief account of Jacobi elliptic functions. For a more extensive exposition, see, for example [Akh1970, WW1990].

An *elliptic integral* is an integral of the form

$$\int R(w, \sqrt{P(w)})\,dw,$$

where R is a rational function, and P a polynomial of degree 3 or 4. In special cases, such an integral can be expressed via elementary functions.

Consider the integral

$$z = \int_0^w \frac{dw}{(1 - w^2)(1 - k^2w^2)}, \tag{3.6}$$

where k is a real constant, $|k| \leq 1$. The function inverse to this integral,

$$w = \operatorname{sn}(z) = \operatorname{sn}(z; k),$$

is called an *elliptic sine-function*, and k is an *elliptic modulus*.

Exercise 3.95. Prove that function sn is doubly periodic with periods:

$$4 \int_0^1 \frac{dt}{\sqrt{(1 - t^2)(1 - k^2t^2)}} \quad \text{and} \quad 2i \int_1^{1/k} \frac{dt}{\sqrt{(t^2 - 1)(1 - k^2t^2)}}.$$

Exercise 3.96. Prove that sn is an odd function:

$$\operatorname{sn}(-z) = -\operatorname{sn}(z), \quad \text{for all } z.$$

We introduce also

$$\operatorname{cn} z = \sqrt{1 - \operatorname{sn}^2 z}, \quad \operatorname{dn} z = \sqrt{1 - k^2 \operatorname{sn}^2 z}.$$

Exercise 3.97. Prove that elliptic Jacobi functions satisfy the following relations:

$$\operatorname{sn}^2 z + \operatorname{cn}^2 z = 1, \quad k^2 \operatorname{sn}^2 z + \operatorname{dn}^2 z = 1.$$

Now, let us derive the formulae for differentiation of an elliptic function.

Proposition 3.98. *It holds that*

$$\frac{d\operatorname{sn} z}{dz} = \operatorname{cn} z \cdot \operatorname{dn} z, \quad \frac{d\operatorname{cn} z}{dz} = -\operatorname{sn} z \cdot \operatorname{dn} z, \quad \frac{d\operatorname{dn} z}{dz} = -k^2 \cdot \operatorname{sn} z \cdot \operatorname{cn} z.$$

Proof. From (3.6),

$$\frac{dw}{dz} = \sqrt{(1 - w^2)(1 - k^2 w^2)}, \quad w = \operatorname{sn} z$$

and the first formula immediately follows. We obtain the other ones by differentiating the identities from Exercise 3.97. □

Exercise 3.99. Prove that functions $w = \operatorname{cn} z$ and $w = \operatorname{dn} z$ are inverse to the integrals

$$z = \int_1^w \frac{dw}{\sqrt{(1 - w^2)(k'^2 + k^2 w^2)}} \quad \text{and} \quad z = \int_1^w \frac{dw}{\sqrt{(1 - w^2)(w^2 - k'^2)}}.$$

Theorem 3.100 (Addition theorem for elliptic Jacobi functions). *Elliptic Jacobi functions satisfy the following addition rules:*

$$\operatorname{sn}(z + \zeta) = \frac{\operatorname{sn} z \operatorname{cn} \zeta \operatorname{dn} \zeta + \operatorname{sn} \zeta \operatorname{cn} z \operatorname{dn} z}{1 - k^2 \operatorname{sn}^2 z \operatorname{sn}^2 \zeta},$$

$$\operatorname{cn}(z + \zeta) = \frac{\operatorname{cn} z \operatorname{cn} \zeta - \operatorname{sn} z \operatorname{sn} \zeta \operatorname{dn} z \operatorname{dn} \zeta}{1 - k^2 \operatorname{sn}^2 z \operatorname{sn}^2 \zeta},$$

$$\operatorname{dn}(z + \zeta) = \frac{\operatorname{dn} z \operatorname{dn} \zeta - \operatorname{sn} z \operatorname{sn} \zeta \operatorname{cn} z \operatorname{cn} \zeta}{1 - k^2 \operatorname{sn}^2 z \operatorname{sn}^2 \zeta}.$$

Proof. Consider the differential equation

$$\frac{dw}{\sqrt{(1 - w^2)(1 - k^2 w^2)}} + \frac{dv}{\sqrt{(1 - v^2)(1 - k^2 v^2)}} = 0. \tag{3.7}$$

Immediately we see that its integral is

$$\int_0^w \frac{dw}{\sqrt{(1 - w^2)(1 - k^2 w^2)}} + \int_0^v \frac{dv}{\sqrt{(1 - v^2)(1 - k^2 v^2)}} = C. \tag{3.8}$$

To get another integral, consider the auxiliary system

$$\frac{dw}{du} = \sqrt{(1-w^2)(1-k^2w^2)},$$

$$\frac{dv}{du} = -\sqrt{(1-v^2)(1-k^2v^2)}.$$

We have

$$\frac{\frac{d}{du}\left(v\frac{dw}{du} - w\frac{dv}{du}\right)}{v\frac{dw}{du} - w\frac{dv}{du}} = -\frac{2k^2wv\left(v\frac{dw}{du} + w\frac{dv}{du}\right)}{1 - k^2w^2v^2},$$

or

$$\frac{d}{du}\ln\left(v\frac{dw}{du} - w\frac{dv}{du}\right) = \frac{d}{du}\ln(1 - k^2w^2v^2),$$

i.e.,

$$v\frac{dw}{du} - w\frac{dv}{du} = \text{const} \cdot (1 - k^2w^2v^2).$$

From there we finally get another integral of (3.7):

$$\frac{w\sqrt{(1-v^2)(1-k^2v^2)} + v\sqrt{(1-w^2)(1-k^2w^2)}}{1 - k^2w^2v^2} = C_1. \qquad (3.9)$$

Now, letting $w = \operatorname{sn} z$, $v = \operatorname{sn} \zeta$, from the integrals (3.8) and (3.9), we get

$$z + \zeta = C, \qquad \frac{\operatorname{sn} z \operatorname{cn} \zeta \operatorname{dn} \zeta + \operatorname{sn} \zeta \operatorname{cn} z \operatorname{dn} z}{1 - k^2 \operatorname{sn}^2 z \operatorname{sn}^2 \zeta} = C_1,$$

C, C_1 being constants. By the uniqueness theorem, $C_1 = \varphi(C) = \varphi(z + \zeta)$. If we take $\zeta = 0$, we get $\varphi(z) = \operatorname{sn} z$, i.e.,

$$\operatorname{sn}(z + \zeta) = \frac{\operatorname{sn} z \operatorname{cn} \zeta \operatorname{dn} \zeta + \operatorname{sn} \zeta \operatorname{cn} z \operatorname{dn} z}{1 - k^2 \operatorname{sn}^2 z \operatorname{sn}^2 \zeta}.$$

Addition theorems for the other Jacobi functions can be analogously derived. $\square$

Cubic curves

We have already seen that cubic curves of the form

$$y^2 = 4x^3 + ax + b$$

are substantially connected with complex tori. This section is devoted to some important properties of cubic curves.

Definition 3.101. The *flex* of a cubic curve is its non-singular point which is a triple intersection of the curve with its tangent.

Exercise 3.102. Let $y^2 = 4x^3 + ax + b$ be a non-singular curve. Show the following:

- Point p of the curve is its flex if and only if $3\mathcal{A}(p) = 0$ on the Jacobian $\mathbf{C}/\Lambda$ of the curve.
- The curve has exactly nine flexes.
- A line that contains two flexes of the curve, contains a third one.

Lemma 3.103. *Let cubic curve $\mathcal{C}$ in $\mathbf{CP}^2$ be given by the equation*

$$P(x, y, z) = 0,$$

and

$$H(x, y, z) = \det \begin{pmatrix} \frac{\partial^2 P}{\partial x^2} & \frac{\partial^2 P}{\partial x \partial y} & \frac{\partial^2 P}{\partial x \partial z} \\ \frac{\partial^2 P}{\partial x \partial y} & \frac{\partial^2 P}{\partial y^2} & \frac{\partial^2 P}{\partial y \partial z} \\ \frac{\partial^2 P}{\partial x \partial z} & \frac{\partial^2 P}{\partial y \partial z} & \frac{\partial^2 P}{\partial z^2} \end{pmatrix}$$

its Hessian. Then flex points p of $\mathcal{C}$ are exactly solutions of the system $P(p) = H(p) = 0$.

Proof. First, let us note that the solutions of the system $P(p) = H(p) = 0$ do not depend on the choice of the coordinate system. Thus, let us choose the coordinate system where the chosen point $p \in \mathcal{C}$ has coordinates $(0, 0, 1)$ and the tangent at p is $y = 0$. In this coordinate system, the equation of $\mathcal{C}$ is

$$P(x, y, z) = y \cdot z^2 + (\alpha x^2 + 2\beta xy + \gamma y^2) \cdot z + p_3(x, y), \quad \deg p_3 = 3.$$

We immediately see that p is a flex if and only if $\alpha = 0$.

On the other hand,

$$H(p) = \det \begin{pmatrix} 2\alpha & 2\beta & 0 \\ 2\beta & 2\gamma & 2 \\ 0 & 2 & 0 \end{pmatrix} = -8\alpha,$$

which proves the lemma. $\square$

Applying the Bézout theorem, we immediately get that any smooth cubic curve has exactly nine flexes. Now, we are ready to prove

Proposition 3.104. *In a suitable chosen affine coordinate system, the equation of any smooth cubic curve can be written in the form*

$$y^2 = 4x^3 + ax + b,$$

where $4x^3 + ax + b$ is a polynomial without multiple roots.

Proof. Take the coordinate system such that the curve has a flex at $(0, 1, 0)$, and that $z = 0$ is the tangent at this point. If $P(x, y, z) = 0$ is the equation of the curve in this coordinate system, then $P(x, 1, 0)$ has $x = 0$ as a triple root, that is,

$$P(x, y, z) = \mu x^3 + zQ(x, y, z), \quad \mu \neq 0, \quad \deg Q = 2.$$

Without losing generality, suppose $\mu = -4$. The affine equation of the curve is

$$p(x, y) = -4x^3 + \alpha x^2 + 2\beta xy + \gamma y^2 + 2\alpha' x + 2\beta' y + \gamma'.$$

Since $(0, 1, 0)$ is a non-singular point, we get $\gamma \neq 0$. Again, without losing generality, suppose $\gamma = 1$. With the transformation

$$x \mapsto x, \quad y \mapsto y + \beta x + \beta',$$

we get

$$p(x, y) = y^2 - (4x^3 + ax + b).$$

Since the curve is smooth, the polynomial $4x^3 + ax + b$ has no multiple roots. □

Corollary 3.105. *Any line that contains two flexes of a smooth cubic curve, passes through a third one.*

Proof. It follows by Proposition 3.104 and Exercise 3.102. □

3.6 Hyperelliptic curves

In this section, we are going to define an important class of curves, namely *hyperelliptic curves*, which share many properties with the class of elliptic curves. After the definition, we proceed with proving some important theorems of algebraic geometry, for hyperelliptic curves.

Definition of hyperelliptic curves

Definition 3.106. A compact Riemann surface of genus $g > 1$ is called *hyperelliptic* if there is a holomorphic mapping of degree 2 from the surface into a Riemann sphere. A curve is called *hyperelliptic* if its normalization is a hyperelliptic Riemann surface.

Exercise 3.107. Prove that the holomorphic mapping of degree 2 from a hyperelliptic surface of genus g to the Riemann sphere has exactly $2g + 2$ branching points.

Example 3.108. Consider the curve in $\mathbf{CP}^2$ with the affine equation

$$\mathcal{C} \;:\; y^2 = P(x),$$

where P is a polynomial of degree $2g + 2$, $g > 1$, with all zeroes mutually distinct. Then the normalization $\widetilde{\mathcal{C}}$ of the curve $\mathcal{C}$ is a hyperelliptic surface of genus g.

Proof. Let $(\widetilde{\mathcal{C}}, \sigma)$ be the normalization of the curve $\mathcal{C}$. Then $x \circ \sigma$ is the function of degree 2 from $\widetilde{\mathcal{C}}$ to a Riemann sphere. It is easy to see that its ramification divisor is

$$R = \sigma^{-1}(0, a_1) + \cdots + \sigma^{-1}(0, a_{2g+2}),$$

where $a_1, \ldots, a_{2g+2}$ are roots of polynomial $P(x)$. Thus, from the Riemann–Hurwitz formula, we immediately obtain that the genus of $\widetilde{\mathcal{C}}$ is equal to g. $\square$

It is possible to prove the opposite statement.

Theorem 3.109. *Each hyperelliptic surface of genus g can be represented as a normalization of a plane algebraic curve of degree $2g + 2$.*

Exercise 3.110. Prove that the normalization of the curve given with the affine equation $y^2 = P(x)$, $\deg P = 2g + 1$, $g > 1$ is a hyperelliptic Riemann surface of genus g.

Properties of hyperelliptic curves

Theorem 3.111. *Let φ be a meromorphic function on the hyperelliptic curve*

$$y^2 = P(x).$$

Then φ is a rational function of x, y.

Proof. Functions $\varphi(x, y) \cdot \varphi(x, -y)$ and $\varphi(x, y) + \varphi(x, -y)$ depend only on x, and their only singular points are poles. Thus, they may be viewed as meromorphic functions on a Riemann sphere. By Proposition 3.10, they are rational functions of x:

$$R_1(x) = \varphi(x, y), \quad R_2(x) = \varphi(x, -y).$$

It follows that $\varphi(x, y)$, $\varphi(x, -y)$ are solutions of the equation

$$\varphi^2 - R_1(x)\varphi + R_2(x) = 0.$$

Rewrite this equation in the form

$$P_0(x)\varphi^2 + P_1(x)\varphi + P_2(x) = 0,$$

where P_0, P_1, P_2 are polynomials. Then:

$$\varphi(x, y) = \frac{P_1(x) + \sqrt{P_1^2(x) - 4P_0(x)P_2(x)}}{2P_0(x)}.$$

Going around each of the roots of $P(x)$ transforms $\varphi(x, y)$ to $\varphi(x, -y)$, as well as going around each of the roots of $P_1^2(x) - 4P_0(x)P_2(x)$. Thus, $P(x)$ and $P_1^2(x) - 4P_0(x)P_2(x)$ have the same set of roots, i.e.,

$$\varphi(x, y) = \frac{P_1(x) + c \cdot y \cdot Q(x)}{2P_0(x)}.$$
$\square$

Let us remark that this theorem will hold in general – for arbitrary curves.

Exercise 3.112. Prove that each of the differentials

$$\frac{dx}{y}, \frac{xdx}{y}, \ldots, \frac{x^{g-1}dx}{y} \tag{3.10}$$

is holomorphic on the curve $y^2 = P(x)$, where P is a polynomial of degree $2g + 2$.

Proposition 3.113. *The dimension of the linear space of holomorphic differentials on a hyperelliptic curve of genus g is equal to g.*

Proof. The g differentials given in Exercise 3.112 are holomorphic and linearly independent. Thus, the dimension of the space of holomorphic differentials is at least g.

Let ω be another holomorphic differential. Then $\dfrac{\omega}{dx/y}$ is a rational function, i.e., $\omega = R(x, y)\dfrac{dx}{y}$. It is easy to calculate

$$\left(\frac{dx}{y}\right) = (g-1)P_\infty^+ + (g-1)P_\infty^-,$$

where P_∞^+ and P_∞^- are two points on the normalization lying over the single intersection point of the curve with the line of infinity. This means that R cannot have any zeros in the affine part, so it is a polynomial. Taking into account that the pole divisors of x and y on the curve are

$$(x)_\infty = P_\infty^+ + P_\infty^-,$$
$$(y)_\infty = (g+1)P_\infty^+ + (g+1)P_\infty^-,$$

we come to the conclusion that R is a polynomial of x of degree at most $g - 1$. Thus ω is a linear combination of the differentials (3.10), which concludes the proof. □

On a hyperelliptic surface, it is possible naturally to define a non-trivial involution.

Let Γ be a hyperelliptic surface, and $f : \Gamma \to \mathbf{C}$ a meromorphic function of degree 2. Let $j : \Gamma \to \Gamma$ be a mapping defined by

$$j(P) = \begin{cases} Q & \text{if } \nu_P(f) = 1, \text{where } f(P) = f(Q), \ P \neq Q; \\ P & \text{if } \nu_P(f) = 2. \end{cases}$$

Exercise 3.114. Prove that j is holomorphic.

Exercise 3.115. Let P, Q be two points on a hyperelliptic surface. Prove that $P + j(P) \sim Q + j(Q)$.

Exercise 3.116. Denote by B the set of branching points of Γ, i.e., the set of fixed points of j. For a set $T \subset B$ of even cardinality, write

$$e_T = \sum_{P \in T} P - \frac{|T|}{2}\left(Q + j(Q)\right),$$

where $Q \in \Gamma$ is an arbitrary point. Show that

- $2e_T \sim 0$;
- $e_{T_1} + e_{T_2} \sim e_{T_1 \Delta T_2}$, where Δ denotes the symmetric set difference;
- $e_{T_1} \sim e_{T_2}$ if and only if $T_1 = T_2$ or $T_1 = B \setminus T_2$;
- for a meromorphic differential ω on Γ, show that $(\omega) \sim (g-1)(Q + j(Q))$, where g is the genus of Γ.

3.7 Abel's theorem

Matrix of periods of a Riemann surface

Suppose Γ is a given, compact Riemann surface of genus g. Denote by $(a_1, \ldots, a_g, b_1, \ldots, b_g)$ a basis of homologies $H_1(\Gamma, \mathbf{Z})$, which is *canonical*, i.e., such that

$$a_i \circ a_j = b_i \circ b_j = 0, \ a_i \circ b_j = \delta_{ij}, \quad i,j = 1, \ldots, g.$$

Denote by $\tilde{\Gamma}$ the fundamental $4g$-angle, with edges

$$a_1 b_1 a_1^{-1} b_1^{-1} \cdots a_g b_g a_g^{-1} b_g^{-1}.$$

The surface Γ can be realized by gluing the edges of $\tilde{\Gamma}$.

Let ω, ω' be closed differentials on Γ, and let

$$A_i = \int_{a_i} \omega, \quad B_i = \int_{b_i} \omega, \quad A_i' = \int_{a_i} \omega', \quad B_i' = \int_{b_i} \omega',$$

for $i = 1, \ldots, g$ be their periods on a canonical basis of cycles. Then

$$\iint_{\Gamma} \omega \wedge \omega' = \sum_{i=1}^{g} (A_i B_i' - A_i' B_i).$$

Let us fix a basis of holomorphic differentials $[\omega_1, \ldots, \omega_g]$ such that

$$\int_{a_j} \omega_k = 2\pi i \delta_{jk}, \quad j,k = 1, \ldots, g.$$

For a basis normalized in that way, denote by B_{jk} the matrix of b-periods,

$$B_{jk} = \int_{b_j} \omega_k, \quad j,k = 1, \ldots, g.$$

Definition 3.117. The matrix B_{jk} is called a period matrix of Riemann surface Γ.

Proposition 3.118 (Riemann bilinear relations). *For the period matrix B_{jk} of a Riemann surface, it holds that*

- *The matrix B is symmetric.*
- *The matrix B has a negatively defined real part.*

Definition 3.119. A matrix B is called a *Riemannian matrix*, if it satisfies properties of the last proposition. The set of such $g \times g$ matrices is called the *Sigel half-plane* and is denoted by $\mathcal{H}_g$.

Thus, every period matrix of a Riemann surface is a Riemannian matrix. The converse question is highly nontrivial: *Which Riemannian matrices are period matrices of some Riemann surface?* This classical and very important problem of XIX century algebraic geometry is known as *the Riemann–Schottky problem* and it was open for more than a century. This problem was solved quite recently, in the middle of the 1980s; using the techniques of *the soliton theory*, Japanese mathematician Shiota proved the so-called *Novikov's conjecture*. We will say something more about this at the and of this Section.

Jacobian of a Riemann surface. The Abel mapping

Denote the standard basis of $\mathbf{C}^g$ by $e = [e_1, \ldots, e_g]$, $(e_i)_k = \delta_{ik}$.

Exercise 3.120. Let B be a Riemannian matrix. Then $2g$ vectors $e_1, \ldots, e_g$, $Be_1, \ldots, Be_g$ are linearly independent over $\mathbf{R}$.

Let us consider an integer-valued lattice Λ_B in $\mathbf{C}^g$ generated by the vectors $2\pi i e_j, Be_k, k, j = 1, \ldots, g$:

$$\Lambda_B: \quad 2\pi i M + BN, \quad M, N \in \mathbf{Z}^g.$$

Then $2g$-dimensional torus $\mathbf{T}^{2g} = \mathbf{T}(B) = \mathbf{C}^g/\Lambda_B$ defines a g-dimensional *Abel variety*, a g-dimensional complex torus.

Definition 3.121. If a matrix B is a period matrix of some Riemann surface Γ of genus g, then $\mathbf{T}(B)$ is called *the Jacobian variety of a surface* Γ, denoted by $\mathbf{T}(B) = \text{Jac}(\Gamma)$.

Let a compact, smooth Riemann surface Γ of genus g be given with some canonical basis of homologies (a, b) and with a corresponding normalized basis of holomorphic differentials $[\omega_1, \ldots, \omega_g]$. Choosing an arbitrary point P_0 on Γ, let us consider g *Abel integrals*

$$u_i(P) = \int_{P_0}^{P} \omega_i, \quad i = 1, \ldots, g, \tag{3.11}$$

assuming one and the same integration path every time.

Together with holomorphic differentials, known also as *Abel differentials of the first kind*, meromorphic differentials play an important role as well.

Definition 3.122. *Abel differentials of the second kind* $\omega_P^{(n)}$ *are meromorphic differentials with a unique pole at a point P of order $n+1$, locally represented by*

$$\omega_P^{(n)} = \frac{dz}{z^{n+1}} + \cdots .$$

Abel differentials of the third kind ω_{PQ} *are determined by unique simple poles P, Q with residua $+1, -1$.*

These differentials are uniquely determined by the conditions:

$$\int_{a_i} \omega_P^{(n)} = 0, \quad \int_{a_i} \omega_{PQ} = 0, \quad i = 1, \ldots, g.$$

Exercise 3.123. Prove the following formulae:

$$\int_{b_i} \omega_P^{(n)} = \frac{1}{n!} \frac{d^{n-1} f_i(Q)}{dz^{n-1}}, \quad i = 1, \ldots, g, \ n \in \mathbf{N}, \tag{3.12}$$

$$\int_{b_i} \omega_{PQ} = \int_Q^P \omega_i, \quad i = 1, \ldots, g, \tag{3.13}$$

where $\omega_i = f_i(z)\,dz$ locally represent a basic holomorphic differential around a point Q.

Exercise 3.124. Given four arbitrary points on a Riemann surface, prove that

$$\int_{Q_1}^{Q_2} \omega_{Q_3 Q_4} = \int_{Q_3}^{Q_4} \omega_{Q_1 Q_2}.$$

Exercise 3.125. Prove that the formula

$$\mathcal{A}(P) = (u_1(P), \ldots, u_g(P)) \tag{3.14}$$

defines a mapping $\mathcal{A} : \ \Gamma \to \mathrm{Jac}\,(\Gamma)$, where u_i are introduced by the formula (3.11).

Definition 3.126. The mapping $\mathcal{A} : \ \Gamma \to \mathrm{Jac}\,(\Gamma)$ defined by formula (3.14) is called *the Abel mapping*.

The natural question is *whether given points $P_1, \ldots, P_n$ and $Q_1, \ldots, Q_n$ represent a divisor of zeroes and poles of some meromorphic function on a surface Γ*. The answer is given in the following

Theorem 3.127 (Abel). *Given points $P_1, \ldots, P_n$ and $Q_1, \ldots, Q_n$ form divisors of zeroes and poles of a meromorphic function on a Riemann surface Γ if and only if the relation*

$$\sum_{i=1}^n \mathcal{A}(P_i) = \sum_{i=1}^n \mathcal{A}(Q_i)$$

takes place on the Jacobian $\mathrm{Jac}\,(\Gamma)$.

Riemann theta-function

An important tool is introduced by the following

Definition 3.128. Let B be an arbitrary $g \times g$ Riemann matrix, $B \in \mathcal{H}_g$. The *Riemann theta-function* $\theta(z, B)$ is defined by the series

$$\theta(z, B) = \sum_{n \in \mathbf{Z}^g} \exp\big((Bn, n) + (n, z)\big). \tag{3.15}$$

Proposition 3.129. *The series* (3.15) *converges uniformly and absolutely on every compact subset of* $\mathbf{C}^g \times \mathcal{H}_g$ *and it defines a holomorphic function.*

Proposition 3.130. *The Riemann theta-function satisfies the following periodic relations:*

$$\theta(z + 2\pi i e_k, B) = \theta(z, B), \quad k = 1, \dots, g,$$
$$\theta(z + B e_k, B) = \exp(-B_{kk}/2 - z_k)\, \theta(z, B), \quad k = 1, \dots, g.$$

Similarly, *Riemann theta-functions with characteristics* can be introduced for arbitrary real vectors $a, b \in \mathbf{R}^g$:

$$\theta[2a, 2b](z) = \exp\Big\{\frac{1}{2}(Ba, a) + (z + 2\pi i b, a)\Big\}\theta(z + 2\pi i b + Ba).$$

The Jacobi inversion problem and the Riemann theorem about zeroes of theta-function

Starting from the case of genus 2 Riemann surfaces, there is no sense in inverting a fixed Abel integral. The following system of equations,

$$\zeta_1 = \int_{P_0}^{P_1} \frac{dz}{\sqrt{P_5(z)}} + \int_{P_0}^{P_2} \frac{dz}{\sqrt{P_5(z)}},$$

$$\zeta_2 = \int_{P_0}^{P_1} \frac{z\,dz}{\sqrt{P_5(z)}} + \int_{P_0}^{P_2} \frac{z\,dz}{\sqrt{P_5(z)}},$$

appeared in mechanical systems, in connection with the Kowalevski case of rigid body motion, see [Kow1889]. The problem is to determine points P_1, P_2 as functions of given values ζ_1, ζ_2. Observing the symmetric appearance of points P_1 and P_2 in the above formulae, the problem can be reduced to finding expressions of symmetric functions of P_1, P_2, through ζ_1, ζ_2. Historically, it was Jacobi who solved this problem in the genus 2 case.

For an arbitrary genus, the corresponding general Jacobi problem of inversion was formulated and solved by Riemann.

Given an arbitrary, smooth Riemann surface Γ of genus g, with a fixed canonical basis of homology cycles and corresponding basis of holomorphic differentials. By using the Abel mapping, we define

$$\mathcal{A}^n :\ S^n(\Gamma) \to \text{Jac}\,(\Gamma), \quad \mathcal{A}^n(P_1,\ldots,P_n) = \sum_{i=1}^{n} \mathcal{A}(P_i),$$

where $S^n(X)$ denotes the symmetric nth degree of a set X.

Proposition 3.131. *Let a nonspecial divisor $D = P_1 + \cdots + P_g$ be given; then in a neighborhood of the point $\mathcal{A}^g(P_1,\ldots,P_g) \in \text{Jac}\,(\Gamma)$ the mapping $\mathcal{A}^g$ is invertible.*

In the general case, the divisor $D = P_1 + \cdots + P_g$ is nonspecial. Thus, the inverse of the mapping $\mathcal{A}^g$ is defined almost everywhere. To find explicitly the inverse, Riemann essentially used theta-functions. Let us present some of their basic properties, necessary for the solution of the Jacobi inversion problem.

Suppose a vector $f \in \mathbf{C}^g$ is given. Consider the function $F(P) = \theta(\mathcal{A}(P) - f)$, where $\theta(z) = \theta(z, B)$ is the theta-function of the surface Γ. Function F is well defined and analytic on the fundamental $4g$-angle $\tilde{\Gamma}$, and for almost all f it is not identically equal to zero.

Proposition 3.132. *If the function F is not identically zero, then it has exactly g zeroes in $\tilde{\Gamma}$.*

Definition 3.133. *A vector $\mathcal{K} = (K_1,\ldots,K_g)$, where*

$$K_j = \frac{2\pi i + B_{jj}}{2} - \frac{1}{2\pi i} \sum_{l \neq j} \left(\int_{a_l} \omega_l(P) \int_{P_0}^{P} \omega_j \right), \quad j = 1,\ldots,g$$

is called the vector of Riemann constants.

Proposition 3.134. *If a function F is not identically zero and if P_1, ..., P_g are its zeroes, then $\mathcal{A}^g(P_1,\ldots,P_g) = f - \mathcal{K}$.*

Theorem 3.135 (Riemann). *Given a vector f such that*

$$F(P) = \theta(\mathcal{A}(P) - \mathcal{K} - f)$$

is not identically zero. Then:

- *the function F has exactly g zeroes $P_1,\ldots,P_g$, giving the solution of the Jacobi problem $u_i(P_1) + \cdots + u_i(P_g) = f_i$, $i = 1,\ldots,g$.*
- *The divisor $P_1 + \cdots + P_g$ is nonspecial.*

The set of zeroes of the theta-function defined on the Jacobian of the Riemann surface Γ is called *the theta divisor* or *the Θ-divisor* of the Riemann surface, and we denote it by Θ_Γ.

The Baker–Akhiezer function

In the theory of integrable systems an important role is played by the notion of the Baker–Akhiezer function.

Definition 3.136. Given n points $P_1, \ldots, P_n$ on a Riemann surface of genus g, with local parameters k_i^{-1}, $i = 1, \ldots, n$, $k_i^{-1}(P_i) = 0$, n polynomials $q_i(k)$ and a nonspecial divisor D, then n-point Baker–Akhiezer function ψ corresponding to the data, is

- meromorphic on $\Gamma \setminus \{P_1, \ldots, P_n\}$,
- for its divisor it holds that $(\psi) + D \geq 0$,
- when P tends to P_i, the function $\psi(P) \exp(-q_i(k_i(P))$ is analytical.

Theorem 3.137. *[DKN2001] Given a nonspecial divisor D of degree N. Then the dimension of the space of Baker–Akhiezer functions is $N - g + 1$.*

Example 3.138. If $N = g$, then the Baker–Akhiezer function ψ is determined uniquely up to a scalar factor. It is given by the formula

$$\psi(P) = a \exp\left(\sum_{j=1}^{n} \int_{Q}^{P} \Omega_{q_j} \right) \frac{\theta\left(\mathcal{A}(P) + \sum_{j=1}^{n} U^{(q_j)} - \mathcal{A}(D) - \mathcal{K} \right)}{\theta\left(\mathcal{A}(P) - \mathcal{A}(D) - \mathcal{K} \right)},$$

where Ω_{q_j} are Abel differentials of the second order, with a principle part around P_j of the form $dq_j(k_j(P))$ normalized by the condition of annulation of the a periods; $2\pi i U^{(q_j)}$ are the vectors of their b-periods.

Riemann–Schottky problem and Novikov's conjecture

We saw that every period matrix of a Riemann surface is a Riemannian matrix. The converse question: *which Riemannian matrices are period matrices of some Riemann surface* is classical and a very important problem of XIX century algebraic geometry known as *the Riemann–Schottky problem*. It was solved quite recently, in the middle of the 1980s, using the techniques of the Baker–Akhiezer functions and the soliton theory, through the so-called *Novikov's conjecture*, see [Dub1981].

It was known after Krichever (see [DKN2001] and references therein) that there exist certain theta-function formulae associated with period matrices which give solutions of the *Kadomtsev–Petviashvili* equation from the soliton theory

$$(u_t + u u_x + u_{xxx})_x + u_{yy} = 0.$$

The Novikov conjecture is a converse statement: that a Riemannian matrix is a period matrix *only if it gives a solution of the Kadomtsec–Petviashvili equation through the Krichever formulae.*

In a weak form the Novikov conjecture has been proven by Dubrovin in 1981 [Dub1981]. The complete solution of Novikov's conjecture and the Riemann–Schottky problem was done by Shiota in 1986 [Shi1986, Mum1983]. The highlight of Shiota's proof was use of a notion of the *tau-function* introduced by Sato's school a few years before, giving an opportunity to involve simultaneously the whole hierarchy of integrable systems associated with the Kadomtsev–Petviashvili equation.

Novikov's conjecture demonstrates the deepest relationship between the theory of integrable dynamical systems and the theory of algebraic curves. It solved a century old, general and important problem of description of period matrices of Jacobians among Riemannian matrices through the solutions of the Kadomtsev-Petviashvili integrable hierarchy.

In this book, we are going to follow the same ideology on a very specialized and simple level. We are going to demonstrate a deep, intimate relationship between general hyperelliptic Jacobians on one hand and integrable billiard systems generated by pencils of quadrics on the other hand.

3.8 Points of finite order on the Jacobian of a hyperelliptic curve

In this section we are going to give the analytical characterization of some classes of finite order divisors on a hyperelliptic curve.

Let the curve $\mathcal{C}$ be given by

$$y^2 = (x - x_1) \ldots (x - x_{2g+1}), \quad x_i \neq x_j \text{ when } i \neq j.$$

It is a regular hyperelliptic curve of genus g, embedded in $\mathbf{CP}^2$. Let $\mathcal{J}(\mathcal{C})$ be its Jacobian variety and

$$\mathcal{A} : \mathcal{C} \to \mathcal{J}(\mathcal{C})$$

the Abel-Jacobi map.

Take E to be the point which corresponds to the value $x = \infty$, and choose $\mathcal{A}(E)$ to be the neutral in $\mathcal{J}(\mathcal{C})$. According to Abel's theorem (Theorem 3.127), $\mathcal{A}(P_1) + \cdots + \mathcal{A}(P_n) = 0$ if and only if there exists a meromorphic function f with zeroes $P_1, \ldots, P_n$ and a pole of order n at the point E. Let $\mathcal{L}(nE)$ be the vector space of meromorphic functions on $\mathcal{C}$ with a unique pole E of order at most n, and $f_1, \ldots, f_k$ a basis of $\mathcal{L}(nE)$. The mapping

$$F : \mathcal{C} \to \mathbf{CP}^{k-1}, \quad X \mapsto [f_1(X), \ldots, f_k(X)]$$

is a projective embedding whose image is a smooth algebraic curve of degree n. Hyperplane sections of this curve are zeroes of functions from $\mathcal{L}(nE)$. Thus, the equality $n\mathcal{A}(P) = 0$ is equivalent to

$$\text{rank} \begin{pmatrix} f_1(P) & f_2(P) & \cdots & f_k(P) \\ f_1'(P) & f_2'(P) & \cdots & f_k'(P) \\ & & \cdots & \\ f_1^{(n-1)}(P) & f_2^{(n-1)}(P) & \cdots & f_k^{(n-1)}(P) \end{pmatrix} < k. \qquad (3.16)$$

Lemma 3.139. *For $n \leq 2g$, there does not exist a point P on the curve $\mathcal{C}$, such that $n\mathcal{A}(P) = 0$ and $P \neq E$.*

Proof. Let P be a point on $\mathcal{C}$, $P \neq E$, and $x = x_0$ its corresponding value. Consider the case of n even. Since E is a branch point of a hyperelliptic curve, its Weierstrass gap sequence is $1, 3, 5, \ldots, 2g - 1$ [Gun1966]. Now, applying the Riemann–Roch theorem, we obtain $\dim \mathcal{L}(nE) = n/2 + 1$. Choosing a basis $1, x, \ldots, x^{n/2}$ for $\mathcal{L}(nE)$, and substituting in (3.16), we come to a contradiction. $\qquad\square$

Lemma 3.140. *Let $P(x_0, y_0)$ be a non-branching point on the curve $\mathcal{C}$. For $n > 2g$, equality $n\mathcal{A}(P) = 0$ is equivalent to*

$$\text{rank} \begin{pmatrix} B_{m+1} & B_m & \cdots & B_{g+2} \\ B_{m+2} & B_{m+1} & \cdots & B_{g+3} \\ & & \cdots & \\ B_{2m-1} & B_{2m-2} & \cdots & B_{m+g} \end{pmatrix} < m - g, \quad \text{when} \quad n = 2m, \qquad (3.17)$$

$$\text{rank} \begin{pmatrix} B_{m+1} & B_m & \cdots & B_{g+1} \\ B_{m+2} & B_{m+1} & \cdots & B_{g+2} \\ & & \cdots & \\ B_{2m} & B_{2m-1} & \cdots & B_{m+g} \end{pmatrix} < m - g + 1, \quad \text{when} \quad n = 2m + 1,$$

and

$$\sqrt{(x - x_1) \ldots (x - x_{2g+1})} = B_0 + B_1(x - x_0) + B_2(x - x_0)^2 + B_3(x - x_0)^3 + \cdots .$$

Proof. This claim follows from previous results by choosing a basis for $\mathcal{L}(nE)$:

$$1, x, \ldots, x^m, y, xy, \ldots, x^{m-g-1}y \text{ if } n = 2m,$$
$$1, x, \ldots, x^m, y, xy, \ldots, x^{m-g}y \text{ if } n = 2m + 1,$$

similarly as in [GH1978a]. $\qquad\square$

In the next lemma, we are going to consider the case when the curve $\mathcal{C}$ is singular, i.e., when some of the values $x_1, x_2, \ldots, x_{2g+1}$ coincide.

Lemma 3.141. *Let the curve $\mathcal{C}$ be given by*

$$y^2 = (x - x_1) \ldots (x - x_{2g+1}), \quad x_1 \cdot x_2 \cdot \cdots \cdot x_{2g+1} \neq 0,$$

P_0 *one of the points corresponding to the value $x = 0$ and E the infinite point on C. Then $2nP_0 \sim 2nE$ is equivalent to (3.17), where*

$$y = \sqrt{(x - x_1)\ldots(x - x_{2g+1})} = B_0 + B_1 x + B_2 x^2 + \cdots$$

is the Taylor expansion around the point P_0.

Proof. Suppose that, among $x_1, \ldots, x_{2g+1}$, only x_{2g} and x_{2g+1} have the same values. Then $(x_{2g}, 0)$ is an ordinary double point on C. The normalization of the curve C is the pair $(\tilde{C}, \pi)$, where $\tilde{C}$ is the curve given by

$$\tilde{C} : \tilde{y}^2 = (\tilde{x} - x_1)\ldots(\tilde{x} - x_{2g-1}),$$

and $\pi : \tilde{C} \to C$ is the projection

$$(\tilde{x}, \tilde{y}) \xmapsto{\pi} (x = \tilde{x}, \; y = (\tilde{x} - x_{2g})\tilde{y}).$$

The genus of $\tilde{C}$ is $g - 1$.

Denote by A and B points on $\tilde{C}$ which are mapped to the singular point $(x_{2g}, 0) \in C$ by the projection π. Any other point on C is the image of a unique point of the curve $\tilde{C}$. Let

$$\pi(\tilde{E}) = E, \quad \pi(\tilde{P}_0) = P_0.$$

The relation $2nP_0 \sim 2nE$ holds if and only if there exists a meromorphic function f on $\tilde{C}$, $f \in \mathcal{L}(2n\tilde{E})$, having a zero of order $2n$ at $\tilde{P}_0$ and satisfying $f(A) = f(B)$.

For $n \leq g - 1$, according to Lemma 3.140, $2n\tilde{E} \sim 2n\tilde{P}_0$ cannot hold. For $n \geq g$, choose the following basis of the space $\mathcal{L}(2n\tilde{E})$:

$$1, \; \tilde{y}, \; f_1 \circ \pi, \; \ldots, \; f_{n-g-1} \circ \pi,$$

where $1, f_1, \ldots, f_{n-g-1}$ is a basis of $\mathcal{L}(2nE)$ as in the proof of Lemma 3.140.

Since $\tilde{y}$ is the only function in the basis which has different values at points A and B, we obtain that the condition

$$2n\tilde{E} \sim 2n\tilde{P}_0$$

is equivalent to (3.17).

Cases when C has more singular points or singularities of higher order, can be discussed in the same way. $\qquad\square$

Lemma 3.142. *Let the curve C be given by*

$$y^2 = (x - x_1)\ldots(x - x_{2g+2}),$$

with all x_i distinct from 0, and Q_+, Q_- the two points on $\mathcal{C}$ over the point $x = 0$.
Then $nQ_+ \sim nQ_-$ is equivalent to

$$\text{rank} \begin{pmatrix} B_{g+2} & B_{g+3} & \cdots & B_{n+1} \\ B_{g+3} & B_{g+4} & \cdots & B_{n+2} \\ \cdots & \cdots & \cdots & \cdots \\ \cdots & \cdots & \cdots & \cdots \\ B_{g+n} & \cdots & \cdots & B_{2n-1} \end{pmatrix} < n - g \quad and \quad n > g, \qquad (3.18)$$

where $y = \sqrt{(x - x_1) \dots (x - x_{2g+2})} = B_0 + B_1 x + B_2 x^2 + \cdots$ is the Taylor expansion around the point Q_-.

Proof. $\mathcal{C}$ is a hyperelliptic curve of genus g. The relation $nQ_+ \equiv nQ_-$ means that there exists a meromorphic function on $\mathcal{C}$ with a pole of order n at the point Q_+, a zero of the same order at Q_- and neither other zeros nor poles. Denote by $\mathcal{L}(nQ_+)$ the vector space of meromorphic functions on $\mathcal{C}$ with a unique pole Q_+ of order at most n. Since Q_+ is not a branching point on the curve, $\dim \mathcal{L}(nQ_+) = 1$ for $n \leq g$, and $\dim \mathcal{L}(nQ_+) = n - g + 1$, for $n > g$. In the case $n \leq g$, the space $\mathcal{L}(nQ_+)$ contains only constant functions, and the divisors nQ_+ and nQ_- can not be equivalent. If $n \geq g + 1$, we choose the following basis for $\mathcal{L}(nQ_+)$:

$$1, f_1, \dots, f_{n-g},$$

where

$$f_k = \frac{y - B_0 - B_1 x - \cdots - B_{g+k-1} x^{g+k-1}}{x^{g+k}}.$$

Thus, $nQ_+ \equiv nQ_-$ if there is a function $f \in \mathcal{L}(nQ_+)$ with a zero of order n at Q_-, i.e., if there exist constants $\alpha_0, \dots, \alpha_{n-g}$, not all equal to 0, such that

$$\begin{array}{rcllcccl} \alpha_0 & + & \alpha_1 f_1(Q_-) & + & \cdots & + & \alpha_{n-g} f_{n-g}(Q_-) & = & 0 \\ & & \alpha_1 f_1'(Q_-) & + & \cdots & + & \alpha_{n-g} f_{n-g}'(Q_-) & = & 0 \\ & & \cdots & & & & & & \\ & & \cdots & & & & & & \\ & & \alpha_1 f_1^{(n-1)}(Q_-) & + & \cdots & + & \alpha_{n-g} f_{n-g}^{(n-1)}(Q_-) & = & 0. \end{array}$$

Existence of a non-trivial solution to this system of linear equations is equivalent to condition (3.18).

When some of the values $x_1, \dots, x_{2g+2}$ coincide, the curve $\mathcal{C}$ is singular. This case can be considered by the procedure of normalization of the curve, as in the previous Lemma 3.141. The condition for the equivalence of the divisors nQ_+ i nQ_-, in the case when $\mathcal{C}$ is singular, is again (3.18). $\square$

Chapter 4

Projective Geometry

4.1 Preliminaries

We will start with an overview of some basic statements from the line projective geometry which are going to be used in the sequel. Projective line $\mathbf{KP}^1$ can be identified with $\widehat{\mathbf{K}} = \mathbf{K} \cup \{\infty\}$: to a pair of homogeneous coordinates (x_1, x_2) we associate $t = x_2/x_1 \in \mathbf{K}$ if $x_1 \neq 0$. To $(0, x_2)$ we associate the symbol ∞. For the standard projective frame of $\mathbf{KP}^1$ we choose the triplet $(\infty, 0, 1)$.

Definition 4.1. Given four points A, B, C, D, the first three of which are distinct, on a projective line ℓ. *The cross-ratio* (A, B, C, D) is equal to the value $H(D) \in \mathbf{KP}^1$, where $H : \ell \to \mathbf{KP}^1$ is the unique projective transformation such that $H(A) = \infty$, $H(B) = 0$ and $H(C) = 1$.

Given four points A, B, C, D, the first three of which are distinct, in the projective space $\mathbf{KP}^1$. The cross-ratio (A, B, C, D) is equal to

$$(A, B, C, D) = \frac{(C - A)(D - B)}{(C - B)(D - A)}.$$

It is assumed that if $C = B$ the cross-ratio is ∞ and if $D = \infty$ the cross-ratio is $(C - A)/(C - B)$.

Proposition 4.2. *For a given four distinct points A, B, C, D on a projective line ℓ and a given four distinct points A_1, B_1, C_1, D_1 on a projective line ℓ_1 there exists a projective transformation $H : \ell \to \ell_1$ mapping A, B, C, D to A_1, B_1, C_1, D_1 respectively if and only if the two cross-ratios are equal,*

$$(A, B, C, D) = (A_1, B_1, C_1, D_1).$$

Given two polynomials $P, Q \in \mathbf{K}[X]$, they define *a rational fraction*

$$R(X) = \frac{P(X)}{Q(X)}.$$

The fraction is reduced if P and Q are relatively prime. It defines a *rational map* $\mathbf{KP}^1 \to \mathbf{KP}^1$ extending the standard evaluation map by mapping the zeroes of Q in the reduced case to ∞.

Proposition 4.3. *If a rational map $\mathbf{KP}^1 \to \mathbf{KP}^1$ is invertible and if its inverse is a rational map, then they are projective transformations.*

We will use a notion of *semi-projective transformation* for the composition of a projective transformation with a field automorphism.

Four points A, B, C, D are in *harmonic division* if

$$(A, B, C, D) = -1.$$

A (1-1)-correspondence of a projective space in which the relation of linear dependence is invariant is called *collineation*. In other words, collineation is transformation of a projective space which maps lines to lines.

Theorem 4.4 (Fundamental Theorem of Projective Geometry). *Every collineation is a semi-projective transformation.*

Proposition 4.5. *If $\mathbf{K}$ is a field of characteristics different from 2, then a bijection $\mathbf{KP}^1 \to \mathbf{KP}^1$ is semi-projective if and only if it preserves harmonic divisions.*

A projective transformation H of a projective line onto itself is *involution* if $H \neq \mathrm{id}$, $H \circ H = \mathrm{id}$. Two points x, x' are *homologous* if they are H-images of each other.

Let us remark that a (1-1)-linear map $G : U \to V$ induces a projective map $\mathbf{P}(G) : \mathbf{P}(U) \to \mathbf{P}(V)$, since it maps one-dimensional linear subspaces onto one-dimensional linear subspaces.

Proposition 4.6. *For a projective transformation H of a projective line into itself, the next three conditions are equivalent:*

(a) *H is an involution;*

(b) *$H = \mathbf{P}(G)$ where G is a traceless linear operator;*

(c) *there exists a point M such that $H(M) \neq M$ and $H^2(M) = M$.*

Notice that a nonidentical projective transformation cannot have more than two fixed points.

Proposition 4.7. *Given a projective transformation H of a projective line l having two distinct fixed points P, Q:*

(a) *The cross-ratio $(P, Q, M, H(M)) = k$ is constant, not depending on $M \in l$.*

(b) *This constant k is equal to the ratio of the eigen-values of the linear operator G such that $H = \mathbf{P}(G)$.*

(c) *H is an involution if and only if $k = -1$.*

Thus, a projective transformation with two fixed points P, Q is involution if and only if the points $P, Q, M, H(M)$ are in harmonic division.

4.2   Conics and quadrics

Given an affine or projective space $\mathcal{A}$ over field $\mathbf{K}$. A quadratic equation $F(x) = 0$ in $\mathcal{A}$ defines hypersurface of zeroes of the quadratic form F. This hypersurface is called *a quadric*. If the field $\mathbf{K}$ is algebraically closed, then the hypersurface of zeroes determines the quadratic form F uniquely up to a scalar multiple. If the space $\mathcal{A}$ is projective, then F is homogeneous and the quadric is *projective*, otherwise if $\mathcal{A}$ is affine, the quadric is *affine*.

If the space $\mathcal{A}$ is two-dimensional, then the quadric is an affine or a projective *conic*. For example, let (x_1, x_2, x_3) be projective coordinates in two-dimensional projective space $\mathcal{A} = \mathbf{KP}^2$. Any quadratic form F is determined by a symmetric 3×3 matrix $A = [a_{ij}]$, $a_{ij} = a_{ji}$, where $i, j = 1, 2, 3$. The corresponding conic is defined by the equation

$$0 = F(x) = \sum_{i,j=1}^{3} a_{ij} x_i x_j. \tag{4.1}$$

If F is irreducible over the algebraic closure of the field $\mathbf{K}$, then the conic is *non-degenerate*. If F is reducible, then the conic is *degenerate*. A degenerate conic consists either of a double line, if F is proportional to the square, or it consists of two different lines if F is a product of two distinct linear forms. The conic determined by equation (4.1) is non-degenerate if and only if $\det A \neq 0$.

As an easy consequence of the Bézout theorem, one can prove

Proposition 4.8. *A projective conic is degenerate if and only if it has at least one multiple point.*

The symmetric matrix A defined by (4.1) up to the nonzero scalar is called a *conic matrix*. The set of conics is thus related to the five-dimensional projective space $\mathbf{KP}^5$.

Another consequence of the Bézout theorem is the following

Proposition 4.9. *If two conics do not have a common line, then they intersect in at most four points, counting multiplicity. If $\mathbf{K}$ is algebraically closed, then they intersect in four points.*

From equation (4.1) we see that the six parameters, the entries of a symmetric matrix A, define a conic. Also, it is easy to see that symmetric matrices A and αA, where $\alpha \in \mathbf{K} \setminus \{0\}$, determine the same conic. Since any condition that a point belongs to a conic leads to a linear equation on coefficients a_{ij}, this suggests that a conic is uniquely determined by five of its points. More precisely, we have

Proposition 4.10. *A conic in the projective plane is uniquely determined by its five points, if no four of them are collinear.*

Proof. We only need to prove uniqueness. If there are two conics $\mathcal{C}_1$ and $\mathcal{C}_2$ which both contain the given five points, then, according to Proposition 4.9, these two conics have a common line ℓ_1. That common line contains at most three of the given points. Thus there are at least two more points common for $\mathcal{C}_1$ and $\mathcal{C}_2$ and they determine another line ℓ_2 common for $\mathcal{C}_1$ and $\mathcal{C}_2$. Then, $\mathcal{C}_1$ and $\mathcal{C}_2$ coincide since both are equal to $\ell_1 + \ell_2$.   □

A line and a non-degenerate conic have at most two points of intersection, counting multiplicity; if they have one double common point A, we say that the line is *the tangent* to the conic at the point A.

Let us consider the real affine case. Since any line intersects a non-degenerate conic in at most two points, that is true for the infinite line as well. In the real affine case we recognize the ellipse, parabola or hyperbola as a non-degenerate conic having 0, 1-double or 2 distinct intersection points with the infinite line.

Circles

Starting from the well-known notion of circle in the Euclidean plane with an equation in an orthonormal frame of the form

$$x^2 + y^2 + ax + by + c = 0,$$

we define a notion of circle in a projective plane $\mathbf{KP}^2$ as the set of points with homogeneous coordinates satisfying an equation of the form

$$a(x_1^2 + x_2^2) + bx_1x_3 + cx_2x_3 + dx_3^2 = 0.$$

We suppose that (a, b, c, d) is not the null-vector. Two equations of the previous form define the same set of points if and only if the equations are proportional, assuming an algebraically closed field $\mathbf{K}$.

One can easily see that a conic is a circle in $\mathbf{CP}^2$ if and only if it contains the points
$$\hat{I} = (1, i, 0), \quad \hat{J} = (1, -i, 0).$$

We call these points *the cyclic points*, because all circles contain them. A line containing a cyclic point in infinity is called *an isotropic line*.

We distinguish three cases of circles:

(a) ordinary circles, such that $a \neq 0$,

(b) the union of the infinity line with some other line, $a = 0$, $(b, c) \neq (0, 0)$,

(c) the double infinity line, when $a = b = c = 0$.

The circles in a plane form a three-dimensional projective subspace of the five-dimensional projective space of conics.

4.3 Projective structure on a conic

Let $\mathcal{C}$ be a non-degenerate conic, A one of its points, and $\mathcal{X}_A$ the pencil of lines containing point A. A projective structure on $\mathcal{C}$ can be induced by the structure of pencil $\mathcal{X}_A$: consider a map

$$i_A : \mathcal{X}_A \to \mathcal{C}, \qquad i_A(\ell) = B \longleftrightarrow \ell \cap \mathcal{C} = \{A, B\}, \tag{4.2}$$

which maps a given line $\ell \ni A$ to the second point of intersection ℓ with the conic $\mathcal{C}$. If a line t is the tangent to the conic $\mathcal{C}$ at the point A, then $i_A(t) = A$. Induced structure does not depend on the choice of the point $A \in \mathcal{C}$, as follows from the next theorem.

Theorem 4.11. *Given a non-degenerate conic $\mathcal{C}$ and two of its points A and B. Then the map*

$$i_A^{-1} \circ i_B : \mathcal{X}_B \to \mathcal{X}_A$$

is a projective transformation of pencils.

Proof. Lines as elements of a pencil of lines are parametrized by their slope. The function which maps the slope of a line ℓ, $A \in \ell$ to the coordinates of the point $i_A(\ell)$ is a rational one. It is related to calculation of the second root of a quadratic equation with one root known. Thus, the function $i_A^{-1} \circ i_B$ and its inverse are rational functions. Now, the proof follows from general considerations.

We can write down explicit formulae for the map. Let the coordinates be chosen such that $A = (0,0,1)$ and $B = (0,1,0)$, Then the equation of the conic is of the form

$$ax^2 + byz + cxy + dxz = 0.$$

The equations of the lines l and s from $\mathcal{X}_A$ and $\mathcal{X}_B$ are of the forms $y = kx$ and $z = k'x$. The relationship between the slope coefficients follows from the condition that $\ell \cap s \in \mathcal{C}$:

$$a + bkk' + ck + dk' = 0. \tag{4.3}$$

$\square$

Given a non-degenerate conic $\mathcal{C}$ and four of its points E, F, G, H. For a fixed fifth point $A \in \mathcal{C}$ we can define a number $\mathrm{in}_A(E, F, G, H)$ as the cross-ratio of the four lines AE, AF, AG and AH as members of the line pencil $\mathcal{X}_A$. It appears that the number does not depend on the choice of the point $A \in \mathcal{C}$. We have

Corollary 4.12 (Chasles Theorem). *Given a non-degenerate conic $\mathcal{C}$, its four points E, F, G, H and two more points $A, B \in \mathcal{C}$. Two cross-ratios $\mathrm{in}_A(E, F, G, H)$ and $\mathrm{in}_A(E, F, G, H)$ calculated in two pencils $\mathcal{X}_A$ and $\mathcal{X}_B$ are equal,*

$$\mathrm{in}_A(E, F, G, H) = \mathrm{in}_B(E, F, G, H).$$

We are going to denote the invariant as

$$\text{in}_{\mathcal{C}}(E, F, G, H) := \text{in}_A(E, F, G, H)$$

because it does not depend on the choice of the fifth point A but only on the conic $\mathcal{C}$.

The proof follows immediately from Theorem 4.11, because the cross-ratio is an invariant of the projective transformations of pencils.

Theorem 4.11 also has an interesting converse.

Theorem 4.13. *Let $f : \mathcal{X}_A \to \mathcal{X}_B$ be a projective transformation of pencils of lines in a projective plane. There exists a conic $\mathcal{C}$ such that when ℓ runs in $\mathcal{X}_A$, then points $\ell \cap f(\ell)$ belong to $\mathcal{C}$. If $f(\ell_{AB}) \neq \ell_{AB}$ the conic $\mathcal{C}$ is non-degenerate. The conic $\mathcal{C}$ contains the points A and B. If $f(\ell_{AB}) = \ell_{AB}$, then the conic $\mathcal{C}$ is degenerate and it has the line ℓ_{AB} as one of its components.*

Proof. Any projective-linear transformation of the pencils of lines is of the form (4.3), $ab - cd \neq 0$. From the proof of Theorem 4.11, we see that it leads to the conic

$$ax^2 + byz + cxy + dxz = 0.$$

The conic contains the points A, B, the basis of the pencils. It does not contain a line $\ell \in \mathcal{X}_A$ different from ℓ_{AB} because of $ab - cd \neq 0$. If it contains the line ℓ_{AB}, then $b = 0$ and ∞ is a fixed point of the projective-linear transformation. Thus, the transformation maps ℓ_{AB} into itself.   □

Theorem 4.14 (Frégier Theorem). *In the projective plane, let a non-degenerate conic $\mathcal{C}$ and a point $F \notin \mathcal{C}$ be given. Define the mapping $I_F : \mathcal{C} \to \mathcal{C}$, such that it maps a point $M \in \mathcal{C}$ to the point $M' \in \mathcal{C}$, the second intersection point of the line ℓ_{FM} with the conic $\mathcal{C}$. Then I_F is an involution on $\mathcal{C}$.*

Conversely, every involution I of the conic $\mathcal{C}$ is of that form, i.e., there exists $F \notin \mathcal{C}$ such that $I = I_F$.

Proof. The mapping I_F obviously satisfies $I_F^2 = id$ and it is a rational function from $\mathcal{C}$ to $\mathcal{C}$. Thus, it is a projective transformation and an involution.

To prove the converse, we consider an arbitrary involution $i : \mathcal{C} \to \mathcal{C}$. By taking two pairs of points, $A, i(A)$ and $B, i(B)$ and intersecting the two lines they determine, we get a point F, $F = \ell_{Ai(A)} \cap \ell_{Bi(B)}$. Now, we compare two involutions, i and I_F and we see that they have two common corresponding pairs of points. Thus, they coincide, $i = I_F$. This concludes the proof.   □

Definition 4.15. *The Frégier point of the involution I is the point which is the intersection of lines $\ell_{Mi(M)}$.*

Exercise 4.16. If the field $\mathbf{K}$ is of characteristic 2, prove that then all the tangents to the conic pass through one and the same point.

Is that point the Frégier point of the transformation $I(t) = -t$?

Proposition 4.17. *Let $\mathcal{C}$ be a non-degenerate conic, ℓ a line intersecting $\mathcal{C}$ in two points F, G and $A, B, A', B' \in \mathcal{C}$ four given points on the conic. There exists a projective transformation from $\mathcal{C}$ to $\mathcal{C}$ with H, G as fixed points and mapping A to A' and B to B' if and only if*

$$\ell_{AB'} \cap \ell_{BA'} \in \ell.$$

Proof. Consider a mapping f from $\mathcal{C}$ to $\mathcal{C}$ defined as follows: for a given point $M \in \mathcal{C}$ construct $P = \ell_{MA'} \cap \ell$ and $M' = \mathcal{C} \cap \ell_{PA}$. Put $f(M) = M'$. Since h is rational having rational inverse, it is a projective transformation from $\mathcal{C}$ to $\mathcal{C}$. It maps, by definition A to A' and it has H, G as fixed points. Thus, $f(B) = B'$ is equivalent to the condition $\ell_{AB'} \cap \ell_{BA'} \in \ell$. $\qquad \square$

Exercise 4.18. Prove the previous proposition in the case when the line ℓ is tangent to the conic $\mathcal{C}$.

Theorem 4.19 (Pascal Theorem). *Let $\mathcal{C}$ be a non-degenerate conic with given six points $M, N, O, P, Q, R \in \mathcal{C}$. Then the points $D = l_{MN} \cap l_{PQ}$, $E = \ell_{NO} \cap \ell_{QR}$ and $F = \ell_{OP} \cap \ell_{RM}$ are collinear.*

Proof. Denote the intersection points of the conic $\mathcal{C}$ with a line ℓ_{DE} as G, H. According to Proposition 4.17, there exists a projective transformation $f : \mathcal{C} \to \mathcal{C}$ such that G, H are its fixed points and $f(N) = Q$, but also $f(P) = M$ and $f(R) = O$. Applying Proposition 4.17 again, we conclude that the lines ℓ_{OP} and ℓ_{RM} intersect on the line ℓ_{DE}. $\qquad \square$

Exercise 4.20. Let a non-degenerate conic $\mathcal{C}$ be given together with its four points M, N, P, Q. The intersection of the tangents to the conic at M and P is collinear with the points of intersection $\ell_{MN} \cap \ell_{PQ}$ and $\ell_{NP} \cap \ell_{MQ}$.

Exercise 4.21 (Simpson's line). Given three points M, N, P on a non-degenerate conic $\mathcal{C}$. Then the intersections of the tangents to the conic at M, N, P with the opposite sides of the triangle MNP are collinear.

Exercise 4.22 (Pappus Theorem). Formulate and prove the version of the Pascal theorem in a case of degenerate conic.

4.4 Pencils of conics

We have already mentioned that the space of all conics in $\mathbf{KP}^2$ is a five-dimensional projective space $\mathbf{KP}^5$. *A linear system of conics* is a projective subspace of the projective space $\mathbf{KP}^5$. If a linear system of conics is one-dimensional, we call it *a pencil of conics*.

Example 4.23. Given two points A, B in a projective plane. Consider all conics which contain the points A and B. By applying a projective transformation which maps the points A and B to the cyclic points I and J, we map the set of conics to the set of circles. The set of circles is a three-dimensional system of conics which contain cyclic points.

Example 4.24. Consider line ℓ and a pencil of lines through the point A, where A does not belong to ℓ. Then the degenerate conics of the form $\mathcal{C}_s = \ell + s$, with $s \in \mathcal{X}_A$ form a pencil of conics. All of them have a common line. Moreover, any pencil of conics having a common line is of that form.

We will consider, from now on, pencils of conics not having a common line. Such a pencil has at most four *base points*, the points that are common for all the conics in the pencil.

A general pencil of conics has four distinct base points P, Q, R, S. No three of them are collinear. Such a pencil contains three degenerate conics

$$\mathcal{C}_1 = \ell_{PQ} + \ell_{RS}, \quad \mathcal{C}_2 = \ell_{PR} + \ell_{QS}, \quad \mathcal{C}_3 = \ell_{PS} + \ell_{QR}.$$

All other conics in the pencil are non-degenerate. By associating a projective frame to the base points we come to the equation of the pencil,

$$\lambda x_1(x_2 - x_3) + \mu x_2(x_1 - x_3) = 0. \tag{4.4}$$

Theorem 4.25. *The set of conics containing four given points is a pencil if no three of the points are collinear.*

Proof. If we associate a projective frame to the four given points, then the conics passing trough the first three of the points are of the form

$$A x_1 x_2 + B x_2 x_3 + C x_3 x_1 = 0.$$

The fourth condition gives the form of conics

$$C x_1(x_2 - x_3) + B x_2(x_1 - x_3) = 0,$$

which is of the form (4.4). $\square$

Theorem 4.26 (Desargues Theorem). *If line ℓ does not contain any of the base points of the pencil $\mathcal{X}$, then $\mathcal{X}$ induces an involution on ℓ.*

Proof. An involution of the line ℓ corresponds to a pencil of divisors of degree 2. A pencil of divisors of degree 2 corresponds to a pencil of quadratic equations. The intersection of a conic with the line ℓ is described by a quadratic equation. Coefficients of such a quadratic equation are linear functions of coefficients of the corresponding conic from the pencil. $\square$

Corollary 4.27. *If line ℓ does not contain any of the base points of the pencil $\mathcal{X}$, then exactly two conics of $\mathcal{X}$ are tangent to ℓ.*

Examples of pencils of conics

Pencil with a double base point. Consider a pencil with three base points P, Q, R, such that conics from the pencil have an intersection of multiplicity 2 at one of the points, say P. The conic

$$\mathcal{C}_1 = \ell_{PQ} + \ell_{PR}$$

is the only one having P as a double point. All other conics share the same tangent t at P. The pencil contains only one more degenerate conic besides $\mathcal{C}_1$:

$$\mathcal{C}_2 = t + \ell_{QR}.$$

Choosing the appropriate projective frame: $P = (0,0,1)$, $Q = (0,1,0)$ and $R = (1,0,0)$ with $(1,1,1) \in t$, we get the equation of the pencil

$$\lambda x_1 x_2 + \mu x_3 (x_2 - x_1) = 0. \tag{4.5}$$

As an example, one can consider a pencil of circles having common tangent t at a given point.

Pencil with two double base points. Let us consider a pencil of conics having intersections of order 2 at two base points P, Q. This pencil contains a degenerate conic

$$\mathcal{C}_1 = 2\ell_{PQ}.$$

We call such a pencil a *bitangent pencil* because all other conics are tangent to a line $t \ni P$ and to a line $s \ni Q$. The pencil contains another degenerate conic,

$$\mathcal{C}_2 = t + s.$$

In the appropriate projective frame: $P = (0,0,1)$, $Q = (0,1,0)$, $(1,0,0) \in t$ and $(1,1,1) \in s$, we get the equation of the bitangent pencil of conics,

$$\lambda x_1 x_2 + \mu x_3^2 = 0. \tag{4.6}$$

A pencil of concentric circles is an example of a bitangent pencil of conics. Two base points are the cyclic points and two common tangents are isotropic lines from the center.

Pencil with a triple base point. Consider the pencil of conics having intersections of order 3 at P and of order 1 at Q. All the conics from the pencil have a common tangent t at P. The only degenerate conic in the pencil is

$$\mathcal{C}_1 = t + \ell_{PQ}.$$

In the appropriate projective frame, the equation of the pencil can be written in the form

$$\lambda(x_1 x_3 + x_2^2) + \mu x_2 x_1 = 0. \tag{4.7}$$

Pencils with a quadruple base point. The last case considers pencils with one base point P. Then the conics have intersections with multiplicity 4 at P. Any degenerate conic from the pencil is a union of two lines intersecting at P.

If such a pencil contains two degenerate conics $\mathcal{C}_1 = F_1(x_1, x_2, x_3)$ and $\mathcal{C}_2 = F_2(x_1, x_2, x_3)$ then all the conics from the pencil are degenerate as unions of two lines which pass through the point P. The equation of the pencil in this case is

$$\lambda F_1(x_1, x_2, x_3) + \mu F_2(x_1, x_2, x_3) = 0.$$

As a subcase, we consider pencils which contain two double lines. If the characteristic of the base field is different from 2, all other conics from the pencil are not double lines.

If a pencil contains only one degenerate conic, then other, non-degenerate conics intersect at the point P with multiplicity 4; they have a common tangent t at P. With a projective frame such that $P = (0, 0, 1)$ and t as x_2-axis, we can derive the equation of the pencil,

$$\lambda(x_1 x_3 + F(x_1, x_2)) + \mu x_1^2 = 0,$$

where F is a homogeneous polynomial in x_1, x_2 of degree 2.

4.5   Quadrics and polarity

Given vector space V over field $\mathbf{K}$. A quadratic form F defines *a quadric* $\mathcal{Q}$ in the projective space $\mathcal{A} = \mathbf{P}(V)$. There is a bilinear form B associated with the quadratic form F:

$$B(x, x) = F(x).$$

The form B is non-degenerate, $\det B \neq 0$, if and only if the quadric $\mathcal{Q}$ does not have double points. In this case, there is an isomorphism f between the vector space V and its dual V^*:

$$f(x)(y) := B(x, y). \tag{4.8}$$

Denote by $\mathbf{P}(V^*)$ the projective space of hyperplanes in $\mathbf{P}(V)$. Then, f induces an isomorphism $\mathbf{P}(f)$ from $\mathbf{P}(V)$ to $\mathbf{P}(V^*)$.

Definition 4.28. The hyperplane $\mathbf{P}(f)(x_0)$ is called *the polar* of the point x_0 with respect to the quadric $\mathcal{Q}$.

The equation of the polar hyperplane is

$$B(x_0, x) = 0.$$

More generally, $\mathbf{P}(f)$ defines a correspondence between a projective subspace $\mathcal{A}_1 = \mathbf{P}(W)$ of $\mathbf{P}(V)$ and a linear system of hyperplanes $\mathcal{A}_1^* = \mathbf{P}(W^\perp)$ of $\mathbf{P}(V^*)$. Since this correspondence is symmetric, we say that spaces $\mathcal{A}_1$ and $\mathcal{A}_1^*$ are *conjugate* to each other with respect to the quadric $\mathcal{Q}$.

Definition 4.29. A point which is the conjugate of a hyperplane with respect to the quadric Q is called *the pole* of the hyperplane with respect to the quadric Q.

Proposition 4.30. *Let Q be a quadric with no singular points and A_1, A_2 two points not belonging to Q. Denote by M, M' the intersection points of the line $l_{A_1 A_2}$ with the quadric Q. Then, the points A_1 and A_2 are conjugate to each other if and only if the pairs of points (A_1, A_2) and (M, M') are harmonic conjugates:*

$$(A_1, A_2, M, M') = -1.$$

Proof. The points M, M' are determined as roots of the equation $F(A_1 + tA_2) = 0$, quadratic in t. Rewrite the equation

$$F(A_1)t^2 + 2B(A_1, A_2)t + F(A_2) = 0. \tag{4.9}$$

Since by the assumption, $F(A_i) \neq 0$, the sum of roots is equal to zero if and only if $B(A_1, A_2) = 0$. This is equivalent to the condition of being harmonic conjugate with 0 and ∞, the parameters corresponding to A_1 and A_2. $\square$

From equation (4.9) we see that in the case $A_1 \in Q$ we have $F(A_1) = 0$ and the equation has a root $t = 0$. In this case, the condition $B(A_1, A_2) = 0$ implies that 0 is a double root. Thus, in the case $A_1 \in Q$ the line $l_{A_1 A_2}$ is tangent to the quadric Q. We have proved the following

Proposition 4.31. *The polar of a point $A \in Q$ is the tangent hyperplane to the quadric Q at the point A.*

Exercise 4.32.

(a) Let x_0 be the pole of hyperplane α with respect to quadric Q. Then, α is the polar of x_0 with respect to Q.

(b) Let hyperplane α be tangent to quadric Q at point x_0. Then the pole of α with respect to Q is the point x_0.

(c) If a hyperplane α contains point x, then the pole of α belongs to the polar of the point x.

(d) If hyperplanes belong to a pencil, then their poles with respect to a given quadric are collinear.

(e) Consider a line containing the pole z of a hyperplane α with respect to quadric Q. If the line intersects α at point p, and Q at y_1, y_2, then the four points $y_1, y_2; z, p$ are harmonically conjugate.

(f) If the infinite hyperplane is not tangent to quadric Q, its pole is the center of the quadric.

4.6   Polarity and pencils of conics

Now, we are going to apply previous considerations of polarity to pencils of conics. However, we know that a pencil of conics contains degenerate conics as well. Thus, we need first to establish the notion of polarity in the case of degenerate conics.

Let

$$\mathcal{C}_1 = \ell_1 + \ell_2$$

be a degenerate conic consisting of two distinct lines, with P as their intersecting point. If we denote by $B(x, y)$ the corresponding bilinear form, then $B(P, x) = 0$ for any x. If $A \neq P$ then the equation $B(A, x) = 0$ determines a line $p_{\mathcal{C}_1}(A)$, *the polar* of the point A with respect the degenerate conic $\mathcal{C}_1$. If $A \in \ell_1$ then $\ell_A = \ell_1$. If $A \notin \mathcal{C}_1$ then there is an extension of Proposition 4.30:

Corollary 4.33. *Given a degenerate conic* $\mathcal{C}_1 = \ell_1 + \ell_2$, *point* $A \notin \mathcal{C}_1$ *and the polar* ℓ_A *of the point* A *with respect to* $\mathcal{C}_1$. *The four lines* $\ell_1, \ell_2, \ell_{PA}, p_{\mathcal{C}_1}(A)$ *are harmonically conjugate in the pencil of lines through the point* $P = \ell_1 \cap \ell_2$.

Thus, the notion of polar is well defined if $A \neq P$, and a polar contains the point P. But, the notion of the pole of a line with respect to a degenerate conic $\mathcal{C}_1$ is not well defined.

In the case of a double line as a degenerate conic

$$\mathcal{C}_2 = 2\ell,$$

for a point A of the line ℓ we have $B(A, x) = 0$ for every point x. If $A \notin \ell$ the solution of the equation $B(A, x) = 0$ is the line ℓ. We conclude that the notion of polar is defined for $A \notin \ell$ and $\ell_A = \ell$. The notion of the pole is not defined.

For a non-degenerate conic $\mathcal{C}$ and a point $P \notin \mathcal{C}$, the polar $p_{\mathcal{C}}(P)$ of the point P with respect to $\mathcal{C}$ intersects the conic $\mathcal{C}$ at the points A, B of contact of tangents to $\mathcal{C}$ from the point P:

$$p_{\mathcal{C}}(P) \cap \mathcal{C} = \{A, B\} \longleftrightarrow t_{\mathcal{C}}(A) \cap t_{\mathcal{C}}(B) = \{P\}, \qquad (4.10)$$

where $t_{\mathcal{C}}(A)$ denotes the tangent to the conic $\mathcal{C}$ at the point $A \in \mathcal{C}$.

Before we pass to pencils of conics, we introduce the notion of a *self-polar triangle* with respect to a conic $\mathcal{C}$: a triangle P, Q, R is self-polar with respect to a conic $\mathcal{C}$ if any two of its vertices are conjugate with respect to the conic $\mathcal{C}$. It is equivalent to say that every vertex is the pole of the opposite side of the triangle.

For a given conic $\mathcal{C}$ and a point $P \notin \mathcal{C}$ one constructs a self-polar triangle by choosing a point Q on the polar $p_{\mathcal{C}}(P)$, and the point R such that the points Q, R are harmonically conjugate with the intersection points of the polar $p_{\mathcal{C}}(P)$ with the conic $\mathcal{C}$.

For a given triangle $T = PQR$, the set of conics to which the triangle T is self-polar form a two-dimensional projective subspace of the five-dimensional projective space of conics. If a projective frame is chosen with vertices P, Q and R then the equation of such conics is in a diagonal form:

$$Ax_1^2 + Bx_2^2 + Cx_3^2 = 0. \tag{4.11}$$

Indeed, if the coordinates of the vertices are $P = (1,0,0)$, $Q = (0,1,0)$ and $R = (0,0,1)$, these are also homogeneous coordinates of lines ℓ_{QR}, ℓ_{PR} and ℓ_{PQ} respectively. The condition of self-polarity is

$$\hat{C}P = \lambda_1 \ell_{QR}, \quad \hat{C}Q = \lambda_2 \ell_{PR}, \quad \hat{C}R = \lambda_3 \ell_{PQ},$$

where $\hat{C}$ denotes a matrix of the conic $\mathcal{C}$ in the chosen frame. Thus, the matrix $\hat{C}$ is diagonal in the frame.

General pencil of conics

Let us consider a general pencil of conics $\mathcal{X}$ having four distinct base points A, B, C, D, no three of them being collinear. Denote by $\mathcal{Q}(t)$ a general conic from the pencil with the equation

$$Q_1(x) + tQ_2(x) = 0,$$

and denote by $p_{\mathcal{Q}(t)}(M)$ the polar of point M with respect to conic $\mathcal{Q}(t)$. Coefficients of polar $p_{\mathcal{Q}}(t)(M)$ are linear functions of pencil parameter t. Thus, when the conic runs through the pencil, the polar describes a pencil of lines or it is fixed. We want to describe the cases of points M such that $p_{\mathcal{Q}(t)}(M)$ is fixed as a function of t.

The pencil $\mathcal{X}$ contains three degenerate conics,

$$\mathcal{Q}_1 = \ell_{AB} + \ell_{CD}, \quad \mathcal{Q}_2 = \ell_{AC} + \ell_{BD}, \quad \mathcal{Q}_3 = \ell_{AD} + \ell_{BC}.$$

Their double points P, Q, R form the diagonal triangle of quadrilateral $ABCD$.

If $M \notin \{P, Q, R\}$ then polars $p_{\mathcal{Q}(t)}(M)$ are well defined with respect to all conics in the pencil. Being fixed, it would pass through the points P, Q, R. But in characteristics different from 2, the points P, Q, R are not collinear.

If $M = P$, then ℓ_{QR} is the polar of the point P with respect to both other degenerate conics. Indeed, one can easily see that the lines $\ell_{BD}, \ell_{AC}, \ell_{QR}, \ell_{QP}$ are harmonically conjugate in the pencil of lines through Q. Similarly, the lines $\ell_{AD}, \ell_{BC}, \ell_{QR}, \ell_{RP}$ are harmonically conjugate in the pencil of lines through the point R. Thus,

$$\ell_{QR} = p_{\mathcal{Q}_2}(P) = p_{\mathcal{Q}_3}(P),$$

and, due to linearity, $\ell_{QR} = p_{\mathcal{Q}(t)}(P)$ for any t.

The same is true for $M = Q$ and $M = R$. Thus there are only three points having a fixed polar with respect to all conics in the pencil $\mathcal{X}$. Moreover, the triangle $T = PQR$ is the only self-polar triangle with respect to all conics in the pencil.

Exercise 4.34.

(a) Let $\mathcal{X}$ be a pencil of conics touching line t at point $A \in t$, and containing two points B, C. Prove that there are only two points M: $M = A$ and $M = \ell_{BC} \cap t$, having the polar fixed. Prove that there are no triangles that are self-polar with respect to all conics in the pencil.

(b) Let $\mathcal{X}$ be a pencil of bitangent conics, touching lines t, s at $A \in t$, $B \in s$ respectively. Prove that $M = t \cap s$ is the only point having fixed polar with respect to all conics in the pencil. Prove that there is an infinity of triangles that are self-polar with respect to all conics in the pencil $\mathcal{X}$.

(c) Let $\mathcal{X}$ be a pencil of conics, having common tangent t with multiplicity 3 at point $A \in t$ and passing through B. Prove that A is the only point that has fixed polar with respect to all conics in the pencil. Prove that no triangle is self-polar with respect to all conics in the pencil $\mathcal{X}$.

(d) Let $\mathcal{X}$ be a pencil of conics, having common tangent t with multiplicity 4 at $A \in t$. Describe the set of points having fixed polar with respect to all conics in the pencil. Prove that there no triangle is self-polar with respect to all conics in the pencil $\mathcal{X}$.

From the previous considerations we have

Theorem 4.35. *Let $\mathcal{X}$ be a pencil of conics containing a non-degenerate conic.*

(a) *Let M be a point different from any double point of any degenerate conic from the pencil $\mathcal{X}$. Then the mapping joining to conic $\mathcal{Q}(t) \in \mathcal{X}$ its polar $p_{\mathcal{Q}(t)}(M)$ is a projective transformation from the pencil $\mathcal{X}$ onto the pencil of lines through point M'.*

(b) *If M is a double point of a degenerate conic from the pencil, then it has fixed polar with respect to all conics in $\mathcal{X}$.*

(c) *If all conics from pencil $\mathcal{X}$ have exactly one common self-polar triangle, then the pencil $\mathcal{X}$ has four distinct base points.*

(d) *If conics from pencil $\mathcal{X}$ have more than one self-polar triangle, then $\mathcal{X}$ is a pencil of bitangent conics.*

Theorem 4.35 shows that a general pencil of conics $\mathcal{X}$ establishes a correspondence between a point M with a point M', the base of pencil of polars of M with respect to conics of $\mathcal{X}$. On line $\ell_{MM'}$, the pencil of conics $\mathcal{X}$ induces involution $i_{\mathcal{X}}$. Pair M, M' is harmonically conjugate with every pair of intersection points of $\ell_{MM'}$ with a conic from pencil $\mathcal{X}$. This means that points M, M' are fixed points of involution $i_{\mathcal{X}}$ of line $\ell_{MM'}$. As a consequence we get the following

Corollary 4.36. *Given pencil of conics $\mathcal{X}$ and line ℓ. Then, according to Theorem 4.35 there is only one pair of points M, M' on ℓ corresponding to each other.*

Exercise 4.37. Conics having a common self-polar triangle T and passing through a given point P either pass through three other fixed points, or, if P is on a side of the triangle T, they have a common tangent at P and another common tangent at another point Q.

Theorem 4.38. *Let $\mathcal{X}$ be a general pencil of conics with four base points and ℓ a line. Then the set of poles $P_{\mathcal{Q}(t)}(\ell)$ of line ℓ with respect to conics $\mathcal{Q}(t) \in \mathcal{X}$ is a conic $\mathcal{C}$ when t runs through $\mathbf{P}_1$. The conic $\mathcal{C}$ contains the vertices of the self-polar triangle T of pencil $\mathcal{X}$. Moreover, $\mathcal{C}$ is the locus of points M' which correspond to points M, when M runs through the line ℓ.*

Proof. Take points A, B on ℓ, such that none of them is a vertex of the triangle T. As conics $\mathcal{Q}(t)$ run through the pencil $\mathcal{X}$, polars $p_{\mathcal{Q}(t)}(A)$ and $p_{\mathcal{Q}(t)}(B)$ run through pencils of lines with base points A' and B'. According to Theorem 4.35, a projective transformation between the last two pencils is established. Intersections of the corresponding lines from the two line pencils are poles of line ℓ. By applying Theorem 4.13, we get the statement. $\square$

Exercise 4.39. Given a quadrilateral in the Euclidean plane, the mid-points of six of its sides and the vertices of the diagonal triangle are on the same conic.

Exercise 4.40 (Euler circle of nine points). Given a triangle ABC in the Euclidean plane. Then there exists a circle containing the mid-points of the sides of the triangle, the foot-points of altitudes, and the mid-points of the segments AH, BH and CH, where H is the orthocenter of the triangle.

Theorem 4.41. *Let triangle $T = PQR$ be self-polar to conic $\mathcal{Q}$ and inscribed in another conic $\mathcal{S}$. Then, any point $A \in \mathcal{S}$ is a vertex of a triangle ABC inscribed in $\mathcal{S}$ and self-polar with respect to $\mathcal{Q}$.*

Proof. Consider polar $p = p_{\mathcal{Q}}(A)$. Denote by B, C its intersection points with $\mathcal{S}$, and by M, M' the intersecting points with $\mathcal{Q}$. Consider pencil of conics $\mathcal{X}$ defined by the set of base points $\{A, P, Q, R\}$ and denote by (J, K), (J_1, K_1), (J_2, K_2) intersection points of line p with degenerate conics $\ell_{AP}+\ell_{QR}$, $\ell_{AQ}+\ell_{PR}$, $\ell_{AR}+\ell_{PQ}$. We have

$$(J, K, M, M') = -1 = (J_1, K_1, M, M') = (J_2, K_2, M, M'). \qquad (4.12)$$

The first relation holds because ℓ_{AP} is the polar of the point K, as it joins the point A, the pole of p and the point P, the pole of ℓ_{QR}. The other two relations follow similarly. According to the Desargues Theorem 4.26, the pencil $\mathcal{X}$ induces an involution on p. By relations (4.12), we see that M, M' are fixed points of the involution. Conic $\mathcal{S}$ belongs to the pencil $\mathcal{X}$ and gives the pair (B, C) on the line p. Thus

$$(B, C, M, M') = -1,$$

and B and C are conjugate with respect to $\mathcal{Q}$. It follows that ABC is a self-polar triangle with respect to $\mathcal{Q}$. $\square$

Theorem 4.41 is an example of a *poristic* statement.

Corollary 4.42. *The six vertices of two triangles self-polar to a conic $\mathcal{Q}$ belong to a conic $\mathcal{S}$. Conversely, if A, B, C, D, E, F are points of conic $\mathcal{S}$, then there is a conic $\mathcal{Q}$ which the triangles ABC and DEF are self-polar to.*

Proof. Let ABC and DEF be two triangles that are self-polar with respect to $\mathcal{Q}$, and let $\mathcal{S}$ be a conic determined by five points A, B, C, D, E. The polar $p_{\mathcal{Q}}(D)$ intersects $\mathcal{S}$ in E and in some other point G. By Theorem 4.41, G is the pole of ℓ_{DE} with respect to $\mathcal{Q}$. But, F is the pole of the same line, by assumption. Thus, $F = G$.

Conversely, consider a two-dimensional system of conics to which the triangle ABC is self-polar. Assumption that D and ℓ_{EF} are pole and polar adds two linear conditions, with conic $\mathcal{C}$ satisfying all the conditions. According to Theorem 4.41, triangle DEF is self-polar. $\square$

Remark 4.43. Theorem 4.41 can be derived from Serret's Theorem.

4.7   Invariants of pairs of conics

Let $\mathcal{C}_1$ and $\mathcal{C}_2$ be two conics in the projective plane, defined by their matrices of quadratic forms C_1 and C_2. We suppose that the conics are in general position, i.e., that they have four distinct intersecting points.

Symmetric 3×3 matrices, as we know, uniquely up to a scalar multiple, define conics. Thus, pairs of conics describe a ten-dimensional projective space. We are interested in invariants of pairs of conics, with respect to projective transformations of the projective plane, which form an eight-dimensional group. The subgroup which leaves a pair fixed is zero-dimensional, which implies the existence of two independent invariants.

We start with the characteristic polynomial of the pair of forms

$$\det(\lambda C_1 + \mu C_2) = \lambda^3 I_1 + \lambda^2 \mu I_2 + \lambda \mu^2 I_3 + \mu^3 I_4,$$

where I_1, I_2, I_3, I_4 are certain polynomials in elements of matrices C_1 and C_2. We will give the expressions later in a special coordinate frame, now notice that $I_1 = \det C_1$ and $I_4 = \det C_2$.

The quantities I_i are invariants of pairs of forms, but not of the pair of conics. A conic determines its form only up to a scalar multiple and the coefficients I_i obviously satisfy the following:

$$I_1(sC_1, tC_2) = s^3 I_1(C_1, C_2),$$
$$I_2(sC_1, tC_2) = s^2 t I_2(C_1, C_2),$$
$$I_3(sC_1, tC_2) = st^2 I_3(C_1, C_2),$$
$$I_4(sC_1, tC_2) = t^3 I_4(C_1, C_2).$$

Thus, true invariants of pairs of conics can be obtained by homogenizing expressions in I_is.

Proposition 4.44. *Two invariants of pairs of conics are given by the formulae*

$$\alpha = \frac{I_1 I_3}{I_2^2}, \quad \beta = \frac{I_2 I_4}{I_3^2}.$$

Now, we pass to the projective frame associated with the common self-polar triangle PQR for conics $\mathcal{C}_1$ and $\mathcal{C}_2$. We know, see equations (4.11), that in that frame, the matrices of conics $\mathcal{C}_1$ and $\mathcal{C}_2$ are simultaneously diagonalized:

$$\mathcal{C}_1 : \quad a_1 x_1^2 + b_1 x_2^2 + c_1 x_3^2 = 0,$$
$$\mathcal{C}_2 : \quad a_2 x_1^2 + b_2 x_2^2 + c_2 x_3^2 = 0.$$

Thus

$$I_1 = a_1 b_1 c_1,$$
$$I_2 = b_1 c_1 a_2 + a_1 c_1 b_2 + a_1 b_1 c_2,$$
$$I_3 = b_2 c_2 a_1 + a_2 c_2 b_1 + a_2 b_2 c_1,$$
$$I_4 = a_2 b_2 c_2.$$

Let us consider matrix $C_2^{-1} C_1$. Its eigen-vectors are coordinates of vertices of the self-polar triangle, see (4.11). The eigen-values are:

$$\lambda_1 = \frac{a_1}{a_2}, \quad \lambda_2 = \frac{b_1}{b_2}, \quad \lambda_3 = \frac{c_1}{c_2}.$$

Ratios of the eigen-values are invariants of the pair of conics and they can be expressed in terms of cross-ratios.

Proposition 4.45. *Let $\{M_1, N_1\} = \ell_{PQ} \cap \mathcal{C}_1$ and $\{M_2, N_2\} = \ell_{PQ} \cap \mathcal{C}_2$. Then*

$$\frac{\lambda_1}{\lambda_2} = (M_1, M_2, P, Q)^2.$$

Proof. The equation of the line l_{PQ} is $x_3 = 0$ and the points of intersection have the following coordinates:

$$M_1 = (\sqrt{b_1}, i\sqrt{a_1}, 0), \quad N_1 = (\sqrt{b_1}, -i\sqrt{a_1}, 0),$$
$$M_2 = (\sqrt{b_2}, i\sqrt{a_2}, 0), \quad N_2 = (\sqrt{b_2}, -i\sqrt{a_2}, 0).$$

Now, it is easy to compute the cross-ratio and to finish the proof. □

From the relations

$$\frac{I_1}{I_4} = \lambda_1 \lambda_2 \lambda_3,$$

$$\frac{I_2}{I_4} = \lambda_1 \lambda_2 + \lambda_2 \lambda_3 + \lambda_1 \lambda_3,$$

$$\frac{I_3}{I_4} = \lambda_1 + \lambda_2 + \lambda_3,$$

we can easily get the following

Proposition 4.46. *There is a connection between the invariants*

$$\alpha = \frac{I_1 I_3}{I_2^2} = \frac{\operatorname{Tr}\,(C_2^{-1}C_1)\,\det C_2}{(\operatorname{Tr}\,(C_1^{-1}C_2))^2\,\det C_1},$$

$$\beta = \frac{I_4 I_2}{I_3^2} = \frac{\operatorname{Tr}\,(C_1^{-1}C_2)\,\det C_1}{(\operatorname{Tr}\,(C_2^{-1}C_1))^2\,\det C_2}.$$

Proof. Follows by direct calculation, taking into account

$$\frac{I_1}{I_4} = \frac{\det C_1}{\det C_2}, \quad \frac{I_2}{I_1} = \operatorname{Tr}\,(C_1^{-1}C_2), \quad \frac{I_3}{I_4} = \operatorname{Tr}\,(C_2^{-1}C_1). \qquad \square$$

The four points in the intersection of the conics determine the cross-ratios k_1 and k_2 on each of the conics $\mathcal{C}_1$ and $\mathcal{C}_2$, according to the Chasles Theorem, see Corollary 4.12. We want now to compare these invariants with the previous ones. Denote by h_1, h_2, h_3 zeroes of the polynomial

$$P(t) = \det(C_1 + tC_2).$$

Obviously, $h_i = -\lambda_i$.

Theorem 4.47. *The pair of invariants (k_1, k_2) can be expressed in the following ways:*

$$k_1 = (0, h_2, h_3, h_1) = \frac{\frac{\lambda_2}{\lambda_1} - 1}{\frac{\lambda_2}{\lambda_3} - 1},$$

$$k_2 = (\infty, h_2, h_3, h_1) = \frac{1 - \frac{\lambda_1}{\lambda_2}}{1 - \frac{\lambda_3}{\lambda_2}}.$$

Proof. Change the projective frame to the four intersection points $A = (1,0,0)$, $B = (0,1,0)$, $C = (0,0,1)$, $D = (1,1,1)$. In the new frame, the equation of conics becomes

$$C_1: \quad A_1 x_1 x_2 + B_1 x_2 x_3 - (A_1 + B_1)x_1 x_3 = 0,$$
$$C_2: \quad A_2 x_1 x_2 + B_2 x_2 x_3 - (A_2 + B_2)x_1 x_3 = 0.$$

Then we easily calculate

$$h_1 = -\frac{B_1}{B_2}, \quad h_2 = -\frac{A_1}{A_2}, \quad h_3 = -\frac{A_1 + B_1}{A_2 + B_2}.$$

Using the following parametrization of the conic $\mathcal{C}_1$,

$$t \mapsto ((B_1(1-t) - A_1 t)t, B_1(1-t) - A_1 t, -A_1 t),$$

we get coordinates of the intersection points A, B, C, D on $\mathcal{C}_1$:

$$(A, B, C, D) = (\infty, 0, \frac{B_1}{A_1 + B_1}, 1).$$

Now one calculates

$$k_1 = (A, B, C, D)_{C_1} = 1 + \frac{A_1}{B_1} = (0, h_2, h_3, h_1).$$

The rest follows by direct calculations. $\qquad\qquad\qquad\qquad\qquad\qquad$ $\square$

Exercise 4.48. Two conics C_1 and C_2 are given such that there exists a triangle inscribed in C_1 and circumscribed about C_2.

(a) Prove that there exists a frame such that the conics have equations of the form

$$
\begin{aligned}
C_1 &: \; Ax_1x_2 + Bx_2x_3 + Cx_1x_3 = 0, \\
C_2 &: \; x_1^2 + x_2^2 + x_3^2 - 2(x_1x_2 + x_1x_3 + x_2x_3) = 0.
\end{aligned}
\tag{4.13}
$$

(b) Prove

$$I_3^2 = 4I_4I_2 \Leftrightarrow \beta = \frac{1}{4}. \tag{4.14}$$

Exercise 4.48 gives the invariant relation ((4.14)) satisfied by a pair of conics C_1, C_2 in order to be C_1 3-*circumscribed* to C_2.

Exercise 4.49. Two conics C_1 and C_2 are given such that two of their common tangents intersect on a common chord.

(a) Prove that there exists a projective transformation mapping the pair of conics onto a pair of congruent circles.

(b) Prove the following invariant relation between such conics:

$$I_1I_3^3 = I_2^3I_4 \Leftrightarrow \alpha = \beta. \tag{4.15}$$

Exercise 4.50. Two conics C_1 and C_2 are given such that there exists a quadrilateral inscribed in C_1 and circumscribed about C_2. Then, they satisfy the invariant relation

$$I_3^2 + 8I_1I_4^2 - 4I_2I_3I_4 = 0. \tag{4.16}$$

Exercise 4.50 provides an invariant relation satisfied by a pair of conics, one of which being 4-*circumscribed* about the other one.

4.8 Duality. Complete conics

Given a non-degenerate quadric Q in $\mathbf{KP}^n$, the polarity correspondence with respect to Q maps a point $P \in Q$ to the tangent hyperplane $p_Q(P)$ to quadric Q at P. In this way, a set of hyperplanes is associated to the set of points of the quadric Q, namely the set of tangent hyperplanes of the quadric Q. We can denote this set as a *dual quadric* Q^v. The dual quadric is really a quadric in the dual space $\mathbf{KP}^*_n$ of hyperplanes of the original space $\mathbf{KP}^n$.

Set $V = \mathbf{K}^{n+1}$ and let B be the bilinear form associated with the quadric $\mathcal{Q}$. It induces an isomorphism

$$f : V \to V^*, \quad f(x)(y) = B(x,y),$$

and defines a bilinear form on V^*:

$$B^v(u,v) = B(f^{-1}(u), f^{-1}(v)).$$

The dual quadric $\mathcal{Q}^v$ is a quadric associated with the bilinear form B^v.

Proposition 4.51. *The quadric $\mathcal{Q}^v$ is dual to the quadric $\mathcal{Q}$ if and only if*

$$Q^v(u) = Q(f^{-1}(u)).$$

Proof. Hyperplane h with the equation $u(x) = 0$ is tangent to $\mathcal{Q}$ if and only if $u \in f(\mathcal{Q})$. This means $f^{-1}(u) \in \mathcal{Q}$ and $Q(f^{-1}(u)) = 0$. $\qquad\square$

Thus a quadric dual to a non-degenerate quadric $\mathcal{Q}$ is also non-degenerate. Its equation $Q^v(u) = 0$ is called *the tangential equation* of the quadric $\mathcal{Q}$ and the dual quadric is sometimes called *the hyperplane envelope of the quadric $\mathcal{Q}$.* The original quadric is, in this context, sometimes called the *locus* or the *point locus* quadric.

If M is the matrix of the form B in some basis e of V, then the matrix of B^v in the dual basis e^* of V^* is the inverse matrix M^{-1}. Obviously, the dual of the dual of the non-degenerate quadric $\mathcal{Q}$ is $\mathcal{Q}$ itself:

$$\mathcal{Q}^{vv} = \mathcal{Q}.$$

Conics and duality

We now pass to study of duality in the case of conics.

Exercise 4.52. Find the tangential equations of the conics

(a) $A_1 x_1^2 + A_2 x_2^2 + A_3 x_3^2 = 0$,
(b) $x_1 x_2 + 3 x_1 x_3 + 5 x_2 x_3 = 0$.

In the next exercises, dual versions of theorems which are previously proved for conics are formulated.

Exercise 4.53. Prove the following statements:

(a) (Dual version of Theorem 4.13.) Let $g : \ell \to \ell_1$ be a projective transformation between lines ℓ and ℓ_1. If the point $T = \ell \cap \ell_1$ is not a fixed point for g, then lines $\ell_{Mg(M)}$ envelope a conic tangent to ℓ and ℓ_1.

If $g(T) = T$, then all lines $\ell_{Mg(M)}$, $M \neq T$, contain a fixed point.

(b) (*Dual version of the Frégier Theorem, see Theorem* 4.14.) Given a non-degenerate conic $\mathcal{C}$ and an involution i on it. The intersection points of tangents to $\mathcal{C}$ at M and $i(M)$ run through a line when M runs through the conic $\mathcal{C}$. This line contains the fixed points of the involution i.

(c) *Brianchon Theorem.* (*Dual of the Pascal theorem.*) Given a non-degenerate conic $\mathcal{C}$ and its six tangents $t_1, \ldots, t_6$. Denote the intersection points by $i = t_i \cap t_{i+1 (\mathrm{mod}\, 6)}$. The lines $\ell_{14}, \ell_{25}, \ell_{36}$ are concurrent.

Complete conics

As we investigated degenerations of a point conic locus, degenerations of envelope conics need to be studied as well. There are two cases:

(a) $\mathcal{C}_1^v = A + B$ represents the sum of two pencils of lines, one with the base point A and the other one with B as the base point.

(b) $\mathcal{C}_2^v = 2A$ consisting of the pencil of lines through the point A counted twice.

Now, we want to extend the notion of dual conic from the non-degenerate case to degenerate conics. Up to now, the duality defines a correspondence in the space $\mathbf{KP}^5 \times \mathbf{KP}^{v5}$ of all conics in the space and all conics in the dual space, determined for non-degenerate conics. Starting from the relation

$$MM^v = E$$

between matrices of bilinear forms of a non-degenerate conic and its dual, we can extend it to the degenerate case by putting

$$MM^v = \rho E,$$

where ρ is a scalar. Analyzing the system of equations as a linear system with unknowns-entries of M^v and coefficients-entries of M we come to the conclusion:

(a) the dual associated to degenerate conic $\mathcal{C}_1 = \ell_1 + \ell_2$ is $\mathcal{C}_1^v = 2A$ where $\{A\} = \ell_1 \cap \ell_2$;

(b) to degenerate conic $\mathcal{C}_2 = 2\ell$, one can associate, as a dual, any conic of the form $A + B$ where $A, B \in \ell$ is a pair of points in ℓ.

(a^v) the dual associated to degenerate dual conic $\mathcal{C} = A + B$ represented by a pair of distinct points A, B, is conic $\mathcal{C}^v = 2\ell_{AB}$ that consists of line ℓ_{AB} counted twice.

(b^v) to degenerate dual conic $\mathcal{C} = 2A$ represented by a double point, any conic of the form $\ell_1 + \ell_2$, where $\ell_1 \cap \ell_2 = \{A\}$, can be associated as a dual one.

Previous consideration leads to the notion of *complete conics*. More precisely, denote D and D^v the sets of degenerate conics in $\mathbf{KP}^5$ and $\mathbf{KP}^{5*}$ respectively and denote by V and V^v the singular sets of D and D^v. V consists of double lines and V^v of double points. Denote by G_0 the graph of the correspondence in $(\mathbf{KP}^5 \setminus D) \times (\mathbf{KP}^{5v} \setminus D^v)$ between non-degenerate conics and their duals. Denote by G the closure of G_0 in $\mathbf{KP}^5 \times \mathbf{KP}^{5v}$.

Definition 4.54. The elements of G are called *complete conics*. For $C \in G$ the image of the first projection $p_1(C) \in \mathbf{KP}^5$ is called *the conic locus* and the image of the second projection $p_2(C) \in \mathbf{KP}^{5v}$ is called *the conic envelope*.

For a non-degenerate conic, the complete conic is uniquely determined either by the locus or the envelope. For a degenerate conic, the situation is different and a complete conic brings more information than just one of the projections.

Remark 4.55. It can be shown that the projection $p_1 : G \to \mathbf{KP}^5$ is the blow-up along V and similarly the projection $p_2 : G \to \mathbf{KP}^{5v}$ can be identified with the blow-up along V^v.

From the general theory, it follows that G is a smooth irreducible projective variety. It also follows that the exceptional varieties $A = p_1^{-1}(V)$ and $A^v = p_2^{-1}(V^v)$ are smooth and irreducible. A and A^v intersect transversally, and their intersection, i.e., the set of complete conics of double lines with double points, is also smooth and irreducible.

4.9   Confocal conics

Confocal family of conics, elliptic coordinates

It is well known that an ellipse with focal points F_1 and F_2 in the Euclidean plane is defined with a positive real number l as the set of all points M such that the sum of distances to focal points is equal to $2l$:

$$|MF_1| + |MF_2| = 2l.$$

By replacing the word "sum" by "difference" in the previous definition, we come to the notion of a hyperbola, as the set of points M satisfying

$$||MF_1| - |MF_2|| = 2l.$$

By varying l, with F_1 and F_2 fixed, we get a whole family of confocal conics. In some orthogonal frame, the family can be represented by the formula

$$\frac{x^2}{A - \lambda} + \frac{y^2}{B - \lambda} = 1, \quad A > B > 0. \tag{4.17}$$

The focal points are

$$F_1 = (c, 0), \quad F_2 = (-c, 0), \quad c = \sqrt{A - B}.$$

The values of l vary together with λ since $l = A - \lambda$.

Each pair (x, y) of Cartesian coordinates uniquely defines a pair (λ_1, λ_2) such that

$$\frac{x^2}{A - \lambda_i} + \frac{y^2}{B - \lambda_i} = 1.$$

Moreover, the inequality $A > \lambda_2 > B > \lambda_1$ is satisfied.

Pairs (λ_1, λ_2) are called *elliptic coordinates*.

Equation $\lambda = \lambda_1$ defines an ellipse, which contains point (x, y), while equation $\lambda = \lambda_2$ defines a hyperbola passing through the same point. The two conics are orthogonal to each other at the intersection point.

It is instructive to see a confocal family in projective settings. Denote by Ω the degenerate envelope conic consisting of the cyclic points

$$\Omega = \hat{I} + \hat{J}.$$

Let $\mathcal{C}$ be the conic given by the equation

$$\mathcal{C} : \quad \frac{x_1^2}{A} + \frac{x_2^2}{B} - x_3^2 = 0.$$

The equation of its dual is

$$\mathcal{C}^v : \quad Au^2 + Bv^2 - w^2 = 0.$$

Consider a pencil of dual conics $\mathcal{C}^v - \lambda\Omega$. The equations are

$$\mathcal{C}^v - \lambda\Omega : \quad (A - \lambda)u^2 + (B - \lambda)v^2 - w^2 = 0.$$

The corresponding point-locus conic equations are

$$\frac{x^2}{A - \lambda} + \frac{y^2}{B - \lambda} = z^2,$$

where we recognize equation (4.17) of a confocal family.

Definition 4.56. For a given algebraic curve γ in the projective plane, its *class* is the number of tangents to γ from a given general point.

Definition 4.57. *Focal points* of a given algebraic curve γ of class s in the projective plane are s^2 intersection points of s tangents to γ from cyclic point $\hat{I}$ with s tangents to γ from the other cyclic point $\hat{J}$.

Conics are obviously curves of class 2. Thus, there are four tangents to a conic from the cyclic points and there are four focal points of a conic.

Definition 4.58. *A confocal pencil of conics* is the set of conics in a projective plane having four common tangents.

The notion of a confocal pencil of conics is obviously dual to the notion of a general pencil of conics which have four distinct points in common, as the base set for the pencil. Going back to the example we started with, we see now that F_1 and F_2 are two of the four focal points of the confocal family. Points F_1 and F_2 are *real focal points*. The other two focal points $G_1 = (0, ic)$, $G_2 = (0, -ic)$ are not real. The family of confocal conics includes three degenerate conics:

$$(2\ell_{F_1 F_2}; F_1, F_2), \quad (2\ell_{G_1 G_2}; G_1, G_2), \quad (2\ell_{\hat{i}\hat{j}}; \hat{I}, \hat{J}) = \Omega.$$

According to Section 4.8, each of these degenerate conics, as a complete conic, consists of a double line with a pair of marked points.

Remark 4.59. The polar with respect to conic $\mathcal{C}$ of one of its foci F is called *the directrix* associated with F. It contains points A_1, A_2 of contact of the isotropic tangents it_1 and it_2 from F to $\mathcal{C}$. The bitangent pencil of conics with tangents it_1, it_2 at points A_1, A_2 contains $\mathcal{C}$ and two degenerate conics: $2\ell_{A_1 A_2}$ and $it_1 + it_2$.

Put the origin at F and in the chosen coordinates write down the equation of the directrix $\ell_{A_1 A_2}$ as $ax + by + c = 0$. Then there exists a parameter s such that the equation of $\mathcal{C}$ is

$$x^2 + y^2 + s(ax + by + c)^2 = 0.$$

From the last equation we get one of the defining properties of a conic:

A conic is the set of points whose distances from the focus F and to the directrix $\ell_{A_1 A_2}$ are in a constant ratio.

This ratio is equal to $\sqrt{s(a^2 + b^2)}$.

Definition 4.60. Two curves γ_1 and γ_2 of class s in the projective plane are *confocal* if they have the same s focal points, neither two of which belong to the same isotropic tangent.

Exercise 4.61. Describe the family of conics in the Euclidean plane that are confocal with a given parabola.

Tangential pencils

We saw that a confocal pencil can be understood as the dual of a general linear pencil of conics, i.e., of a pencil with four distinct base points. Now, we are going to consider briefly other possible tangential pencils, as dual to linear pencils we have already studied. Thus, *a tangential pencil of conics* is a set of conics with tangential equations of the form:

$$\mathcal{Q}_\lambda : \quad Q_0^v(u, v, w) + \lambda Q_\infty^v(u, v, w) = 0,$$

where $\lambda \in \mathbf{KP}^1$ is a parameter of the pencil; $Q_0^v = 0$ and $Q_\infty^v = 0$ are tangential equations of two given conics. We will consider only pencils which contain a non-degenerate conic.

Example 4.62.

(a) The most general example is a pencil with four distinct common tangents. The case of confocal conics we studied before belongs to this case. Such a pencil contains three degenerate conics.

(b) The following and the next example represent pencils which are at the same time pencils of locus conics. Now, we consider a pencil of conics with two given tangents ℓ_1, ℓ_2 at the two given points A_1, A_2. The degenerate conics from the pencil are $A_1 + A_2$ and $2\ell_1 \cap \ell_2$.

(c) The set of conics super-osculating with the common tangent t at the point A.

In a dual manner, we can easily formulate basic properties of tangential pencils, starting from the facts proven for locus pencils before.

Proposition 4.63. *Given a tangential pencil of conics $\mathcal{X}$.*

(a) *Given a line ℓ not joining the points of degenerate conics from the pencil. Then the poles of ℓ with respect to conics from the pencil form line ℓ'. In this way, a projective transformation from $\mathcal{X}$ to ℓ' is established.*

(b) *If the base set of the pencil is a quadrilateral, and P is a given point, then all polars of P with respect to non-degenerate conics from $\mathcal{X}$ are tangents to some conic $\mathcal{C}$. The three diagonals of the base quadrilateral are tangent to $\mathcal{C}$ as well.*

4.10 Quadrics, their pencils and linear subsets

Let us start with two well-known examples of quadrics in the three-dimensional Euclidean space.

Example 4.64 (One-sheeted hyperboloid). Let us consider a hyperbola $xy = 1$ in the Euclidean plane. It has two symmetry axes: the line $x = y$ and the line $x = -y$. By rotation about the first one, a two-sheeted hyperboloid is obtained. Rotating about the second symmetry axes, we get a one-sheeted hyperboloid.

We are interested here in its *line structure*. A one-sheeted hyperboloid is a double line surface: through an arbitrary point of the hyperboloid there are two lines which are completely contained in the hyperboloid. (See Figure 4.1.)

One can easily imagine the line structure of a one-sheeted hyperboloid having in mind the following model: two congruent circles connected by ropes of the same length. After stretching the ropes we get a model of a cylinder. But after rotations of one of the circles about the axes of the cylinder, we come to a model of a one-sheeted hyperboloid.

Exercise 4.65. Every one-sheeted hyperboloid can be presented in appropriate coordinates by an equation of the form

$$\frac{x^2}{a^2} + \frac{y^2}{b^2} - \frac{z^2}{c^2} = 1,$$

where x, y, z are rectangular coordinates and a, b, c are positive constants.

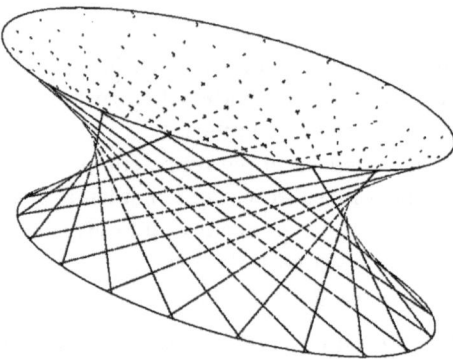

Figure 4.1: One-sheeted hyperboloid

Line $x = 0$, $y = 0$ is *the axis* of the hyperboloid. The intersection with the xy-plane,

$$\frac{x^2}{a^2} + \frac{y^2}{b^2} = 1,$$

is called *the middle ellipse*.

Given point P_0 of the hyperboloid, with coordinates (x_0, y_0, z_0), one can easily check the existence of two *generatrices*, the lines that are completely contained in the hyperboloid, passing through P_0. The equations of the lines are

$$x = x_0 - \sigma \frac{a}{b} y_1 t, \quad y = y_0 + \sigma \frac{b}{a} x_1 t, \quad x = z_0 + ct, \quad t \in \mathbf{R}, \qquad (4.18)$$

with $\sigma = 1$ and $\sigma = -1$. Point (x_1, y_1) belongs to the middle ellipse and is given by the formula

$$x_1 = \frac{x_0 + \sigma \frac{a}{b} \frac{z_0}{c} y_0}{\frac{x_0^2}{a^2} + \frac{y_0^2}{b^2}}, \quad y_1 = \frac{y_0 - \sigma \frac{b}{a} \frac{z_0}{c} x_0}{\frac{x_0^2}{a^2} + \frac{y_0^2}{b^2}}.$$

The generatrices are divided into two families corresponding to $\sigma = 1$ and $\sigma = -1$. The basic properties of these two families of lines are collected in the following

Exercise 4.66. Prove:

(a) Every point of a hyperboloid belongs to exactly one generatrix from each family of lines.

(b) Each pair of generatrices from different families is contained in a plane.

(c) For a given generatrix, there exists exactly one other generatrix parallel to it.

(d) Any two generatrices from one family are skew.

(e) Given three generatrices from the same family. The set of lines not skew to any of three lines coincides with the other family.

Observe at the end of Exercise 4.66 that a one-sheeted hyperboloid can be obtained by rotations of any of the generators about the axes.

Example 4.67 (Hyperbolical paraboloid). This is a surface given in a chosen rectangular frame by an equation of the form

$$\frac{x^2}{a^2} - \frac{y^2}{b^2} = 2z,$$

with positive constants a, b. For given point $M(x_0, y_0, z_0)$ of the surface, there are two lines passing through it, which are contained in the surface

$$x = x_0 + at, \quad y = y_0 + \sigma bt, \quad z = z_0 + t\left(\frac{x}{a} - \sigma\frac{y}{b}\right),$$

where $\sigma = \pm 1$. According to the sign of σ, one divides the lines into two families.

Basic properties of the two families of lines are given in the next

Exercise 4.68. Prove:

(a) Any point of a hyperbolic paraboloid is contained in exactly one generatrix from each family.

(b) Every two generatrices from distinct families are intersecting each other.

(c) Every two generatrices from the same family are skew.

(d) There exists a plane parallel to all the lines from one family.

Previous examples clearly manifest the importance of the structure of linear subsets of quadrics. We are going to study such structures in general settings.

For given quadric $\mathcal{Q}$ in $\mathbf{KP}^n$, denote by Q the associated $(n+1) \times (n+1)$ matrix. Then, the quadric is non-degenerate if and only if $\det Q \neq 0$.

If $\mathcal{Q}$ is degenerate, denote by r the rank of the matrix Q, $r \leq n$. Then the quadric $\mathcal{Q}$ is a cone with $\mathbf{KP}^{n-r}$ as the vertex over a smooth quadric in $\mathbf{KP}^{r-1}$.

We will mostly deal with *general quadrics*, which means quadrics which are either non-degenerate or have co-rank 1.

Proposition 4.69. *Let $\mathcal{Q}$ be a general quadric in $\mathbf{KP}^n$.*

(a) *If $n = 2k + 1$ then maximal linear subspaces of $\mathcal{Q}$ have dimension k.*

(b) *If $n = 2k$ then maximal linear subspaces of $\mathcal{Q}$ have dimension $k - 1$.*

(c) *The collection of all maximal linear subsets, denoted by $S(\mathcal{Q})$, is an algebraic variety of dimension $n(n + 1)/2$.*

(d) *If rank of $\mathcal{Q}$ is even, then $S(\mathcal{Q})$ has two components, otherwise it has one component. Each of the components is a unirational variety.*

Pencils of quadrics

The theory of pencils of quadrics is in many questions just a straightforward generalization of the theory of pencils of conics. Given two quadrics $\mathcal{Q}_1$ and $\mathcal{Q}_2$ in $\mathbf{KP}^n$, they generate a pencil of quadrics of the form

$$\mathcal{Q}(\lambda) \ : \quad Q_1 + \lambda Q_2,$$

where $\lambda \in \mathbf{KP}^1$ is a projective parameter.

Recall that by $p_{\mathcal{Q}}(P)$ we denoted the polar of the point P with respect to $\mathcal{Q}$. When $\mathcal{Q}$ runs through pencil $\mathcal{X}$, the polars of the fixed point P with respect to $\mathcal{Q}$, regarded as points of the dual space, describe a line or a fixed point.

Lemma 4.70. *For a given pencil $\mathcal{X}$ of quadrics*

$$\mathcal{Q}(\lambda) \ : \quad Q_1 + \lambda Q_2,$$

in $\mathbf{KP}^n$ the next three conditions are equivalent:

(a) *The polynomial $P(\lambda) = \det(Q(\lambda))$, of degree $n+1$ in λ has $n+1$ distinct roots $\lambda_1, \dots, \lambda_{n+1}$.*

(b) *Each quadric $\mathcal{Q}_\lambda$ is general, and the only degenerate ones are $\mathcal{Q}_{\lambda_1}, \dots, \mathcal{Q}_{\lambda_{n+1}}$.*

(c) *The base of the pencil $X = \mathcal{Q}_1 \cap \mathcal{Q}_2$ is smooth.*

Definition 4.71. A pencil of quadrics is *generic* if it satisfies any of the three equivalent conditions from Lemma 4.70.

In such a case, the vertices $P_1, \dots, P_{n+1}$ of degenerate quadrics $\mathcal{Q}_{\lambda_1}, \dots, \mathcal{Q}_{\lambda_{n+1}}$ do not belong to the base set X. They form *the self-polar simplex* and they correspond to eigen-vectors of matrices

$$Q_1^{-1} Q_2.$$

Thus the matrices $Q(\lambda)$ are simultaneously diagonalized in the corresponding frame.

Lemma 4.72. *If $n = 2k$, then the base set X contains maximal linear subsets of dimension $k-1$. If $n = 2k+1$ then X does not contain any linear subsets of dimension k.*

Theorem 4.73 (Generalized Desargues Theorem). *A pencil of quadrics $\mathcal{X}$ induces an involution on a general line.*

Exercise 4.74. Consider a pencil of quadrics $\mathcal{X}$ in three-dimensional space.

(a) The polar planes of fixed point P with respect to quadrics of the pencil form a pencil with a line v_P as its base set. Such a line v_P is an axis of the pencil.

(b) Let ℓ be a given line. Then the lines $p_\lambda(\ell)$, that are polar to ℓ with respect to quadrics $\mathcal{Q}(\lambda)$ of the pencil $\mathcal{X}$, and the axes v_P of points $P \in \ell$, form the two systems of generators of some quadric $\mathcal{S}$. $\mathcal{S}$ contains the vertices of the degenerate quadrics of $\mathcal{X}$. (Compare with Theorem 4.35).

(c) If $\ell = v_P$ then $\mathcal{S}$ is a cone with vertex P.

(d) Given a plane π, its poles with respect to quadrics from the pencil are contained in a twisted cubic passing through the vertices of the four degenerate quadrics of the pencil.

There is a great variety of *non-generic* pencils of quadrics but we are not going to be more specific on this occasion. As a classical reference one can use [Tod1947].

4.11 Confocal quadrics

In this section, we are going to define families of confocal quadrics in the d-dimensional Euclidean space $\mathbf{E}^d$. Having in mind the situation with confocal conics, we need to take into account that in higher dimensions there is no such clear notion of focal points. Thus, we start from the analytical expression of a confocal family.

Definition 4.75. *A family of confocal quadrics* in the d-dimensional Euclidean space $\mathbf{E}^d$ is a family of the form

$$\mathcal{Q}_\lambda \; : \; \frac{x_1^2}{a_1 - \lambda} + \cdots + \frac{x_d^2}{a_d - \lambda} = 1 \qquad (\lambda \in \mathbf{R}), \tag{4.19}$$

where $a_1, \ldots, a_d$ are real constants.

Let us notice that a family of confocal quadrics in the Euclidean space is determined by only one quadric.

From now on, we are going to consider the non-degenerate case when $a_1, \ldots, a_d$ are all distinct.

Theorem 4.76 (Jacobi). *Any point of the d-dimensional Euclidean space is the intersection of exactly d quadrics of the confocal family* (4.19). *The quadrics are perpendicular to each other at the intersecting points. (See Figure 4.2.)*

Proof. In terms of the dual space, the statement is that hyperplane $(s, x) = 1$ is tangent to n quadrics of a Euclidean pencil, with an additional condition: radius-vectors of the contact points are mutually orthogonal. The main axes of the quadric $(Sx, x) = 2(s, x)^2$ correspond to these vectors. $\qquad\square$

From here, it follows that to every point $x \in \mathbf{E}^d$ we may associate a d-tuple of distinct parameters $(\lambda_1, \ldots, \lambda_d)$, such that x belongs to quadrics $\mathcal{Q}_{\lambda_1}, \ldots, \mathcal{Q}_{\lambda_d}$.

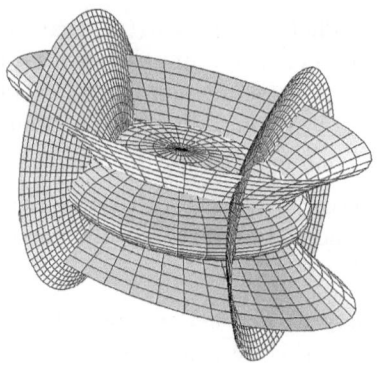

Figure 4.2: Three confocal quadrics in $\mathbf{E}^3$

Additionally, if we order $a_1 < a_2 < \cdots < a_d$ and $\lambda_1 < \lambda_2 \cdots < \lambda_d$, then

$$\lambda_1 \leq a_1 < \lambda_2 \leq a_2 < \cdots < \lambda_d \leq a_d.$$

Definition 4.77. *Jacobi elliptic coordinates* of point $x = (x_1, \ldots, x_d) \in \mathbf{E}^d$ are values $(\lambda_1, \ldots, \lambda_d)$ that satisfy

$$\frac{x_1^2}{a_1 - \lambda_i} + \cdots + \frac{x_d^2}{a_d - \lambda_i} = 1, \quad 1 \leq i \leq d.$$

Theorem 4.78 (Chasles). *Any line in $\mathbf{E}^d$ is tangent to exactly $d-1$ quadrics from a given confocal family. Tangent hyperplanes to these quadrics, constructed at the points of tangency with the line, are orthogonal to each other.*

Exercise 4.79.

(a) Shadows of quadrics from a confocal family form a confocal family of quadrics in the target hyperplane.

(b) The intersection of a pencil of quadrics with a hyperplane is a pencil of quadrics.

(c) The intersection of a family of confocal quadrics with a hyperplane is not necessarily a confocal family of quadrics in the hyperplane.

Proof of Chasles theorem. Let us project, parallel to a given line, the given confocal family. The statement follows from Exercise 4.79 and Theorem 4.76.   □

Theorem 4.80 (Jacobi–Chasles). *Given a geodesic line on a given quadric in $\mathbf{E}^n$. Then the tangents of the quadric are tangent to $n-2$ other quadrics, which are confocal to the given one, and which are the same for all tangents.*

Example 4.81 (Confocal quadrics in $\mathbf{R}^3$). Let a family be given by the equation

$$\frac{x^2}{A-\lambda} + \frac{y^2}{B-\lambda} + \frac{z^2}{C-\lambda} = 1, \quad A > B > C > 0.$$

Every triplet (x, y, z) of Cartesian coordinates uniquely determines a triplet $(\lambda_1, \lambda_2, \lambda_3)$ such that

$$\frac{x^2}{A-\lambda_i} + \frac{y^2}{B-\lambda_i} + \frac{z^2}{C-\lambda_i} = 1.$$

We assume $A \geq \lambda_3 > B \geq \lambda_2 > C \geq \lambda_1$.

Consider the equation $\lambda = \lambda_1$; it determines an ellipsoid, which contains the point (x, y, z).

Exercise 4.82.

(a) What kind of quadric is one defined by the equation $\lambda = \lambda_2$? And what by $\lambda = \lambda_3$?

(b) Analyze the cases when some of the coordinates λ_i is equal to some of the semi-axes A, B, C.

   One should have in mind that the triplet $(\lambda_1, \lambda_2, \lambda_3)$ of elliptical coordinates does not determine uniquely a point in the space.

(c) Describe points which share the same elliptical coordinates.

   We introduce the notation

$$L(z) := \prod_{j=1}^{n}(z - \lambda_j), \quad A(z) := \prod_{j=1}^{n}(z - a_j).$$

The following identities are very useful in dealing with elliptic coordinates.

Exercise 4.83. Prove:

(a) $\dfrac{a_n + \lambda_2}{a_1 + \lambda_2} - \dfrac{a_n + \lambda_1}{a_1 + \lambda_1} = \dfrac{(a_1 - a_n)(\lambda_2 - \lambda_1)}{(a_1 + \lambda_2)(a_1 + \lambda_1)};$

(b) $\dfrac{\partial}{\partial q_i} = 2 \sum_{k=1}^{n} \dfrac{A(\lambda_k)}{L'(\lambda_k)} \dfrac{q_i}{a_i - \lambda_k} \dfrac{\partial}{\partial \lambda_k};$

(c) $\sum_{i=1}^{n} \dfrac{q_i}{(z - a_i)} \dfrac{\partial}{\partial q_i} = 2 \dfrac{L(z)}{A(z)} \sum_{k=1}^{n} \dfrac{A(\lambda_k)}{L'(\lambda_k)} \dfrac{1}{z - \lambda_k} \dfrac{\partial}{\partial \lambda_k};$

(d) $\dfrac{\partial q_i}{\partial \lambda_k} = \dfrac{q_i}{2(\lambda_k - a_i)}.$

Example 4.84 (The Jacobi problem for geodesics on an ellipsoid). Given an ellipsoid

$$\frac{x^2}{A} + \frac{y^2}{B} + \frac{z^2}{C} = 1, \quad A > B > C > 0$$

corresponding to the equation $\lambda_1 = 0$ in elliptic coordinates. The Hamiltonian of the motion of a material point of the unit mass under the inertia on the ellipsoid, is expressed in the elliptic coordinates by formula

$$H = \frac{2}{\lambda_3 - \lambda_2} \left(\frac{(A - \lambda_3)(\lambda_3 - B)(\lambda_3 - C)}{\lambda_3} p_1^2 + \frac{(A - \lambda_2)(B - \lambda_2)(\lambda_2 - C)}{\lambda_2} p_2^2 \right).$$

The problem of integration of the Hamiltonian equations of motion is *separable* in elliptic coordinates. A complete solution of the Hamilton–Jacobi equation can be obtained in the form

$$S(\lambda_2, \lambda_3; a, E) = \sqrt{E/2} \left(\int \frac{\lambda_2 - a}{\sqrt{R(\lambda_2)}} + \int \frac{\lambda_3 - a}{\sqrt{R(\lambda_3)}} \right),$$

where E is the value of the energy integral, the constant a belongs to the interval (C, A) and

$$R(x) = -\frac{(x - a)(x - A)(x - B)(x - C)}{x}.$$

Integrable potential perturbations of this problem will be discussed in Section 7.5 (see Theorem 7.33).

As in the case of confocal conics, one can see a confocal pencil of quadrics as a tangential hyperplane pencil of quadrics. Let a quadric $\mathcal{Q}$ be given:

$$\mathcal{Q} : \quad \frac{x_1^2}{a_1} + \cdots + \frac{x_d^2}{a_d} = 1.$$

Its dual tangential hyperplane equation is

$$\Sigma : \quad a_1 u_1^2 + \cdots + a_d u_d^2 = 1.$$

Denote by Ω the quadric with tangential hyperplane equation

$$\Omega : \quad u_1^2 + \cdots + u_d^2 = 1.$$

Then, the confocal family of quadrics $\mathcal{Q}(\lambda)$ is the set of quadrics forming a tangential hyperplane pencil of quadrics

$$\Sigma + \lambda\Omega$$

with tangential equations of the form

$$\Sigma(\lambda) = \Sigma - \lambda\Omega : \quad (a_1 - \lambda)u_1^2 + \cdots + (a_d - \lambda)u_d^2 = 1.$$

Example 4.85. In the three-dimensional case there are four degenerate quadrics in a confocal pencil, corresponding to the values $\lambda_i = a_i$ and $\lambda = \infty$. The first three cases lead to three *focal conics*: a virtual conic in the plane $x_1 = 0$, a hyperbola in the plane $x_2 = 0$ and an ellipse in the plane $x_3 = 0$ (see Figure 4.3). The fourth degenerate quadric is Ω.

As in the case of conics, there are also confocal families consisting entirely of paraboloids.

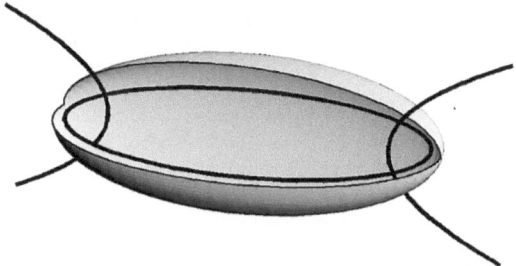

Figure 4.3: Ellipsoid and confocal degenerate quadrics

4.12   (2-2)-correspondences

(1-1)- and (2-1)-correspondences

Together with a projective transformation

$$g : \mathbf{P}^1 \to \mathbf{P}^1$$

given in some affine coordinates by the equation

$$v := g(u) = \frac{au+b}{cu+d}, \quad ad \neq bc,$$

one can consider its graph Γ_g as a subset of $\mathbf{P}^1 \times \mathbf{P}^1$. Moreover, the graph $\Gamma_g \in \mathbf{P}^1 \times \mathbf{P}^1$ is a curve in $\mathbf{P}^1 \times \mathbf{P}^1$ defined by the equation in affine coordinates

$$F(u,v) = buv + dv - au - c = 0.$$

The degree of the equation F is one in both variables u and v. Generalizing the notion of projective transformation to multivalued maps we come to the notion of *correspondence* as a curve $\mathcal{C}$ in $\mathbf{P}^1 \times \mathbf{P}^1$ given by polynomial equation

$$F_{\mathcal{C}}(u,v) = 0.$$

If the polynomial $F_{\mathcal{C}}$ is of degree m in u and of degree n in v we say that the correspondence $\mathcal{C}$ is of type $(m-n)$. Thus, projective transformations correspond to (1-1)-correspondences.

Let us note that the product $\mathbf{P}^1 \times \mathbf{P}^1$ differs from $\mathbf{P}^2$ and can be embedded into $\mathbf{P}^3$ as a quadric in the following way.

Denote by $(u_1 : u_2)$ projective coordinates of the first component and $(v_1 : v_2)$ projective coordinates of the second component. Coordinates of $\mathbf{P}^3$ are denoted by $(z_1 : z_2 : z_3 : z_4)$. It was proven by Segre that formulae

$$z_1 = u_1v_1,$$
$$z_2 = u_1v_2,$$
$$z_3 = u_2v_1,$$
$$z_4 = u_2v_2$$

define embedding

$$s \; : \mathbf{P}^1 \times \mathbf{P}^1 \to \mathcal{S} \in \mathbf{P}^3$$

onto a quadric $\mathcal{S}$ defined by the equation

$$\mathcal{S} \; : \; z_1z_4 = z_2z_3.$$

The quadric $\mathcal{S}$ will be called *the Segre quadric*. In a similar way, one can construct embedding of $\mathbf{P}^m \times \mathbf{P}^n$ into $\mathbf{P}^{mn+m+n}$. The image is called a *Segre variety*.

The graph Γ_g of the projective transformation g is the intersection of the plane

$$bz_1 + dz_3 - az_2 - cz_4 = 0$$

with the Segre quadric $\mathcal{S}$. Thus, a (1-1)-correspondence is a conic.

The next case is (2-1). It is given by the equations of the form

$$F_{\mathcal{B}}(u, v) := R(u)v + P(u) = 0,$$

where P, R are polynomials of degree 2. The correspondence $\mathcal{B}$ is a cubic with rational parametrization $(-uP(u), uR(u), -P(u), R(u))$. From the Bézout theorem, it follows that $\mathcal{B}$ is not a complete intersection of the Segre quadric $\mathcal{S}$ with some other algebraic surface. The cubic $\mathcal{B}$ is not a plane cubic, but *a twisted* one. To summarize, we can say that a (2-1)-correspondence is a rational, twisted cubic.

(2-2)-correspondences, definition and properties

A (2-2)-correspondence $\mathcal{E}$ is defined by a biquadratic equation of the form

$$F_{\mathcal{E}}(u, v) = au^2v^2 + bu^2v + b'uv^2 + cuv + du^2 + d'v^2 + eu + e'v + f = 0.$$

It is obvious that it is obtained as the intersection of the Segre quadric $\mathcal{S}$ with a quadric $\mathcal{Q}$:

$$Q(z_1, z_2, z_3, z_4)$$
$$= az_1^2 + bz_1z_2 + b'z_1z_3 + cz_1z_4 + dz_2^2 + d'z_3^2 + ez_2z_4 + e'z_3z_4 + fz_4^2 = 0.$$

Conversely, any quadric $\mathcal{Q}$ intersecting with the Segre quadric defines a (2-2)-correspondence.

For such a curve $\mathcal{E}$, which represents a (2-2)-correspondence, we say that it is a *biquadratic* in the Segre quadric $\mathcal{S}$. This curve is of degree 4. It intersects any plane in four points of intersection of two conics which are obtained as intersections of the plane with quadrics $\mathcal{S}$ and $\mathcal{Q}$ respectively.

Possible degenerate cases are the following:

(1) $\mathcal{E}$ is the union of two distinct non-degenerate conics: $\mathcal{E} = \mathcal{C}_1 + \mathcal{C}_2$ when the pencil of quadrics generated with $\mathcal{S}$ and $\mathcal{Q}$ contains the sum of two planes.

(2) $\mathcal{E}$ is equal to a double non-degenerate conic: $\mathcal{E} = 2\mathcal{C}$ when the pencil contains a double plane.

(3) $\mathcal{E} = \mathcal{B} + \ell$, where $\mathcal{B}$ is a twisted cubic and ℓ is a line, a common linear generator of $\mathcal{Q}$ and $\mathcal{S}$.

(4) $\mathcal{E} = \ell_1 + \ell_2 + \mathcal{C}_2$, with ℓ_1, ℓ_2 being lines, and $\mathcal{C}_2$ a non-degenerate cubic, when the pencil generated with $\mathcal{Q}$ and $\mathcal{S}$ contains a degenerate quadric, one component of which is a plane tangent to $\mathcal{S}$.

(5) $\mathcal{E}$ decomposes into four lines: $\mathcal{E} = \ell_1 + \ell_2 + \ell_3 + \ell_4$, when the sum of two planes tangent to $\mathcal{S}$ is contained in the pencil.

Symmetric correspondences

We will say that correspondence $\mathcal{E}$ defined in the product of $\mathbf{P}^1$ with itself by the equation $F_{\mathcal{E}}(u, v) = 0$ is *symmetric* if

$$F_{\mathcal{E}}(u, v) = F_{\mathcal{E}}(v, u).$$

A symmetric non-degenerate (1-1)-correspondence is an involution:

$$F(u, v) := a(u + v) + buv + c = 0.$$

Obviously, symmetric correspondence is of degree $(n - n)$.

A general symmetric (2-2)-correspondence $\mathcal{E}$ is of the form

$$F_{\mathcal{E}}(u, v) = au^2v^2 + buv(u + v) + c(u^2 + v^2) + duv + e(u + v) + f = 0. \quad (4.20)$$

The form of the last equation remains symmetric if we apply the same projective transformation on both factors of the product $\mathbf{P}^1 \times \mathbf{P}^1$. We will also use the notion of *Euler–Chasles correspondence* as a synonym for symmetric (2-2)-correspondences.

A (2-2)-correspondence can be not symmetric. This can happen, for example when the two factors of the product $\mathbf{P}^1 \times \mathbf{P}'^1$ are not identical. It can be proved that there exists a projective map $f : \mathbf{P}^1 \to \mathbf{P}'^1$ which symmetrizes non-degenerate nonsymmetric (2-2)-correspondence $\mathcal{E}_1$: the (2-2)-correspondence $\mathcal{E}_2 \subset \mathbf{P}^1 \to \mathbf{P}^1$ defined by

$$(u, v) \in \mathcal{E}_2 \iff (u, f(v)) \in \mathcal{E}_1$$

is symmetric.

Suppose that (2-2)-correspondence $\mathcal{E}$ is defined by the equation

$$F_{\mathcal{E}}(u,v) := P(u)v^2 + Q(u)v + R(u) = 0, \qquad (4.21)$$

where P, Q, R are polynomials of degree 2. We may assume in general that $\mathcal{E}$ is a correspondence between different projective spaces $\mathbf{P}^1$ and $\mathbf{P}'^1$. As a *critical point* in the first space we will assume a point with coordinate u to which corresponds a double point in the second space. In other words, critical points of the first space are zeros of the discriminant $D(u)$ of the previous equation (4.21) understood as a quadratic equation in v:

$$D = D(u) := Q^2(u) - 4P(u)R(u). \qquad (4.22)$$

The discriminant $D(u)$ is a polynomial in u of degree 4. Thus, the divisor of critical points of the first space $D_1^c(\mathcal{E})$ is of degree 4. In the same way we can define critical points of the second space and their divisor $D_2^c(\mathcal{E})$ is again of degree 4. Moreover, if the correspondence $\mathcal{E}$ is symmetrizable, then there exists a projective map f which maps $D_1^c(\mathcal{E})$ to $D_2^c(\mathcal{E})$. Further, if points of the divisor $D_1^c(\mathcal{E})$ are all distinct, then the cross-ratio of the points of the first divisor is equal to the cross-ratio of the points of the second divisor.

The divisor $D_1^c(\mathcal{E})$ of critical points of given (2-2)-correspondence $\mathcal{E}$ may consist of

(1) four distinct points,
(2) one double point and two other distinct points,
(3) one triple and one simple point,
(4) two double points,
(5) one point of order 4.

In the non-degenerate case, there are three possibilities: the case (1) if $\mathcal{E}$ does not have multiple points; the case (2) if it has an ordinary double point; the case (3) occurs if the correspondence $\mathcal{E}$ has a cusp.

Geometric interpretation of a symmetric (2-2)-correspondence. Poncelet theorem

Let us start with the situation of the Poncelet theorem. Suppose conics Γ and $\mathcal{K}$ are given. Consider (2-2)-correspondence on Γ induced by $\mathcal{K}$ in the following way. To a point $M \in \Gamma$ correspond points M_1 and M_1' such that the lines ℓ_{MM_1} and $\ell_{MM_1'}$ are tangent to the conic $\mathcal{K}$. In this way, a symmetric (2-2)-correspondence is defined. Moreover, every symmetric (2-2)-correspondence on a conic is defined in this way. Indeed, suppose the conic Γ is rationally parametrized $(s^2 : s : 1)$ and suppose the conic $\mathcal{K}$ is given by its tangential equation

$$F(u,v,w) = Au^2 + Bv^2 + Cw^2 + Dvw + Ewu + Fuv = 0.$$

The condition that points M with parameter s_1 and M_2 with parameter s_2 are such that the line ℓ_{MM_1} is tangent to the conic $\mathcal{K}$ is

$$F(1, -(s_1 + s_2), s_1 s_2) = 0.$$

The last equation is equivalent to the general symmetric (2-2)-correspondence:

$$A + B(s_1 + s_2)^2 + Cs_1^2 s_2^2 - Ds_1 s_2(s_1 + s_2) + Es_1 s_2 - F(s_1 + s_2) = 0.$$

One can easily see

Proposition 4.86. *The divisor $D^c(\mathcal{E})$ of critical points of the last symmetric (2-2)-correspondence $\mathcal{E}$ generated by the conics Γ and $\mathcal{K}$ is equal to the intersection divisor*

$$D^c(\mathcal{E}) = \Gamma \cdot \mathcal{K}.$$

Exercise 4.87. Prove that the symmetric (2-2)-correspondence $\mathcal{E}$ determined by conics Γ and $\mathcal{K}$

(a) is without multiple points if and only if Γ and $\mathcal{K}$ have four distinct common points,

(b) has an ordinary double point if and only if Γ and $\mathcal{K}$ are tangent at one point,

(c) has a cusp if and only if Γ and $\mathcal{K}$ osculate.

Exercise 4.88. Describe the structure of the symmetric (2-2)-correspondence $\mathcal{E}$ determined by the conics Γ and $\mathcal{K}$ in the following cases:

(a) when Γ and $\mathcal{K}$ are bitangent;

(b) Γ and $\mathcal{K}$ super-osculate.

For a point M on Γ we say that it is a *fixed point* of the correspondence $\mathcal{E}$ induced by Γ and $\mathcal{K}$ if the pair (M, M) belongs to $\mathcal{E}$. In other words the parameter $s_1 = s_2$ of the point M should satisfy the degree 4 equation of $\mathcal{E}$. This shows that the divisor of fixed points $D^f(\mathcal{E})$ is of degree 4. If we denote by $\mathcal{D}$ the diagonal correspondence, then

$$D^f(\mathcal{E}) = \mathcal{E} \cdot \mathcal{D}.$$

Geometrically, it is obvious that M is a fixed point of $\mathcal{E}$ if and only if M is the contact point with Γ of a bitangent of the conics Γ and $\mathcal{K}$. It is well known that there are four bitangents for two given conics.

In order to approach the Poncelet theorem we are going to study compositions of symmetric (2-2)-correspondences.

Let us recall the general definition of the composition of two correspondences $F = F(u, v) \subset A \times B$ and $G = G(v, w) \subset B \times C$. The composition $F \circ G \subset A \times C$ is defined by the condition $(u, w) \in F \circ G$ if and only if there exists $v \in B$ such that $(u, v) \in F$ and $(v, w) \in G$.

To simplify the exposition we will assume that Γ and $\mathcal{K}$ intersect in four distinct points. Denote by $\mathcal{E} = \mathcal{E}_1$ the symmetric (2-2)-correspondence induced by Γ and $\mathcal{K}$. We will consider correspondence $\mathcal{E}_k$ defined in the following way: start from a point $M = M_0 \in \Gamma$. Apply $\mathcal{E}$ and obtain M_1 and M_{-1}. Then apply again $\mathcal{E}$ to points M_1 and M_{-1} by constructing new tangents to conic $\mathcal{K}$, different from $L_{M_0 M_1}$ and $L_{M_0 M_{-1}}$. We obtain new points M_2 and M_{-2}. Repeating the procedure k times, we come to the points M_k and M_{-k}, thus associated to the starting point M_0. In this way a symmetric (2-2)-correspondence is defined on Γ. We will denote it as $\mathcal{E}_k$. We know that every symmetric (2-2)-correspondence on Γ is induced by a conic. Denote by $\mathcal{K}_n$ the conic associated to $\mathcal{E}_n$.

Theorem 4.89. *The conics $\mathcal{K}_n$ belong to a pencil generated by Γ and $\mathcal{K}$.*

Proof. The proof follows from the consideration of critical points. If A_0 is a critical point of $\mathcal{E}$ then $A_1 = A_{-1}$, $A_2 = A_{-2}$ and so on $A_n = A_{-n}$. This means that A_0 is also a critical point of $\mathcal{E}_n$. Thus the sets of critical points of all correspondences $\mathcal{E}_n$, $n \in \mathbf{N}$ coincide with the critical set of $\mathcal{E}$. On the other hand, the critical set is equal to the intersection set of the conic Γ with the conic $\mathcal{K}_n$. This means that intersection sets of Γ and $\mathcal{K}_n$ coincide with the intersection set of Γ and $\mathcal{K}$. Thus, conics $\mathcal{K}_n$, $\mathcal{K}$ and Γ belong to the same pencil. $\square$

Exercise 4.90. Prove
$$\mathcal{E}_k \circ \mathcal{E}_l = \mathcal{E}_{k+l} \circ \mathcal{E}_{|k-l|}.$$

Now, we come to the Poncelet theorem.

Theorem 4.91 (Poncelet). *Assume that there exists a point M_0 on the conic Γ such that the polygon $M_0 M_1 \ldots M_{n-1} M_0$ of $n \geq 3$ sides is inscribed in the conic Γ and circumscribed about conic $\mathcal{K}$ and that lines $\ell_{M_0 M_i}$, $i = 2, \ldots, n-2$, are not tangent to $\mathcal{K}$. Then, for every $N_0 \in \Gamma$ there exists a polygon of n sides $N_0 N_1 \ldots N_{n-1} N_0$ inscribed in Γ and circumscribed about $\mathcal{K}$.*

Proof. We consider the (2-2)-correspondence $\mathcal{E}_n$ obtained by iterations from the correspondence $\mathcal{E}$ generated with the conics Γ and $\mathcal{K}$. By the conditions of the theorem, the point (M_0, M_0) is a double fixed point of the correspondence $\mathcal{E}_n$. Moreover, all other vertices of the polygon induce double fixed points (M_i, M_i) of the correspondence $\mathcal{E}_n$. Thus, it has more than four fixed points, and as a consequence, it contains the diagonal $\mathcal{D}$. By further analysis of fixed points we see that another component of $\mathcal{E}_n$ has to be the diagonal $\mathcal{D}$ again. Thus

$$\mathcal{E}_n = 2\mathcal{D}.$$

Taking any other point $N_0 \in \Gamma$ we now have that it is also a double fixed point of $\mathcal{E}_n$. This means, that repeating the application of $\mathcal{E}$ on N_0, then on N_1 and so on, in n steps we come to the initial point N_0. In this way the polygon $N_0 N_1 \ldots N_{n-1} N_0$ inscribed in Γ and circumscribed about K is constructed, finishing the proof of the Poncelet theorem. $\square$

This line of investigation of the Poncelet theorem and its proof in this manner originated in the works [Tru1853, Tru1863] of Italian mathematician Trudi, of around 1853.

We will finish this chapter by stating a theorem which goes back even further to the past, to Euler and which denoted the beginning of the study of elliptic functions and related addition theorems (see [Eul1766]).

Theorem 4.92 (Euler theorem). *For the general symmetric (2-2)-correspondence (4.20) there exists an even elliptic function ϕ of the second degree and a constant shift c such that*

$$u = \phi(z), \quad v = \phi(z \pm c).$$

The proof of the Euler theorem will be presented at the beginning of Chapter 10.

From the last theorem one can easily deduce another proof of the Poncelet theorem.

Chapter 5

Poncelet Theorem and Cayley's Condition

We have just seen one proof of the Poncelet Theorem (see Theorems 2.14 and 4.91). Now, we present a proof of the Poncelet theorem for circles, which was given by Jacobi.

Theorem 5.1 (Poncelet theorem for circles). *Let $\mathcal{C}$, $\mathcal{D}$ be two circles in $\mathbf{E}^2$. If there is a closed polygon inscribed in $\mathcal{C}$ and circumscribed about $\mathcal{D}$, then there are infinitely many such polygons. Moreover, each point of $\mathcal{C}$ is a vertex of such a polygon, and all the polygons have the same number of sides.*

Proof. Denote the centers of $\mathcal{C}$, $\mathcal{D}$ by M, m, their radii by R, r, distance between the centers by a $(a + r < R)$, and by O the intersection point of the line through the centers with $\mathcal{C}$. Let $P_0 P_1 P_2 \cdots$, be a polygonal line inscribed in $\mathcal{C}$ and circumscribed about $\mathcal{D}$. Let us set $\angle AMP_i = 2\varphi_i$, $i \in \{0, 1, 2, \ldots\}$.

It is elementary to get

$$R \cos(\varphi_1 - \varphi_0) + a \cos(\varphi_1 + \varphi_0) = r,$$

and then

$$\cos \varphi_0 \cos \varphi_1 + \frac{R - a}{R + a} \sin \varphi_0 \sin \varphi_1 = \frac{r}{R + a}. \qquad (5.1)$$

Now, let

$$\varphi_0 = \operatorname{am} u, \quad \varphi_1 = \operatorname{am}(u + c), \quad \varphi\chi = \operatorname{am} c,$$

with am being the inverse of

$$u = \int_0^{\varphi} \frac{dt}{\sqrt{1 - k^2 \sin^2 t}}.$$

It is easy to prove that $\operatorname{cn} = \cos \operatorname{am}$, $\operatorname{sn} = \sin \operatorname{am}$.

Now, from addition formulae for Jacobi elliptic functions (Theorem 3.100), we obtain

$$\cos \varphi_0 \cos \varphi_1 + \sin \varphi_0 \sin \varphi_1 \cdot \sqrt{1 - k^2 \sin^2 \chi} = \cos \chi.$$

Comparing this with equation (5.1), we have

$$\cos \chi = \frac{r}{R + a}, \quad 1 - k^2 \sin^2 \chi = \frac{(R - a)^2}{(R + a)^2},$$

and finally we get

$$k^2 = \frac{4aR}{(R + a)^2 - r^2}.$$

To summarize, we have

$$\varphi_n = \operatorname{am}(u + nc).$$

The condition for the polygonal line to be closed, i.e., $P_n = P_0$, is then equivalent to

$$\varphi_n = \varphi_0 + s\pi$$

$$\Leftrightarrow$$

$$\operatorname{am}(u + nc) = \operatorname{am}(u + 2sK), \quad K = \int_0^{\pi/2} \frac{dt}{\sqrt{1 - k^2 \sin^2 t}}$$

$$\Leftrightarrow$$

$$u + nc = u + 2sK$$

$$\Leftrightarrow$$

$$c = \frac{2sK}{n}.$$

The obtained condition does not depend on P_0, thus the theorem follows.   □

In the next section we prove a more general statement, namely the *Full Poncelet theorem*. This theorem was also proved by Poncelet [Pon1822]. In our exposition, we will follow Lebesgue's presentation from [Leb1942].

Let us consider polygons inscribed in a conic Γ, whose sides are tangent to $\Gamma_1, \ldots, \Gamma_k$, where $\Gamma, \Gamma_1, \ldots, \Gamma_k$ all belong to a pencil of conics. In the dual plane, such polygons correspond to billiard trajectories having caustic Γ^* with bounces on $\Gamma_1^*, \ldots, \Gamma_k^*$.

Theorem 5.2 (Full Poncelet theorem). *Let conics Γ, Γ_1, $\ldots$, Γ_n belong to a pencil $\mathcal{F}$. If a polygon inscribed in Γ and circumscribed about Γ_1, $\ldots$, Γ_n exists, then infinitely many such polygons exist.*

To determine such a polygon, it is possible to give arbitrarily:

(i) *the order which its sides touch Γ_1, $\ldots$, Γ_n in; let the order be: Γ_1', $\ldots$, Γ_n',*

(ii) *a tangent to Γ_1' containing one side of the polygon,*

(iii) *the intersecting point of this tangent with Γ which will belong to the side tangent to Γ_2'.*

The main object in the proof is the cubic Cayley curve, which parametrizes contact points of tangents drawn from a given point to all conics of the pencil. After completing the proof of the Full Poncelet theorem in Section 5.1, in Section 5.2 we present Cayley's conditions which give an analytical characterization of pairs of conics satisfying the Poncelet theorem conditions: namely those allowing finite polygonal lines inscribed in one conic and circumscribed about the other one.

5.1 Full Poncelet theorem

Basic lemma

The next lemma is the main step in the proof of the Full Poncelet theorem. If one Poncelet polygon is given, this lemma enables us to construct every Poncelet polygon with given initial conditions. Also, the lemma is used in deriving a geometric condition for the existence of a Poncelet polygon.

Lemma 5.3. [Leb1942] *Let $\mathcal{F}$ be a pencil of conics in the projective plane and Γ a conic from this pencil. Then there exist quadrangles whose vertices A, B, C, D are on Γ such that three pairs of its non-adjacent sides AB, CD; AC, BD; AD, BC are tangent to three conics of $\mathcal{F}$. Moreover, the six contact points all lie on a line Δ. Any such quadrangle is determined by two sides and the corresponding contact points.*

Let Γ, Γ_1, Γ_2, Γ_3 be conics of a pencil and ABC a Poncelet triangle corresponding to these conics, such that its vertices lie on Γ and sides AB, BC, CA touch Γ_1, Γ_2, Γ_3 respectively. This lemma gives us a possibility to construct triangle ABD inscribed in Γ whose sides AB, BD, DA touch conics Γ_1, Γ_3, Γ_2 respectively. In a similar fashion, for a given Poncelet polygon, we can, applying Lemma 5.3, construct another polygon which corresponds to the same conics, but its sides are tangent to them in different order.

Circumscribed and Tangent Polygons

Let a triangle ABC be inscribed in a conic Γ and sides BC, AC, AB touch conics Γ_1, Γ_2, Γ_3 of the pencil $\mathcal{F}$ at points M, N, P respectively. According to the lemma, there are two possible cases: either points M, N, P are collinear, when we will say that the triangle is *tangent* to Γ_1, Γ_2, Γ_3; or the line MN intersects AB at a point S which is a harmonic conjugate to P with respect to the pair A, B, then we say that the triangle is *circumscribed* about Γ_1, Γ_2, Γ_3.

Let $ABCD\ldots KL$ be a polygon inscribed in Γ whose sides touch conics Γ_1, $\ldots$, Γ_n of the pencil $\mathcal{F}$ respectively. Denote by (AC) a conic such that $\triangle ABC$ is circumscribed about Γ_1, Γ_2, (AC), by (AD) a conic such that $\triangle ACD$ is circumscribed about (AC), Γ_3, (AD). Similarly, we find conics (AE), $\ldots$, (AK). The triangle AKL can be tangent to the conics (AK), Γ_{n-1}, Γ_n or circumscribed about them, and we will say that $ABCD\ldots KL$ is *tangent* or, respectively, *circumscribed* about conics Γ_1, $\ldots$, Γ_n.

Further, we will be interested only in circumscribed polygons. The Poncelet theorem does not hold for tangent triangles nor, hence, for tangent polygons with greater numbers of vertices.

Proof of the Full Poncelet theorem

Step 1.   First, we are going to show the assertion for the case of a triangle. Let $\triangle ABC$ be inscribed in Γ and circumscribed about Γ_1, Γ_2, Γ_3 and α, β, γ be the contact points on the sides BC, CA, AB respectively. Let $B'C'$ be another tangent to Γ_1, with points B', C' on Γ and α' as the contact point.

We are going to apply Lemma 5.3 to the quadrangle $BCB'C'$. Here, the line Δ is $\alpha\alpha'$. It intersects BB' in the point b, CC' in c and there exists a conic S of the pencil $\mathcal{F}$ which touches BB' in b and CC' in c.

Next, construct a quadrangle inscribed in Γ which is determined by sides CA, CC' and contact points β and c. Let A' be the fourth vertex of the quadrangle, and β', a intersecting points of βc with $C'A'$ and AA'. It holds that $\beta' \in \Gamma_2$, $a \in S$.

The quadrangle $BB'AA'$ is inscribed in Γ, and the corresponding line Δ is ab. It intersects AB, $A'B'$ in points γ_1, γ' which belong to the same conic of the pencil $\mathcal{F}$. We are going to show that this conic is Γ_3, i.e., that γ_1 coincides with γ and not with the point which is conjugate to γ with respect to the pair A, B. Since the lines AA', BB', CC' touch S, they are not concurrent and pairs of triangles ABC, abc and $A'B'C'$, abc are not homological. Thus, triples of points α, β, γ_1 and α', β', γ' are not collinear. It follows that $\gamma_1 = \gamma$ and triangle $A'B'C'$ is circumscribed about Γ_1, Γ_2, Γ_3.

Step 2.   To prove this case completely, we need to show that a triangle is uniquely determined by conditions (i), (ii), (iii).

Let $\triangle ABC$ be inscribed in Γ and circumscribed about Γ_1, Γ_2, Γ_3, with a, b, c being the contact points of sides BC, AC, AB with these conics respectively. According to Lemma 5.3, construct the quadrangle $ABCD$ determined by sides AB, BC and contact points c, a, $\Delta = ac$. Let $\Delta \cap AD = a'$, $\Delta \cap CD = c'$, $\Delta \cap AC = M$. It holds that $M \neq b$.

Suppose there exists a triangle $\triangle AXC$ inscribed in Γ with the side AX tangent to Γ_3 at a point c'' and CX tangent to Γ_1 at a''. The point c'' does not belong to Δ, because, otherwise, this line would have three common points with the conic Γ_3. Apply Lemma 5.3 to the quadrangle $AXDC$ and contact points $c' \in CD$, $c'' \in AX$. The line $\Delta' = c'c''$ intersects AC in the point of contact with a conic from the pencil $\mathcal{F}$, i.e., in b or M. Since $\Delta' \neq \Delta$, this point is b. Then $b' = \Delta' \cap XD$ is a contact point of XD with conic Γ_2.

Similarly, we obtain that line $\Delta'' = a'a''$ intersects AC and XD in points b, b'. From there, we have $\Delta' = \Delta'' = \Delta$, which is not true.

Step 3. Now, we are going to consider the case of a polygon with an arbitrary number of vertices.

Let $ABC\ldots KL$ be a polygon inscribed in Γ and circumscribed about Γ_1, Γ_2, $\ldots$, Γ_n. Triangles ABC, ACD, $\ldots$, AKL are, according to the notation, circumscribed about conics Γ_1, Γ_2, (AC); (AC), Γ_3, (AD); $\ldots$; (AK), Γ_{n-1}, Γ_n.

Let $A'B'$ be a line touching Γ_1. Construct triangles $A'B'C'$, $A'C'D'$, $\ldots$, $A'K'L'$ circumscribed about conics Γ_1, Γ_2, (AC); (AC), Γ_3, (AD); $\ldots$; (AK), Γ_{n-1}, Γ_n. Polygon $A'B'C'\ldots K'L'$ is circumscribed about $\Gamma_1, \ldots, \Gamma_n$ respectively.

It is left to prove that it is possible to change the order in which sides touch the conics. Let B_1 be the fourth vertex of the quadrangle from Lemma 5.3, determined by sides AB, BC and their contact points with Γ_1 and Γ_2. Polygon $AB_1C\ldots KL$ is circumscribed about Γ_2, Γ_1, Γ_3, $\ldots$, Γ_n. Since any permutation can be represented by transpositions of successive elements, the theorem is proved.

5.2 Cayley's condition

Representation of conics of a pencil by points on a cubic curve

Let pencil $\mathcal{F}$ of conics be determined by the curves $C = 0$ and $\Gamma = 0$. The equation of an arbitrary conic of the pencil is $C + \lambda\Gamma = 0$.

Let $P + \lambda\Pi = 0$ be the equation of the corresponding polar lines from point $A \in \Gamma$. The geometric place of contact points of tangents from A with conics of the pencil is the cubic $\mathcal{C} : C\Pi - \Gamma P = 0$. On this cubic, any conic of $\mathcal{F}$ is represented by two contact points, which we will call *representative points* of the conic. The line determined by these two points passes through point $Z : P = 0$, $\Pi = 0$. There exist exactly four conics of the pencil whose representative points coincide: the conic Γ and three degenerate conics with representative points A, α, β, γ. Lines ZA, $Z\alpha$, $Z\beta$, $Z\gamma$ are tangents to $\mathcal{C}$ constructed from Z. The tangent line to cubic $\mathcal{C}$ at point Z is a polar of point A with respect to the conic of the pencil which contains Z.

Condition for existence of a Poncelet triangle

If triangles inscribed in Γ and circumscribed about Γ_1, Γ_2, Γ_3 exist, we will say that the conics Γ_1, Γ_2, Γ_3 are *joined* to Γ. In this case, the Full Poncelet theorem states there are six such triangles with the vertex $A \in \Gamma$. Let ABC be one of them. Side AB, denote it by *1*, touches Γ_1 in point m_1. Also, it touches another conic of the pencil, denote it by (I), in point M. Side AC, denote it by *2′*, touches Γ_2 in m_2'. Consider the quadrangle $ABCD$ determined by AB, AC and contact points M, m_2'. Line Mm_2' meets BC at μ_3, its point of tangency to Γ_3, and meets AD (which we will denote by *3′*) at the point m_3' of tangency to Γ_3. Triangle ABD is circumscribed about Γ_1, Γ_2, Γ_3.

Similarly, triangle ACE can be obtained by construction of quadrangle $ABCE$ determined by AB, BC and contact points m_1, μ_3. Line AE touches Γ_3 at point $m_3 \in m_1\mu_3$. Denote this line by *3*. Triangles with sides *3′*, *2* and *3*, *1′* are constructed analogously.

There are exactly six tangents from A to conics Γ_1, Γ_2, Γ_3. We have divided these six lines into two groups: *1,2,3* i *1′, 2′, 3′*. Two tangents enumerated by different numbers and do not belong to the same group, determine a Poncelet triangle.

Cubic $\mathcal{C}$ and the cubic consisting of lines m_1M, m_2m_2', m_3m_3' have simple common points m_1, M, m_2, m_2', m_3, m_3', A, and point Z as a double one. A pencil determined by these two cubics contains a curve that passes through a given point of line Mm_2', different from M, m_2', m_3'. This cubic has four common points with line Mm_2', so it decomposes into the line and a conic. Thus, m_1, m_2, m_3 *are intersection points, different from A and Z, of a conic which contains A and touches cubic $\mathcal{C}$ at Z.*

The converse also holds.

Let an arbitrary conic that contains point A and touches cubic $\mathcal{C}$ at Z be given. Denote by m_1, m_2, m_3 the remaining intersection points of the curve $\mathcal{C}$ with this conic. Each of the lines m_1Z, m_2Z, m_3Z has another common point with the cubic $\mathcal{C}$; denote them by m_1', m_2', m_3' respectively. By definition of the curve $\mathcal{C}$, we have that m_1, m_1'; m_2, m_2'; m_3, m_3' are pairs of representative points of some conics Γ_1, Γ_2, Γ_3 from the pencil $\mathcal{F}$. Line Am_1, besides being tangent to Γ_1 at m_1, has to touch another conic from the pencil $\mathcal{F}$. Assume that it is tangent to a conic (I) at M.

Now, in a similar fashion as before, we can conclude that points M, m_2', m_3' are collinear. Applying Lemma 5.3, it is easily deduced that conics Γ_1, Γ_2, Γ_3 are joined to Γ.

So, we have shown the following: *systems of three joined conics are determined by systems of three intersecting points of cubic $\mathcal{C}$ with conics that contain point A and touch the curve $\mathcal{C}$ at Z.*

Cayley's cubic

Let $D(\lambda)$ be the discriminant of conic $C + \lambda\Gamma = 0$. We will call the curve

$$\mathcal{C}_0 : Y^2 = D(X)$$

Cayley's cubic. Representative points of conic $C + \lambda\Gamma = 0$ on Cayley's cubic are two points that correspond to the value $X = \lambda$.

The polar conic of the point Z with respect to cubic $\mathcal{C}$ passes through the contact points of the tangents ZA, $Z\alpha$, $Z\beta$, $Z\gamma$ from Z to $\mathcal{C}$. Thus, points α, β, γ are representative points of three joined conics from the pencil $\mathcal{F}$. Those three conics are obviously the decomposable ones. Corresponding values λ diminish $D(\lambda)$, and these three representative points on Cayley's cubic $\mathcal{C}_0$ lie on the line $Y = 0$.

Using Sylvester's theory of residues, we will show the following:

Let three representative points of three conics of pencil $\mathcal{F}$ be given on the Cayley's cubic $\mathcal{C}_0$. The condition for these conics to be joined to the conic Γ is that their representative points are collinear.

Sylvester's theory of residues

When considering algebraic curves of genus 1, as Cayley's cubic is here, Abel's theorem can always be replaced by application of this theory.

Proposition 5.4. *Let a given cubic and an algebraic curve of degree $m + n$ meet at $3(m + n)$ points. If there are $3m$ points among them which are placed on a curve of degree m, then the remaining $3n$ points are placed on a curve of degree n.*

Proof. Suppose these curves of degrees $3, n, m + n$ respectively are given by the equations:

$$C_3 = 0, \quad C_n = 0, \quad C_{m+n} = 0.$$

We can assume that C_3 contains x^3. Consider C_n and C_{m+n} as polynomials of x and divide them by

$$C_3 = x^3 + px^2 + qx + r.$$

Let the obtained remainders be

$$\Gamma_n = ax^2 + bx + c,$$
$$\Gamma_{m+n} = Ax^2 + Bx + C,$$

where $p, q, r, a, b, c, A, B, C$ are polynomials of y. Curves $\Gamma_n = 0$ and $\Gamma_{m+n} = 0$ meet C_3 at the same $3n$, respectively $3(m+n)$, points as the curves C_n and C_{m+n}. So, replace Γ_n, Γ_{m+n} by C_n, C_{m+n}.

We are going to show that it is possible to find curves

$$\Gamma_m = \alpha x^2 + \beta x + \gamma,$$
$$\Gamma_{m+n-3} = ux + v,$$

of degrees m and $m + n - 3$ respectively, such that

$$\Gamma_{m+n} = \Gamma_n \Gamma_m + C_3 \Gamma_{m+n-3}.$$

$(\alpha, \beta, \gamma, u, v$ are polynomials of y.) This will prove the proposition.

Consider the system

$$
\begin{array}{rclclclclcl}
0 & = & a\alpha & + & 0\beta & + & 0\gamma & + & u & + & 0v \\
0 & = & b\alpha & + & a\beta & + & 0\gamma & + & pu & + & v \\
A & = & c\alpha & + & b\beta & + & a\gamma & + & qu & + & pv \\
B & = & 0\alpha & + & c\beta & + & b\gamma & + & ru & + & qv \\
C & = & 0\alpha & + & 0\beta & + & c\gamma & + & 0u & + & rv.
\end{array}
$$

Its solution $\alpha, \beta, \gamma, u, v$ is a quintuple of rational functions with denominators equal to the determinant Δ of the system. We are going to prove that these functions are polynomials. Assign to the variable y one of the values y_i which diminishes Δ. The numerator α is equal to

$$N = 0(1) + 0(2) + A(3) + B(4) + C(5),$$

where (1), (2), (3), (4), (5) are corresponding minors of Δ. If y_i is the ordinate of only one intersection point of C_3 and C_n, say (x_i, y_i), then all solutions to the system

$$
\begin{array}{rclclclclcl}
0 & = & az_4 & + & bz_3 & + & cz_2 & + & 0z_1 & + & 0z_0 \\
0 & = & 0z_4 & + & az_3 & + & bz_2 & + & cz_1 & + & 0z_0 \\
0 & = & 0z_4 & + & 0z_3 & + & az_2 & + & bz_1 & + & cz_0 \\
0 & = & z_4 & + & pz_3 & + & qz_2 & + & rz_1 & + & 0z_0 \\
0 & = & 0z_4 & + & z_3 & + & pz_2 & + & qz_1 & + & rz_0
\end{array}
$$

fulfill

$$z_0 : z_1 : z_2 : z_3 : z_4 = 1 : x_i : x_i^2 : x_i^3 : x_i^4.$$

Corresponding minors of a homogeneous system of linear equations with zero determinant form also a solution to the system. Hence, (1), (2), (3), (4), (5) are proportional to $x_i^4, x_i^3, x_i^2, x_i, 1$. Since (x_i, y_i) is a point of curve Γ_{m+n}, we have $N = 0$.

It follows that N is divisible by Δ, i.e., α is a polynomial. It is easy to check that its degree is at most $m - 2$. Similarly, this can be shown for β, γ, u, v.

The proposition can be also demonstrated in the case when C_3 and C_n have multiple points of intersection. $\square$

If the union of two systems of points is the complete intersection of a given cubic and some algebraic curve, then we will say that these two systems are *residual* to each other. Now, the following holds:

Proposition 5.5. *If systems $\mathcal{A}$ and $\mathcal{A}'$ of points on a given cubic curve have a common residual system, then they share all residual systems.*

Proof. Suppose $\mathcal{B}$ is a system residual to both $\mathcal{A}$, $\mathcal{A}'$ and $\mathcal{B}'$ is residual to $\mathcal{A}$. Then $\mathcal{A} \cup \mathcal{A}'$ is residual to $\mathcal{B} \cup \mathcal{B}'$, i.e., the system $\mathcal{A} \cup \mathcal{A}' \cup \mathcal{A} \cup \mathcal{A}'$ is a complete intersection of the cubic with an algebraic curve. Since $\mathcal{A} \cup \mathcal{B}$ is also such an intersection, it follows, by the previous proposition, that $\mathcal{A}'$ and $\mathcal{B}'$ are residual to each other.   □

Let us note that this proposition can be derived as a consequence of Abel's theorem, for a plane algebraic curve of arbitrary degree. However, if the degree is equal to 3, i.e., the curve is elliptic, Proposition 5.5 is equivalent to Abel's theorem.

Condition for existence of a Poncelet polygon

Let conics Γ, Γ_1, ..., Γ_n be from a pencil. If there exists a polygon inscribed in Γ and circumscribed about $\Gamma_1, \ldots, \Gamma_n$, we are going to say that conics $\Gamma_1, \ldots, \Gamma_n$ are *joined* to Γ. Then, similarly as in the case of the triangle, it can be proved that tangents from the point $A \in \Gamma$ to $\Gamma_1, \ldots, \Gamma_n$ can be divided into two groups such that any Poncelet n-polygon with vertex A has exactly one side in each of the groups.

This division of tangents gives a division of characteristic points of conics Γ_1, $\ldots, \Gamma_n$ into two groups on $\mathcal{C}$ and, therefore, a division into two groups on Cayley's cubic $\mathcal{C}_0$: *1,2,3,* ... and *1′, 2′, 3′,* ....

Let $ABCD \ldots KL$ be a Poncelet polygon, and let $(AC), (AD), \ldots$ be conics determined as in the definition of a circumscribed polygon. Let c, γ; d, δ; ... be a corresponding characteristic points on $\mathcal{C}_0$, such that triples *1, 2, c*; γ, *3, d*; δ, *4, e* are characteristic points of the same group with respect to corresponding conics.

Points *1, 2, c* are collinear, as γ, *3, d* are. Thus, *1, 2, 3, d* are residual with c, γ. Line $c\gamma$ contains point Z, so system c, γ is residual with Z, too. It is possible to show that Z is a triple point of curve $\mathcal{C}_0$ and it follows that it is residual with system Z, Z. This implies that points *1,2,3, d, Z, Z* are placed on a conic.

If we take a coordinate system such that the tangent line to $\mathcal{C}_0$ at Z is the infinite line and the axis Oy is line AZ, we will have:

four conics are joined to Γ if and only if their characteristic points of the same group are on a parabola with the asymptotic direction Oy.

Continuing deduction in the same manner, we can conclude: $3n - p$ *points of the cubic $\mathcal{C}_0$ are characteristic points of same group for $3n - p$ conics ($1 \leq p \leq 3$) joined to Γ, if and only if these points are placed on a curve of degree n which has Oy as an asymptotic line of the order p.*

Cayley's condition

Let $\mathcal{C}_0 : y^2 = D(x)$ be the Cayley's cubic, where $D(x)$ is the discriminant of the conic $C + x\Gamma = 0$ from pencil $\mathcal{F}$. A system of n conics joined to Γ is determined by n values x if and only if these n values are abscissae of intersecting points of

$\mathcal{C}_0$ and some algebraic curve. Plugging $D(x)$ instead of y^2 in the equation of this curve, we obtain

$$P(x)y + Q(x) = 0,$$

that is

$$P(x)\sqrt{D(x)} + Q(x) = 0.$$

From there:

$$\sqrt{D(x)}(a_0 x^{p-2} + a_1 x^{p-3} + \cdots + a_{p-2}) + (b_0 x^p + b_1 x^{p-1} + \cdots + b_p) = 0, \quad n = 2p;$$

$$\sqrt{D(x)}(a_0 x^{p-1} + a_1 x^{p-2} + \cdots + a_{p-1}) + (b_0 x^p + b_1 x^{p-1} + \cdots + b_p) = 0, \quad n = 2p+1.$$

If $\lambda_1, \ldots, \lambda_k$ denote parameters corresponding to $\Gamma_1, \ldots, \Gamma_k$ respectively, then existence of a Poncelet polygon inscribed in Γ and circumscribed about $\Gamma_1, \ldots, \Gamma_k$ is equivalent to:

$$\begin{vmatrix} 1 & \lambda_1 & \lambda_1^2 & \cdots & \lambda_1^p & \sqrt{D(\lambda_1)} & \lambda_1\sqrt{D(\lambda_1)} & \cdots & \lambda_1^{p-2}\sqrt{D(\lambda_1)} \\ \cdots \\ \cdots \\ 1 & \lambda_k & \lambda_k^2 & \cdots & \lambda_k^p & \sqrt{D(\lambda_k)} & \lambda_k\sqrt{D(\lambda_k)} & \cdots & \lambda_k^{p-2}\sqrt{D(\lambda_k)} \end{vmatrix} = 0,$$

for $k = 2p$;

$$\begin{vmatrix} 1 & \lambda_1 & \lambda_1^2 & \cdots & \lambda_1^p & \sqrt{D(\lambda_1)} & \lambda_1\sqrt{D(\lambda_1)} & \cdots & \lambda_1^{p-1}\sqrt{D(\lambda_1)} \\ \cdots \\ \cdots \\ 1 & \lambda_k & \lambda_k^2 & \cdots & \lambda_k^p & \sqrt{D(\lambda_k)} & \lambda_k\sqrt{D(\lambda_k)} & \cdots & \lambda_k^{p-1}\sqrt{D(\lambda_k)} \end{vmatrix} = 0,$$

for $k = 2p + 1$.

There exists an n-polygon inscribed in Γ and circumscribed about C if and only if it is possible to find coefficients $a_0, a_1, \ldots$; $b_0, b_1, \ldots$ such that function $P(x)\sqrt{D(x)} + Q(x)$ has $x = 0$ as a root of the multiplicity n.

For $n = 2p$, this is equivalent to the existence of a non-trivial solution of the following system:

$$\begin{array}{ccccccc} a_0 C_3 & + & a_1 C_4 & + & \cdots & + & a_{p-2} C_{p+1} & = & 0 \\ a_0 C_4 & + & a_1 C_5 & + & \cdots & + & a_{p-2} C_{p+2} & = & 0 \\ & & \cdots \\ a_0 C_{p+1} & + & a_1 C_{p+2} & + & \cdots & + & a_{p-2} C_{2p-1} & = & 0, \end{array}$$

where

$$\sqrt{D(x)} = A + Bx + C_2 x^2 + C_3 x^3 + \cdots.$$

Finally, for $n = 2p$, we obtain Cayley's condition

$$\begin{vmatrix} C_3 & C_4 & \cdots & C_{p+1} \\ C_4 & C_5 & \cdots & C_{p+2} \\ & & \cdots \\ C_{p+1} & C_{p+2} & \cdots & C_{2p-1} \end{vmatrix} = 0.$$

Similarly, for $n = 2p + 1$, we obtain

$$\begin{vmatrix} C_2 & C_3 & \cdots & C_{p+1} \\ C_3 & C_4 & \cdots & C_{p+2} \\ & & \cdots & \\ C_{p+1} & C_{p+2} & \cdots & C_{2p} \end{vmatrix} = 0.$$

These results can be directly applied to the billiard system within an ellipse: to determine whether a billiard trajectory with a given confocal caustic is periodic, we need to consider the pencil determined by the boundary and the caustic curve.

Exercise 5.6 ([GZ1976]). Let $\mathcal{C}$ be a circle with radius R and $\mathcal{E}$ be an ellipse inside $\mathcal{C}$ with semi-minor axis b. If d_1, d_2 are distances from the center of $\mathcal{C}$ to the foci of $\mathcal{E}$, then there exists a triangle inscribed in $\mathcal{C}$ and circumscribed about $\mathcal{E}$ if and only if

$$(R^2 - d_1^2)(R^2 - d_2^2) = 4b^2 R^2.$$

Example 5.7. Exercise 5.6 gives a generalization of the Chapple–Euler formula for triangles (see Theorem 2.9). Full generalizations of Theorems 2.9 and 2.10 will be, in Cayley's approach,

$$n = 3: \quad C_2 = 0;$$
$$n = 4: \quad C_3 = 0.$$

Some applications of Lebesgue's results

Now, we are going to apply Lebesgue's results to billiard systems within several confocal conics in the plane.

Consider the dual plane. The case with two ellipses, when the billiard trajectory is placed between them and a particle bounces to one and another of them alternately, is of a special interest.

Corollary 5.8. *The condition for the existence of a $2m$-periodic billiard trajectory which bounces exactly m times to the ellipse $\Gamma_1^* = C^*$ and m times to $\Gamma_2^* = (C + \gamma \Gamma)^*$, having Γ^* for the caustic, is*

$$\det \begin{pmatrix} f_0(0) & f_1(0) & \cdots & f_{2m-1}(0) \\ f_0'(0) & f_1'(0) & \cdots & f_{2m-1}'(0) \\ & & \cdots & \\ f_0^{(m-1)}(0) & f_1^{(m-1)}(0) & \cdots & f_{2m-1}^{(m-1)}(0) \\ f_0(\gamma) & f_1(\gamma) & \cdots & f_{2m-1}(\gamma) \\ f_0'(\gamma) & f_1'(\gamma) & \cdots & f_{2m-1}'(\gamma) \\ & & \cdots & \\ f_0^{(m-1)}(\gamma) & f_1^{(m-1)}(\gamma) & \cdots & f_{2m-1}^{(m-1)}(\gamma) \end{pmatrix} = 0,$$

where $f_j = x^j$, $(0 \le j \le m)$, $f_{m+i} = x^{i-1} \sqrt{D(x)}$, $(1 \le i \le m - 1)$.

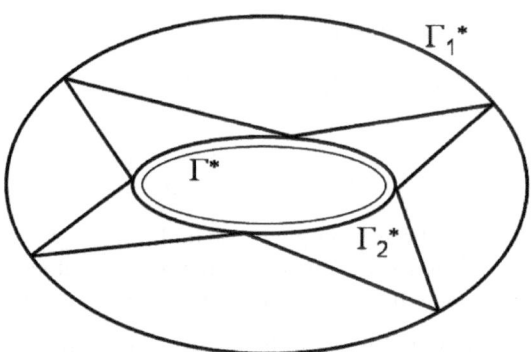

Figure 5.1: A closed billiard trajectory in the domain
bounded by two confocal ellipses

We consider a simple example with four bounces on each of the two conics,
see Figure 5.1.

Example 5.9. The condition on a billiard trajectory placed between ellipses Γ_1^*
and Γ_2^*, to be closed after four alternate bounces to each of them is

$$\det X = 0,$$

where the elements of the 3×3 matrix X are

$$X_{11} = -4B_0 + B_1\gamma + 4C_0 + 3C_1\gamma + 2C_2\gamma^2 + C_3\gamma^3$$
$$X_{12} = -3B_0 + B_1\gamma + 3C_0 + 2C_1\gamma + C_2\gamma^2$$
$$X_{13} = -2B_0 + B_1\gamma + 2C_0 + C_1\gamma$$
$$X_{21} = -6B_0 + B_2\gamma^2 + 6C_0 + 6C_1\gamma + 4C_2\gamma^2 + 3C_3\gamma^3$$
$$X_{22} = -6B_0 + B_1\gamma + B_2\gamma^2 + 6C_0 + 4C_1\gamma + 3C_2\gamma^2$$
$$X_{23} = -5B_0 + 2B_1\gamma + B_2\gamma^2 + 5C_0 + 3C_1\gamma$$
$$X_{31} = -4B_0 + B_3\gamma^3 + 4C_0 + 4C_1\gamma + 4C_2\gamma^2 + 3C_3\gamma^3$$
$$X_{32} = -4B_0 + B_2\gamma^2 + B_3\gamma^3 + 4C_0 + 4C_1\gamma + 3C_2\gamma^2$$
$$X_{33} = -4B_0 + B_1\gamma + B_2\gamma^2 + B_3\gamma^3 + 4C_0 + 3C_1\gamma,$$

with C_i, B_i being coefficients in the Taylor expansions around $x = 0$ and $x = \gamma$
respectively:

$$\sqrt{D(x)} = C_0 + C_1 x + C_2 x^2 + \ldots,$$
$$\sqrt{D(x)} = B_0 + B_1(x - \gamma) + B_2(x - \gamma)^2 + \cdots.$$

On Figure 5.2, we see a Poncelet octagon inscribed in Γ and circumscribed
about Γ_1 and Γ_2. In the dual plane, the billiard trajectory that corresponds to
this octagon has the dual conic Γ^* as the caustic (see Figure 5.1).

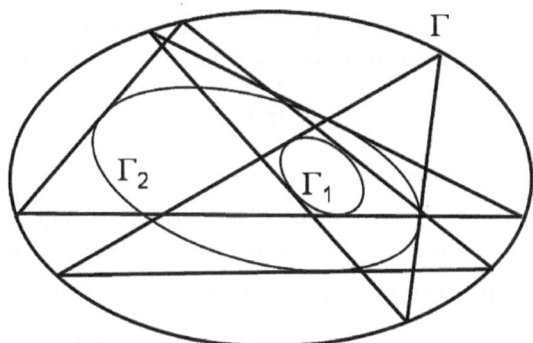

Figure 5.2: A Poncelet octagon whose sides touch
the conics Γ_1 and Γ_2 alternately

5.3 Another proof of Poncelet's theorem and Cayley's condition

In [GH1977, GH1978a], Griffiths and Harris approached classical Poncelet and Cayley theorems using modern algebro-geometric tools and proved them in a very elegant and short manner. Also, they gave an interesting generalization of Poncelet's theorem to the three-dimensional case. They consider polyhedral surfaces which are simultaneously inscribed and circumscribed in two quadrics in the space. They prove, in the same spirit as in the plane, that the existence of one such closed surface implies that infinitely many polyhedra with the same property exist.

In this section, we are going to present these results of Griffiths and Harris.

Let $\mathcal{C}$ and $\mathcal{D}$ be two conics in the complex projective plane $\mathbf{P}^2$, intersecting in points x_0, x_1, x_2, x_3. The dual conic $\mathcal{D}^*$ consists of tangents to $\mathcal{D}$. Consider the configuration

$$E = \{(x, \xi) | x \in \xi\} \subset \mathcal{C} \times \mathcal{D}^*.$$

E is an algebraic curve. We can define its involutions:

$$i(x, \xi) = (x', \xi),$$
$$i'(x', \xi) = (x', \xi'),$$

whose composition $j = i \circ i'$ is given by $j(x, \xi) = (x', \xi')$. It follows that the Poncelet's configuration starting at $p = (x, \xi)$ gives a closed polygon with n sides if and only if

$$j^n(p) = p.$$

By the mapping

$$E \to \mathcal{C} : (x, \xi) \mapsto x,$$

E becomes a two-folded covering of the Riemann sphere $\mathbf{P}^1$. The branching points are x_0, x_1, x_2, x_3. Applying the Hurwitz formula, we calculate the Euler characteristic of the surface E:

$$\chi(E) = 2\chi(\mathbf{P}^1) - 4 = 0.$$

Thus, E is a surface of genus 1, i.e., an elliptic curve.

Choose the point (x_0, ξ_0) to be the neutral of the group on the elliptic curve E, and let $p = (\bar{x}, \bar{\xi})$, where $\bar{x} \in \mathcal{C} \cap \xi_0$ and $\bar{\xi}$ is the other tangent to $\mathcal{D}$ from $\bar{x}$.

Then, the Poncelet theorem can be formulated as follows:

Theorem 5.10. *The Poncelet's construction gives a closed n-polygon with an arbitrary initial condition $q = (x, \xi)$ if and only if $np = 0$ on the elliptic curve E.*

Proof. On the universal covering $\mathbf{C}$ of the Riemann surface E, $E \simeq \mathbf{C}/\Lambda$, the involutions i, i' induce automorphisms

$$\tilde{i}(u) = \alpha u + \tau,$$
$$\tilde{i}'(u) = \alpha' u + \tau'.$$

Since $i^2 = 1$ and i has fixed points, we have $\alpha = -1$. Similarly, $\alpha' = -1$. From $i'(0) = 0$, we obtain

$$\tilde{i}(u) \equiv -u - w \quad \mathrm{mod}\ \Lambda,$$
$$\tilde{i}'(u) \equiv -u \quad \mathrm{mod}\ \Lambda.$$

Thus $\tilde{j}(u) \equiv u + w \quad \mathrm{mod}\ \Lambda$, i.e.,

$$j^n(q) = q \Longleftrightarrow nw \equiv 0 \quad \mathrm{mod}\ \Lambda.$$

Since $p = j(0)$, $w = \tilde{j}(0)$, the statement is proved. □

Let the pencil of conics containing the points x_0, x_1, x_2, x_3 be given by $D_t : t\mathcal{C}(x) + \mathcal{D}(x) = 0$. The determinant $\det(tC + D)$ is a polynomial of the third degree in t with roots t_1, t_2, t_3, all different from zero. For $t \neq t_i$, construct the tangent line to D_t through x_0. Let $x(t)$ be the other intersecting point of the tangent with the conic $\mathcal{C}$. The values $t = t_i$ are mapped to x_i and $t = \infty$ to x_0, because $D_\infty = \mathcal{C}$. Also, $x(0) = \bar{x}$. Thus, the following statement is proved:

Proposition 5.11. *The elliptic curve E is birationally equivalent to the Riemann surface of the algebraic function $\sqrt{\det(tC + D)}$ with the origin at $t = \infty$ and the point $p = (\bar{x}, \bar{\xi})$ corresponding to one of the two points lying over $t = 0$.*

The Cayley condition is obtained by applying the following

Proposition 5.12. *Let an elliptic curve be given by the equation*

$$E \ : \ y^2 = (x - a)(x - b)(x - c),$$

where a, b, c are distinct and non-zero. Choose the point corresponding to the value $x = \infty$ to be the origin on E and let p be one of the points corresponding to $x = 0$. Then p is of finite order n if and only if

$$
\begin{vmatrix}
C_3 & C_4 & \cdots & C_{m+1} \\
C_4 & C_5 & \cdots & C_{m+2} \\
\hdotsfor{4} \\
C_{m+1} & C_{m+2} & \cdots & C_{2m-1}
\end{vmatrix} = 0, \quad n = 2m,
$$

$$
\begin{vmatrix}
C_2 & C_3 & \cdots & C_{m+1} \\
C_3 & C_4 & \cdots & C_{m+2} \\
\hdotsfor{4} \\
C_{m+1} & C_{m+2} & \cdots & C_{2m}
\end{vmatrix} = 0, \quad n = 2m+1,
$$

where $\sqrt{(x-a)(x-b)(x-c)} = A + Bx + C_2 x^2 + C_3 x^3 + \cdots$.

Proof. Denote by P_∞ the point on E corresponding to $x = \infty$.

Let $u_1, \ldots, u_n$ be points on E. We need to find a condition for $u_1 + \cdots + u_n = 0$. When all these points coincide with p, we will obtain the condition for $np = 0$. By Abel's theorem, $u_1 + \cdots + u_n = 0$ is satisfied if and only if a meromorphic function on the curve E exists which has a pole of order n at P_∞ and zeroes $u_1, \ldots, u_n$.

Denote by $\mathcal{L}(nP_\infty)$ the linear space of meromorphic functions on E having a unique pole of order at most n at P_∞. By the Riemann–Roch theorem,

$$
\dim \mathcal{L}(nP_\infty) = n - g + 1 + \dim \mathcal{K}^1(nP_\infty),
$$

where $\mathcal{K}^1(nP_\infty)$ is a space of holomorphic differentials on E which have a zero of order at least n at P_∞ and g is the genus of surface E. Since E is elliptic, $g = 1$.

By the Poincaré–Hopf formula, for any meromorphic differential $\omega \neq 0$ on E we have

$$
\sum_{P \in E} \nu_P(\omega) = -\chi(E) = 0,
$$

where $\nu_P(\omega)$ is the index of form ω at P. Furthermore, for $\omega \in \mathcal{K}^1(nP_\infty)$ we have $\sum_{P \in E} \nu_P(\omega) \geq n$. Thus $\mathcal{K}^1(nP_\infty)$ contains only the zero differential.

It follows that $\dim \mathcal{L}(nP_\infty) = n$. Let $f_1, \ldots, f_n$ be a basis of this space. The mapping

$$
E \to \mathbf{P}^{n-1} : u \mapsto [f_1(u), \ldots, f_n(u)]
$$

is a projective embedding whose image is a smooth algebraic curve of degree n. Intersections of this curve with hyperplanes are zeroes of functions $\mathcal{L}(nP_\infty)$. So, the equality $u_1 + \cdots + u_n = 0$ is equivalent to

$$
\begin{vmatrix}
f_1(u_1) & f_2(u_1) & \cdots & f_n(u_1) \\
f_1(u_2) & f_2(u_2) & \cdots & f_n(u_2) \\
\hdotsfor{4} \\
f_1(u_n) & f_2(u_n) & \cdots & f_n(u_n)
\end{vmatrix} = 0.
$$

For $u_1 = \cdots = u_n = p$, we obtain

$$W(p) = \begin{vmatrix} f_1(p) & f_2(p) & \cdots & f_n(p) \\ f_1'(p) & f_2'(p) & \cdots & f_n'(p) \\ \multicolumn{4}{c}{\cdots\cdots\cdots\cdots\cdots\cdots\cdots\cdots\cdots} \\ f_1^{(n-1)}(p) & f_2^{(n-1)}(p) & \cdots & f_n^{(n-1)}(p) \end{vmatrix}.$$

Take for the basis $f_1, \ldots, f_n$ the following:

$$1, x, \ldots, x^m, y, xy, \ldots, x^{m-1}y, \quad n = 2m+1,$$
$$1, x, \ldots, x^m, y, xy, \ldots, x^{m-2}y, \quad n = 2m.$$

Consider the case $n = 2m$.

The Wronskian $W(p)$ is of the form

$$\begin{vmatrix} 1 & 0 & 0 & \cdots & 0 & - & - & - & - & - & - & - & - \\ 0 & 1! & 0 & \cdots & 0 & - & - & - & - & - & - & - & - \\ 0 & 0 & 2! & \cdots & 0 & - & - & - & - & - & - & - & - \\ \multicolumn{5}{c}{\cdots\cdots\cdots\cdots\cdots\cdots} & - & - & - & - & - & - & - & - \\ \multicolumn{5}{c}{\cdots\cdots\cdots\cdots\cdots\cdots} & - & - & - & - & - & - & - & - \\ 0 & 0 & 0 & \cdots & m! & - & - & - & - & - & - & - & - \\ \\ 0 & 0 & & \cdots & 0 & - & - & - & - & - & - & - & - \\ 0 & 0 & & \cdots & 0 & - & - & - & - & - & - & - & - \\ \multicolumn{5}{c}{\cdots\cdots\cdots\cdots\cdots\cdots} & - & - & - & - & - & - & - & - \\ \multicolumn{5}{c}{\cdots\cdots\cdots\cdots\cdots\cdots} & - & - & - & - & - & - & - & - \\ 0 & 0 & & \cdots & 0 & - & - & - & - & - & - & - & - \end{vmatrix}.$$

Thus, the condition $W(p) = 0$ is equivalent to

$$\begin{vmatrix} \dfrac{d^{m+1}y}{dx^{m+1}} & \dfrac{d^{m+1}(xy)}{dx^{m+1}} & \cdots & \dfrac{d^{m+1}(x^{m-2}y)}{dx^{m+1}} \\[3mm] \dfrac{d^{m+2}y}{dx^{m+2}} & \dfrac{d^{m+2}(xy)}{dx^{m+2}} & \cdots & \dfrac{d^{m+2}(x^{m-2}y)}{dx^{m+2}} \\[3mm] \multicolumn{4}{c}{\cdots\cdots\cdots\cdots\cdots\cdots\cdots\cdots\cdots\cdots\cdots} \\[2mm] \dfrac{d^{2m-1}y}{dx^{2m-1}} & \dfrac{d^{2m-1}(xy)}{dx^{2m-1}} & \cdots & \dfrac{d^{2m-1}(x^{m-2}y)}{dx^{2m-1}} \end{vmatrix} = 0,$$

where d^k/dx^k denotes $d^k/dx^k|_{x=0}$. The previous expression is equivalent to:

$$\begin{vmatrix} (m+1)!C_{m+1} & (m+1)!C_m & \cdots & (m+1)!C_3 \\ (m+2)!C_{m+2} & (m+2)!C_{m+1} & \cdots & (m+2)!C_4 \\ \multicolumn{4}{c}{\cdots\cdots\cdots\cdots\cdots\cdots\cdots\cdots\cdots\cdots\cdots} \\ (2m-1)!C_{2m-1} & (2m-1)!C_{2m-2} & \cdots & (2m-1)!C_{m+1} \end{vmatrix} = 0,$$

and the statement is proved.   □

5.4 One generalization of the Poncelet theorem

At first glance, it seems that polyhedra inscribed in one and circumscribed in another surface would be a natural generalization of the Poncelet's configuration to the three-dimensional space. But, in general, there are infinitely many planes containing a point and touching an algebraic surface. Nevertheless, it is still possible to define more precisely polyhedra corresponding to two quadrics. The analogue for the Poncelet theorem is based on properties of elliptic curves, similarly as in the Griffiths' and Harris' proof of the Poncelet theorem in the plane.

The following lemma gives some properties of lines contained in a second-order surface, which will be necessary for the definition of Poncelet's configuration. Its proof can be found in [GH1977]. See also Proposition 4.69.

Lemma 5.13. [GH1977] *Let S be a quadric in the complex projective three-dimensional space.*

  (i) *The intersection of S and the plane tangent to S at a point $P \in S$ consists of two lines. Any line lying on S and passing through P is one of them.*

  (ii) *Lines on the surface S can be divided into two disjoint families A and B. Each A-line L meets each B-line L'. The plane tangent to S at their intersection point is determined by L and L'. Two distinct lines from the same family are not coplanar.*

  (iii) *Through each line not lying on S, there are exactly two planes tangent to S. Each plane through a line $L \subset S$ touches the surface S at a point of L.*

Now, we are going to describe the analogue of the Poncelet's configuration. Let S and S' be non-degenerate quadrics intersecting in the projective space. Each plane T tangent to both S and S' meets $S \cup S'$ in a quadrilateral with edges L_A, L_B contained in S and L'_A, L'_B in S'. The plane T touches the quadrics at points $P = L_A \cap L_B$ and $P' = L'_A \cap L'_B$. By Lemma 5.13, there exists a plane $\tilde{T}$, distinct from T, which contains L_A and touches S'. By (iii), this plane is tangent also to S at a point of L_A. Let

$$\tilde{T} \cap S = \tilde{L}_A + \tilde{L}_B,$$
$$\tilde{T} \cap S' = \tilde{L}'_A + \tilde{L}'_B,$$
$$\tilde{L}'_A = L_A.$$

Note that from (ii) follows that L'_B, L'_A meet L_A at same points as the lines $\tilde{L}'_A$, $\tilde{L}'_B$.

If we denote by S^* and S'^* the dual quadrics (consisting of tangent planes to S and S'), we can define an involution of the curve $E = S^* \cap S'^*$ by $i_A(T) = \tilde{T}$. We are going to show that E is an elliptic curve.

Lemma 5.14. [GH1977] *The genus of the curve $E = S^* \cap S'^*$ is equal to 1.*

Proof. Instead for E, we are going to show the assertion for the dual curve $C = S \cap S'$. Let L be a B-line on S. Join to the point $P \in C$ the A-line on S which passes through P and, then, the meeting point of that line with L. In this manner, the mapping $\pi_A : C \to \mathbf{P}^1$ is defined. Since L meets S' at two points, π_A is a two-folded covering with branching points corresponding to those A-lines on S which are tangent to S'. Now, it is sufficient to prove that there are exactly four branching points.

Similarly, define function π_B which maps points of the curve C into points of an A-line from S. This is also a two-folded covering of $\mathbf{P}^1$ with the number of branching points the same as π_A. So, we need to prove that there are exactly eight lines on S which are tangent to S'.

If line $L \subset S$ touches S' at point P, then $P \in S \cap S'$ and, by (iii) of Lemma 1, $T_P(S') \in S^*$. Conversely, if $P \in S \cap S'$ and $T_P(S') = T_{P'}(S)$ for some $P' \in S$, then line PP' lies on S and touches S'. Denote by x the coordinates of point $P \in S \cap S'$. Then $T_P(S') \in S^*$ if and only if $Q'x = Qy$ for some y satisfying $(Qy, y) = 0$.

(Here, Q and Q' are symmetric non-singular matrices, such that S, S' have equations $(Qx, x) = 0$, $(Q'x, x) = 0$. The tangent plane at a point on S with the coordinates x_0 is given by the equation $(Qx_0, x) = 0$. The dual quadric S^* is given by $(Q^{-1}x, x) = 0$.)

From here, it follows that points $P \in S \cap S'$, for which $T_P(S') \in S^*$, are given by

$$(Qx, x) = 0, \quad (Q'x, x) = 0, \quad (Q'x, Q^{-1}Q'x) = 0,$$

i.e., they are common points of three quadrics which meet transversely in $\mathbf{P}^3$. There are eight such points, which finishes the proof. $\qquad\square$

Similarly as i_A, we can define mappings i_B, i'_A, i'_B, which we will call *re-flections*. For a given bitangent plane T_0, polyhedral surface $\Pi(T_0)$ is constructed by consecutive application of all possible reflections to T_0. Sides of the polyhedral surface are bitangent to S and S', vertices are intersection points of these two quadrics and edges lie on S and S' alternately.

At this point, it is natural to ask when the configuration $\Pi(T_0)$ is going to be finite.

Theorem 5.15. *There is a finite polyhedron inscribed and circumscribed in S and S' at the same time if and only if there are infinitely many such polyhedra.*

Proof. The curve E is elliptic, so can be represented as $\mathbf{C}/\Lambda$ and the reflections as

$$i_A(u) = -u + \tau_1, \quad i_B(u) = -u + \tau_2,$$
$$i'_A(u) = -u + \tau_3, \quad i'_B(u) = -u + \tau_4.$$

Obviously, the polyhedral surface $\Pi(T_0)$ is finite if and only if $\tau_i - \tau_j \in \mathbf{Q}\Lambda$. Since this condition does not depend on T_0, the theorem is proved. $\qquad\square$

5.5   Poncelet theorem on Liouville surfaces

In this section, we are going to give the presentation and comments to the Darboux results from [Dar1914] on generalization of the Poncelet theorem to Liouville surfaces.

Liouville surfaces and families of geodesic conics

First, we need to define geodesic conics on an arbitrary surface, derive some important properties of theirs and finally to obtain an important characterization of Liouville surfaces via families of geodesic conics.

Let $\mathcal{C}_1$ and $\mathcal{C}_2$ be two fixed curves on a given surface $\mathcal{S}$. *Geodesic ellipses and hyperbolae* on $\mathcal{S}$ are curves given by the equations

$$\theta + \sigma = \text{const},$$
$$\theta - \sigma = \text{const},$$

where θ, σ are geodesic distances from $\mathcal{C}_1$, $\mathcal{C}_2$ respectively.

A coordinate system composed of geodesic ellipses and hyperbolae joined to two fixed curves is orthogonal. In the following proposition, we are going to describe all orthogonal coordinate systems with coordinate curves that can be regarded as a family of geodesic ellipses and hyperbolae.

Proposition 5.16. *Let*

$$ds^2 = A^2 du^2 + C^2 dv^2 \tag{5.2}$$

be the surface element corresponding to an orthogonal system of coordinate curves. Then the coordinate curves represent a family of geodesic ellipses and hyperbolae if and only if the coefficients A, C satisfy a relation of the form

$$\frac{U}{A^2} + \frac{V}{C^2} = 1,$$

with U and V being functions of u, v respectively.

Proof. By assumption, equations of coordinate curves are

$$\theta + \sigma = \text{const}, \quad \theta - \sigma = \text{const},$$

with θ, σ representing geodesic distances from a point of the surface to two fixed curves.

Thus

$$u = F(\theta + \sigma), \qquad v = F_1(\theta - \sigma).$$

Solving these equations with respect to θ and σ, we obtain

$$\theta = \phi(u) + \psi(v), \qquad \sigma = \phi(u) - \psi(v).$$

As geodesic distances, θ and σ need to satisfy the characteristic partial differential equation

$$\frac{A^2(\frac{\partial \xi}{\partial v})^2 + C^2(\frac{\partial \xi}{\partial u})^2}{A^2 C^2} = 1. \tag{5.3}$$

From there, we deduce the desired relation with $U = (\phi'(u))^2$, $V = (\psi'(v))^2$.

The converse is proved in a similar manner.   □

As a straightforward consequence, the following interesting property is obtained:

Corollary 5.17. *If an orthogonal system of curves can be regarded as a system of geodesic ellipses and hyperbolae in two different ways, then it can be regarded as such a system in infinitely many ways.*

Now, let us concentrate on Liouville surfaces, i.e., on surfaces with the surface element of the form

$$ds^2 = (U - V)(U_1^2 du^2 + V_1^2 dv^2), \tag{5.4}$$

where U, U_1 and V, V_1 depend only on u and v respectively.

Now, we are ready to present the characterization of Liouville surfaces via geodesic conics.

Theorem 5.18. *An orthogonal system on a surface can be regarded in two different manners as a system of geodesic conics if and only if it is of the Liouville form.*

Proof. Consider a surface with the element (5.2). If coordinate lines $u = $ const, $v = $ const can be regarded as geodesic conics in two different manners then, by Proposition 5.16, A, C satisfy two different equations:

$$\frac{U}{A^2} + \frac{V}{C^2} = 1, \qquad \frac{U_1}{A^2} + \frac{V_1}{C^2} = 1.$$

Solving them with respect to A, C, we obtain that the surface element is of the form

$$ds^2 = \left(\frac{U}{U - U_1} + \frac{V}{V_1 - V} \right) ((U - U_1)du^2 + (V_1 - V)dv^2).$$

Conversely, consider the Liouville surface with the element (5.4) and the following solutions of equation (5.3):

$$\theta = \int U_1 \sqrt{U - a}\, du + \int V_1 \sqrt{a - V}\, dv,$$

$$\sigma = \int U_1 \sqrt{U - a}\, du - \int V_1 \sqrt{a - V}\, dv.$$

The equations $\theta + \sigma = $ const, $\theta - \sigma = $ const will define the coordinate curves.   □

Generalization of Graves and Poncelet theorems to Liouville surfaces

We learned from Darboux [Dar1914] that Liouville surfaces are exactly those having an orthogonal system of curves that can be regarded in two or, equivalently, infinitely many different ways, as geodesic conics. Now, we are going to present how to make a choice, among these infinitely many presentations, of the most convenient one, which will enable us to show the generalizations of theorems of Graves and Poncelet. All these ideas of enlightening beauty and profoundness belong to Darboux [Dar1914].

Consider a curve γ on a surface S. The *involute* of γ with respect to a point $A \in \gamma$ is the set of endpoints M of all geodesic segments TM, such that:

$T \in \gamma$,

TM is tangent to γ at T,

the length of TM is equal to the length of the segment $TA \subset \gamma$; and these two segments are placed at the same side of the point T.

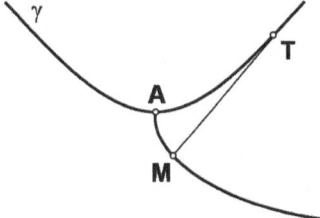

Figure 5.3: Involute

Involutes have the following important property, which follows immediately from the definition:

Lemma 5.19. *The geodesic segments TM are orthogonal to the involute, and the involute itself is orthogonal to γ at A.*

Now, we are going to find explicitly the equations of involutes of coordinate curves on a Liouville surface S with the surface element (5.4).

Lemma 5.20. *The curves on S given by the equations*

$$\theta = \int U_1 \sqrt{U - a} \, du + \int V_1 \sqrt{a - V} \, dv = \text{const},$$

$$\sigma = \int U_1 \sqrt{U - a} \, du - \int V_1 \sqrt{a - V} \, dv = \text{const} \tag{5.5}$$

are involutes of the coordinate curve whose parameter satisfies the equation

$$(U - a)(V - a) = 0.$$

Proof. Fix the parameter a. The equations of geodesics normal to the curves $\theta = $ const, $\sigma = $ const are obtained by differentiating (5.5) with respect to a:

$$\frac{U_1 du}{\sqrt{U-a}} \pm \frac{V_1 dv}{\sqrt{a-V}} = 0. \tag{5.6}$$

Let u_0 be a solution of the equation $U - a = 0$. Then, the geodesic line (5.6) will satisfy $du = 0$ at the point of intersection with the curve $u = u_0$, i.e., it will be tangent to this coordinate curve. The statement now follows from Lemma 5.19. $\square$

Proposition 5.21. *Coordinate curves on a Liouville surface are geodesic conics with respect to any two involutes of one of them.*

Proof. Follows from Lemma 5.20 and the proof of Theorem 5.18. $\square$

Now, we are ready to prove the generalization of Graves' theorem.

Theorem 5.22. *Let $\mathcal{E}_0 : u = u_0$ and $\mathcal{E}_1 : u = u_1$ be coordinate curves on the Liouville surface $\mathcal{S}$. For a point $M \in \mathcal{E}_1$, denote by MP and MP' geodesic segments that touch $\mathcal{E}_0$ at Q, Q'. Then the expression*

$$\ell(MP) + \ell(MP') - \ell(PP')$$

is constant for all M, where $\ell(MP)$, $\ell(MP')$, and $\ell(PP')$ denote lengths of geodesic segments MP, MP', and of the segment $PP' \subset \mathcal{E}_0$ respectively.

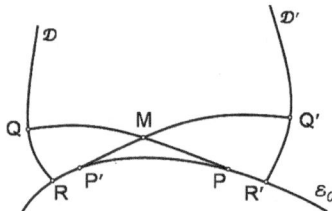

Figure 5.4: Theorem 5.22

Proof. Let $\mathcal{D}$, $\mathcal{D}'$ be involutes of the curve $\mathcal{E}_0$ with respect to points $R, R' \in \mathcal{E}_0$, and Q, Q' intersections of geodesics MP, MP' with these involutes.

Both $\mathcal{E}_0$ and $\mathcal{E}_1$ are geodesic ellipses with base curves $\mathcal{D}$, $\mathcal{D}'$, thus the sum $\ell(MQ) + \ell(MQ')$ remains constant when M moves on $\mathcal{E}_1$.

Since $\ell(PR) = \ell(MP) + \ell(MQ)$, $\ell(P'R') = \ell(MP') + \ell(MQ')$, we have

$$\ell(MQ) + \ell(MQ') = \ell(PR) + \ell(P'R') - \ell(MP) - \ell(MP')$$
$$= \ell(RR') - \big(\ell(MP) + \ell(MP') - \ell(PP')\big),$$

and the theorem is proved. $\square$

From here, the complete analogue of the Poncelet theorem can be derived:

Theorem 5.23. *Let us consider a polygon on the Liouville surface $\mathcal{S}$, with all sides being geodesics tangent to a given coordinate curve, and each vertex but one moving on a coordinate curve. Then the last vertex also remains on a fixed coordinate curve.*

5.6 Poncelet theorem in projective space

In this section, following [CCS1993], a proof of the Full Poncelet theorem in the d-dimensional projective space over an arbitrary field of characteristic not equal to 2 will be presented.

First, we need to define reflection projectively, without metrics.

Let $\mathcal{Q}_1$ and $\mathcal{Q}_2$ be two quadrics. Denote by u the tangent plane to $\mathcal{Q}_1$ at point x and by z the pole of u with respect to $\mathcal{Q}_2$. Suppose lines ℓ_1 and ℓ_2 intersect at x, and the plane containing these two lines meet u along ℓ.

Definition 5.24. If lines ℓ_1, ℓ_2, xz, ℓ are coplanar and harmonically conjugated, we say that rays ℓ_1 and ℓ_2 *obey the reflection law* at the point x of the quadric $\mathcal{Q}_1$ with respect to the confocal system which contains $\mathcal{Q}_1$ and $\mathcal{Q}_2$.

If we introduce a coordinate system in which quadrics $\mathcal{Q}_1$ and $\mathcal{Q}_2$ are confocal in the usual sense, reflection defined in this way is the same as the standard one.

Theorem 5.25 (One Reflection Theorem). *Suppose rays ℓ_1 and ℓ_2 obey the reflection law at x of $\mathcal{Q}_1$ with respect to the confocal system determined by quadrics $\mathcal{Q}_1$ and $\mathcal{Q}_2$. Let ℓ_1 intersect $\mathcal{Q}_2$ at y_1' and y_1, u be a tangent plane to $\mathcal{Q}_1$ at x, and z its pole with respect to $\mathcal{Q}_2$. Then lines $y_1'z$ and y_1z respectively contain intersecting points y_2' and y_2 of ray ℓ_2 with $\mathcal{Q}_2$. The converse is also true.*

Corollary 5.26. *Let rays ℓ_1 and ℓ_2 obey the reflection law of $\mathcal{Q}_1$ with respect to the confocal system determined by quadrics $\mathcal{Q}_1$ and $\mathcal{Q}_2$. Then ℓ_1 is tangent to $\mathcal{Q}_2$ if and only if ℓ_2 is tangent to $\mathcal{Q}_2$; ℓ_1 intersects $\mathcal{Q}_2$ at two points if and only if ℓ_2 intersects $\mathcal{Q}_2$ at two points.*

The next assertion is crucial for proof of the Poncelet theorem.

Theorem 5.27 (Double Reflection Theorem). *Suppose that $\mathcal{Q}_1$, $\mathcal{Q}_2$ are given quadrics and ℓ_1 a line intersecting $\mathcal{Q}_1$ at the point x_1 and $\mathcal{Q}_2$ at y_1. Let u_1, v_1 be tangent planes to $\mathcal{Q}_1$, $\mathcal{Q}_2$ at points x_1, y_1 respectively, and z_1, w_1 their with respect to $\mathcal{Q}_2$ and $\mathcal{Q}_1$. Denote by x_2 second intersecting point of the line w_1x_1 with $\mathcal{Q}_1$, by y_2 intersection of y_1z_1 with $\mathcal{Q}_2$ and by ℓ_2, ℓ_1', ℓ_2' lines x_1y_2, y_1x_2, x_2y_2. Then pairs ℓ_1, ℓ_2; ℓ_1, ℓ_1'; ℓ_2, ℓ_2'; ℓ_1', ℓ_2' obey the reflection law at points x_1 (of $\mathcal{Q}_1$), y_1 (of $\mathcal{Q}_2$), y_2 (of $\mathcal{Q}_2$), x_2 (of $\mathcal{Q}_1$) respectively.*

Proof. Let u_2 and v_2 be tangent planes to quadrics $\mathcal{Q}_1$, $\mathcal{Q}_2$ at points x_2, y_2, and z_2, w_2 their poles with respect to $\mathcal{Q}_2$, $\mathcal{Q}_1$. Since points w_1, x_1, x_2 are collinear, their polar planes v_1, u_1, u_2 with respect to $\mathcal{Q}_1$ belong to a pencil. Poles y_1, z_1, z_2 of these planes with respect to $\mathcal{Q}_2$ are on a line which contains y_2, too. Hence planes v_1, u_1, u_2, v_2 are in a pencil, and it follows that points w_1, x_2, x_2, w_2 are collinear. The assertion follows from the one reflection theorem. $\square$

Corollary 5.28. *If the line ℓ_1 is tangent to a quadrics $\mathcal{Q}'$ confocal with $\mathcal{Q}_1$ and $\mathcal{Q}_2$, then rays ℓ_2, ℓ_1', ℓ_2' also touch $\mathcal{Q}'$.*

Using the Double Reflection theorem, we can prove the Full Poncelet theorem in d-dimensional space. Let $\mathcal{Q}_1$, ..., $\mathcal{Q}_m$, $\mathcal{Q}$ be confocal quadrics and ℓ_1, ..., ℓ_m, $(\ell_{m+1} = \ell_1)$ lines such that pairs ℓ_i, ℓ_{i+1} obey the reflection law at point x_i of $\mathcal{Q}_i$. Let u_i be a plane tangent to $\mathcal{Q}_i$ at x_i, z_i its pole with respect to $\mathcal{Q}$ and $z_i \notin \mathcal{Q}$. If line ℓ_1 intersects $\mathcal{Q}$ at y_1, y_1', then, by the One Reflection theorem, the second intersection point y_2 of line $y_1 z_1$ with quadrics $\mathcal{Q}$ belongs to ℓ_2. Similarly, having point y_i, we construct the point $y_{i+1} \in \ell_{i+1}$. It follows that $y_{m+1} = y_1$ or $y_{m+1} = y_1'$. Suppose $y_{m+1} = y_1$. It can be proved that, if, for a given polygon $x_1 \ldots x_m$ and quadric $\mathcal{Q}$, $y_{m+1} = y_1$ holds, then $y_{m+1} = y_1$ for any surface $\mathcal{Q}$ from the confocal family.

Suppose ℓ_i' are rays ℓ_i reflected from $\mathcal{Q}$ at points y_i. By the Double Reflection theorem, ℓ_i' and ℓ_{i+1}' meet at point $x_i' \in \mathcal{Q}_i$ and obey the reflection law. In this way, we obtained a new polygon $x_1' \ldots x_m'$. If ℓ_1 touches quadrics $\mathcal{Q}^1$, ..., $\mathcal{Q}^{d-1}$ confocal with $\{\mathcal{Q}_1, \mathcal{Q}\}$, then all sides of both polygons touch them. In this way, a $(d-1)$-parameter family of Poncelet polygons is obtained.

5.7   Virtual billiard trajectories

Apart from the real motion of the billiard particle in $\mathbf{E}^d$, it is of interest to consider *virtual reflections*. These reflections were discussed by Darboux in [Dar1914] (see Chapter XIV of Book IV in Volume 2). In this section, we review some results from [DR2006b, DR2008], proving and generalizing a property of virtual reflections formulated by Darboux in the three-dimensional case in the footnote [Dar1914] on p. 320–321:

> "... Il importe de remarquer: le théorème donné dans le texte suppose essentiellement que les côtés du polygone soient formés par les parties **réelles** et non **virtuelles** du rayon réfléchi. Il existe des polygones fermés d'une tout autre nature. Étant donnés, par exemple, deux ellipsoïdes homofocaux (E_0), (E_1), si, par une droite quelconque, on leur mène des plans tangents, on aura quatre points de contact a_0, b_0 sur (E_0), a_1, b_1 sur (E_1). Le quadrilatère $a_0 a_1 b_0 b_1$ sera tel que les bissectrices des angles a_1, b_1 soient les normales de (E_1), et les bissectrices des angles a_0, b_0 les normales de (E_0), mais il ne constituera pas une route **réelle** pour un

rayon lumineux; deux de ses côtés seront formés par les parties virtuelles des rayons réfléchis. De tels polygones mériteraient aussi d'être étudiés, leur théorie offre les rapports les plus étroits avec celle de l'addition des fonctions hyperelliptiques et de certaines surfaces algébriques ..."

Formally, in the Euclidean space, we can define the *virtual reflection* at the quadric $\mathcal{Q}$ as a map of a ray ℓ with the endpoint P_0 ($P_0 \in \mathcal{Q}$) to the ray complementary to the one obtained from ℓ by the real reflection from $\mathcal{Q}$ at the point P_0. On Figure 5.5, ray ℓ_R is obtained from ℓ by the real reflection on the quadric surface $\mathcal{Q}$ at point P_0. Ray ℓ_V is obtained from ℓ by the virtual reflection. Line n is the normal to $\mathcal{Q}$ at P_0.

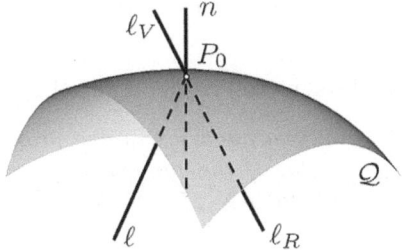

Figure 5.5: Real and virtual reflection

Let us remark that, in the case of real reflections, exactly one elliptic coordinate, the one corresponding to the quadric $\mathcal{Q}$, has a local extreme value at the point of reflection. On the other hand, on a virtual reflected ray, this coordinate is the only one not having a local extreme value at the point of reflection. In the 2-dimensional case, the virtual reflection can easily be described as the real reflection from the other confocal conic passing through the point P_0. In higher-dimensional cases, the virtual reflection can be regarded as the real reflection from the line normal to $\mathcal{Q}$ at P_0 (see Figure 5.5).

The notions of real and virtual reflection cannot be straightforwardly extended to the projective space, since there we essentially use that the field of real numbers is naturally ordered. Nevertheless, it turns out that it is possible to define a certain configuration connected with real and virtual reflection, such that its properties remain in the projective case, too.

Let points $X_1, X_2; Y_1, Y_2$ belong to quadrics $\mathcal{Q}_1, \mathcal{Q}_2$ in $\mathbf{P}^d$.

Definition 5.29. We will say that the quadruple of points X_1, X_2, Y_1, Y_2 constitutes a *virtual reflection configuration* if pairs of lines $X_1Y_1, X_1Y_2; X_2Y_1, X_2Y_2; X_1Y_1, X_2Y_1; X_1Y_2, X_2Y_2$ satisfy the reflection law at points X_1, X_2 of $\mathcal{Q}_1$ and Y_1, Y_2 of $\mathcal{Q}_2$ respectively, with respect to the confocal system determined by $\mathcal{Q}_1$ and $\mathcal{Q}_2$.

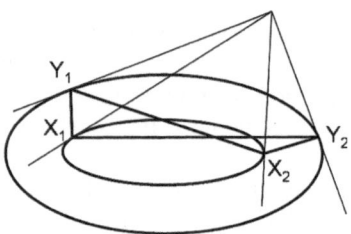

Figure 5.6: Double reflection configuration

If, additionally, the tangent planes to $\mathcal{Q}_1, \mathcal{Q}_2$ at X_1, X_2; Y_1, Y_2 belong to a pencil, we say that these points constitute a *double reflection configuration* (see Figure 5.6).

Now, the Darboux statement can be generalized and proved as follows:

Theorem 5.30. *Let $\mathcal{Q}_1$, $\mathcal{Q}_2$ be two quadrics in the projective space $\mathbf{P}^d$, X_1, X_2 points on $\mathcal{Q}_1$ and Y_1, Y_2 on $\mathcal{Q}_2$. If the tangent hyperplanes at these points to the quadrics belong to a pencil, then X_1, X_2, Y_1, Y_2 constitute a virtual reflection configuration.*

Furthermore, suppose that the projective space is defined over the field of reals. Introduce a coordinate system, such that $\mathcal{Q}_1$, $\mathcal{Q}_2$ become confocal ellipsoids in the Euclidean space. If $\mathcal{Q}_1$ is placed inside $\mathcal{Q}_2$, then the sides of the quadrilateral $X_1Y_1X_2Y_2$ obey the real reflection from $\mathcal{Q}_2$ and the virtual reflection from $\mathcal{Q}_2$.

Proof. Denote by ξ_1, ξ_2 and η_1, η_2 the tangent hyperplanes to $\mathcal{Q}_{\lambda_1}$ at X_1, X_2 and to $\mathcal{Q}_{\lambda_2}$ at Y_1, Y_2, respectively. All these hyperplanes belong to a pencil, thus their poles with respect to any quadric will be collinear – particularly, the pole P of ξ_1 lies on line Y_1Y_2. If $Q = Y_1Y_1 \cap \xi_1$, then pairs P, Q and Y_1, Y_2 are harmonically conjugate. It follows that the lines X_1Y_1, X_1Y_2 obey the reflection law from ξ_1. The rest of the proof can be done in the same way. $\square$

We are going to conclude this section with a statement converse to the previous theorem.

Proposition 5.31. *Let pairs of points X_1, X_2 and Y_1, Y_2 belong to confocal ellipsoids $\mathcal{Q}_1$ and $\mathcal{Q}_2$ in Euclidean space $\mathbf{E}^d$, and let α_1, α_2, β_1, β_2 be the corresponding tangent planes. If a quadruple X_1, X_2, Y_1, Y_2 is a virtual reflection configuration, then planes α_1, α_2, β_1, β_2 belong to a pencil.*

Proof. Consider the pencil determined by α_1 and β_1. Let α_2', β_2' be planes of this pencil, tangent respectively to $\mathcal{Q}_1$, $\mathcal{Q}_2$ at points X_2', Y_2', and distinct from α_1 and β_1. By Theorem 5.30, the quadruple X_1, X_2', Y_1, Y_2' is a virtual reflection configuration. Moreover, if we denote by λ_1, λ_2 parameters of $\mathcal{Q}_1$, $\mathcal{Q}_2$ and assume $\lambda_1 < \lambda_2$, then the sides of the quadrangle obey the reflection law at points X_1, X_2'

and the virtual reflection at Y_1, Y_2'. Since the ray obtained from X_1Y_1 by the virtual reflection of $\mathcal{Q}_2$ at Y_1, has only one intersection with $\mathcal{Q}_1$, we have $X_2 = X_2'$. Points Y_2 and Y_2' coincide, being the intersection of rays obtained from Y_1X_1 and Y_1X_2, by the reflection at the quadric $\mathcal{Q}_1$. Now, the four tangent planes are all in one pencil. $\qquad\square$

5.8 Towards generalization of proof of Cayley's condition

In this section, the analysis of possibility of generalization of the inspiring Lebesgue's procedure from [Leb1942], that is given in Section 5.2, to higher-dimensional cases will be given. Results presented here are obtained in [DR2006b].

A higher-dimensional analogue of the crucial lemma from [Leb1942], which is Lemma 5.3 of this book, is the following:

Lemma 5.32. *Let $\mathcal{Q}_1$, $\mathcal{Q}_2$ be quadrics of a confocal system and let lines ℓ_1, ℓ_2 satisfy the reflection law at point X_1 of $\mathcal{Q}_1$ and ℓ_2, ℓ_3 at Y_2 of $\mathcal{Q}_2$. Then line ℓ_1 meets $\mathcal{Q}_2$ at point Y_1 and ℓ_3 meets $\mathcal{Q}_1$ at point X_2 such that pairs of lines ℓ_1, Y_1X_2 and Y_1X_2,ℓ_3 satisfy the reflection law at points Y_1, X_2 of quadrics $\mathcal{Q}_2$, $\mathcal{Q}_1$ respectively. Moreover, tangent planes at X_1, X_2, Y_1, Y_2 of these two quadrics are in the same pencil.*

This statement can be proved by the direct application of Theorem 5.30 on virtual reflections. Nevertheless, there is no complete analogy between Lemma 5.32 and the corresponding assertion in the plane. Lines ℓ_1, ℓ_3 and ℓ_2, Y_1Y_2 are generically skew. Hence we do not have the third pair of planes tangent to the quadric, containing intersection points of these two pairs of lines.

Nevertheless, a complete generalization of the Basic Lemma, can be formulated as follows:

Theorem 5.33. *Let $\mathcal{F}$ be a dual pencil of quadrics in the three-dimensional space.*

For a given quadric $\Gamma_0 \in \mathcal{F}$, there exist quadruples α, β, γ, δ of planes tangent to Γ_0, and quadrics $\Gamma_1, \Gamma_2, \Gamma_3 \in \mathcal{F}$ touching the pairs of intersecting lines $\alpha\beta$ and $\gamma\delta$, $\alpha\gamma$ and $\beta\delta$, $\alpha\delta$ and $\beta\gamma$ respectively, with the tangent planes to Γ_1, Γ_2, Γ_3 at points of tangency with the lines, all being in one pencil Δ. Moreover, the six intersecting lines are in one bundle. (See Figure 5.7.)

Every such a configuration of planes α, β, γ, δ and quadrics Γ_0, Γ_1, Γ_2, Γ_3 is determined by two of the intersecting lines, and the tangent planes to Γ_1, Γ_2, or Γ_3 corresponding to these two lines.

Proof. Suppose first that two adjacent lines $\alpha\beta$ and $\alpha\gamma$ are given. There exist two quadrics in $\mathcal{F}$ touching $\alpha\beta$ – let Γ_1 be the one tangent to the given plane $\mu \supset \alpha\beta$. Γ_2 denotes the quadric that touches $\alpha\gamma$ and the given plane $\pi \supset \alpha\gamma$. The pencil

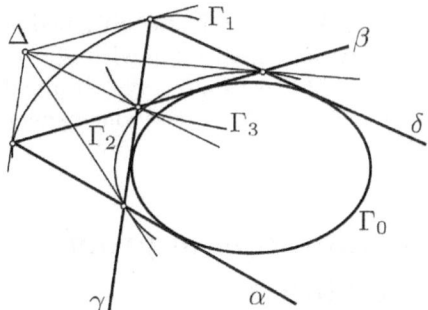

Figure 5.7.

Δ is determined by its planes μ and π. All three lines $\alpha\beta$, $\alpha\gamma$, Δ are in one bundle $\mathcal{B}$, seé Figure 5.8.

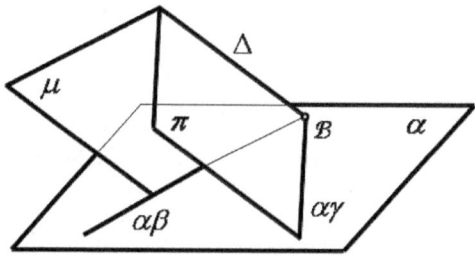

Figure 5.8: Theorem 5.33

Note that lines $\alpha\beta$ and $\alpha\gamma$ determine plane α and that α touches a uni-que quadric Γ_0 from $\mathcal{F}$. Thus, β and γ are determined as tangent planes to Γ_0, containing lines $\alpha\beta$ and $\alpha\gamma$ respectively and being different from α.

Let ν be the plane from Δ, other than μ, that is touching Γ_1. We are going to prove that the point of tangency ν with Γ_1 belongs to γ.

Denote by Φ the dual pencil determined by quadric Γ_1 and the degenerate quadric which consists of lines $\alpha\beta$ and $\gamma\nu$. Since $\gamma\nu \in \mathcal{B}$, these two lines are coplanar.

Dual pencils $\mathcal{F}$ and Φ determine the same involution of the pencil $\alpha\gamma$, because they both determine the pair α, γ and the quadric Γ_1 is common for both pencils. Since quadric $\Gamma_2 \in \mathcal{F}$ determines the pair π, π of coinciding planes, a quadric $\mathcal{Q}$ determining the same pair has to exist in Φ. This quadric, as well as all other quadrics in Φ, touches ν and μ. Since ν, μ, π all belong to the pencil Δ, $\mathcal{Q}$ is degenerate and contains Δ. Since any quadric of Φ touches μ, π at points of lines $\alpha\beta$, $\alpha\gamma$ respectively, the other component of $\mathcal{Q}$ also has to be Δ, thus $\mathcal{Q}$ is the double Δ.

It follows that all quadrics of Φ, and particularly Γ_1, are tangent to ν at a point of $\gamma\nu$.

Similarly, if $\kappa \neq \pi$ is the other plane in Δ tangent to Γ_2, the touching point belongs to β and to δ, the plane, other than γ, tangent to Γ_0 and containing the line $\gamma\nu$.

Now, let us note that $\mathcal{F}$ and Φ determine the same involution on pencil $\alpha\delta$, because they both determine that pair α, δ and Γ_1 belongs to both pencils. Thus, the common plane ρ of pencils Δ and $\alpha\delta$ is tangent to a quadric of $\mathcal{F}$ at a point of $\alpha\delta$. Denote this quadric by Γ_3. Similarly as before, we can prove that Γ_3 is touching the line $\beta\gamma$ and the corresponding tangent plane σ is common to pencils Δ and $\beta\gamma$.

Now, suppose that two non-adjacent lines $\alpha\beta$, $\gamma\delta$, both tangent to a quadric $\Gamma_1 \in \mathcal{F}$, with the corresponding tangent planes μ, ν, are given. Similarly as above, we can prove that the plane ρ is tangent to a quadric from $\mathcal{F}$ at a point of $\alpha\delta$. So, we can consider the configuration as determined by adjacent lines $\alpha\beta$, $\alpha\delta$ and planes μ, ρ. In this way, we reduced it to the previous case. $\square$

Chapter 6

Poncelet–Darboux Curves and Siebeck–Marden Theorem

Although it is impossible to distinguish the nicest mathematical result, probably many mathematicians would agree that the Marden theorem of geometry of polynomials and the great Poncelet theorem from projective geometry of conics by their classical beauty occupy very special places. Our main aim is to present a strong and unexpected relationship between the two theorems obtained very recently in [Dra2008, Dra2010b]. A dynamical equivalence between the full Marden theorem and the Poncelet–Darboux theorem was established. By introducing a class of *isofocal deformations*, a morphism between the Marden curve and the Poncelet–Darboux curve was constructed, using the Moser trick and the Flaschka coordinates. As a byproduct, a complete description of cyclic-symmetric n-correspondences of $\mathbf{P}^1$ was done. Then an effective criterion is presented for complete decomposition of a transversal Poncelet–Darboux curve of degree $n - 1$ on $(n - 1)/2$ conics if n is odd and on $(n - 2)/2$ conics and a line if n is even in terms of a pair of polynomials which defines the Poncelet–Darboux curve. This is an important question in the study of special or 'tHooft instanton bundles (see [Tra1988, Tra1997, HN1982, Har1978, NT1990] and references therein).

6.1 Preliminaries

The Siebeck–Marden theorem

One of the basic theorems in geometric theory of polynomials and rational functions has a long history and is usually referred as *Marden's* after appearance of the book [Mar1966]. The earliest version of this theorem, to the best of our knowledge, goes back to 1864 when Siebeck (see [Sie1864]) formulated and proved it for the case of polynomials with simple roots:

Theorem 6.1 (Siebeck). [Sie1864] *Let $P(z)$ be a polynomial of degree $n \geq 3$ with complex coefficients, such that the zeros $\alpha_1, \ldots, \alpha_n$ are simple and every three noncollinear. There exists a curve $\mathcal{C}$ of class $n - 1$ tangent to every line segment $[\alpha_i, \alpha_j]$ at the midpoint. The foci of the curve $\mathcal{C}$ are zeros of the derivative polynomial $DP(z) = P'(z)$.*

The case $n = 3$.   Even in the simplest case $n = 3$ when the curve $\mathcal{C}$ is a conic, inscribed in the triangle formed by the zeros of a polynomial of degree 3 and tangent to the sides of the triangle at their midpoints, the result of Siebeck's theorem is nontrivial and interesting and attracted a lot of attention not only in the past but also nowadays (see for example [Kal2008]). We are going to present main points of the proof of Siebeck's theorem for $n = 3$ following mostly [Kal2008] and references therein.

The first observation in the proof is that the statement is invariant to applications of affine transformations. Now, we assume that a real plane is identified with the field of complex numbers and that numbers $\alpha_1, \alpha_2, \alpha_3$ correspond to numbers $-1, 1, w$ where w is in the upper half-plane.

The polynomial P gets the form

$$P(z) = (z - 1)(z + 1)(z - w) = z^3 - wz^2 - z + w,$$

and the derivative is

$$DP(z) = 3\left(z^2 - \frac{2}{3}wz - \frac{1}{3}\right).$$

Denote the zeros of the derivative as z_4, z_5. After some analysis of the formulae

$$z_4 + z_5 = \frac{2}{3}w,$$

$$z_4 z_5 = -\frac{1}{3},$$

we conclude that both of the points z_4 and z_5 are in the upper half-plane. Then, from the second formula we conclude that the sum of their arguments θ_4 and θ_5 is equal to π. Denote the line connecting z_4 and the origin O as $L_{z_4 O}$ and the line connecting z_5 and the origin O as $L_{z_5 O}$. As a result, we come to the conclusion that the angle between the line $L_{z_4 O}$ and negative real semi-axis is equal to the angle of the line $L_{z_5 O}$ and the positive real semi-axis. By application of focal properties of ellipses, Proposition 2.3 and Proposition 2.4, we see that there is an ellipse $\mathcal{E}_1$, with z_4, z_5 as foci, such that it touches the real axis at the origin. In other words, it touches the segment $[\alpha_1, \alpha_2]$ at the midpoint.

By symmetry, we see that there are confocal ellipses $\mathcal{E}_1$, $\mathcal{E}_2$ and $\mathcal{E}_3$ with focal points at the zeros of the derivative of the polynomial, each of which touches one of the sides of the triangle at the midpoint.

Next, we need to show that these three ellipses coincide. Again, we apply affine transformation and transform the triangle to a new one with vertices $0, 1, w$, where w is again in the upper half-plane.

The polynomial P gets the form

$$P(z) = z(z-1)(z-w) = z^3 - (1+w)z^2 + wz,$$

and the zeros z_4, z_5 of the derivative

$$DP(z) = 3\left(z^2 - \frac{2}{3}(1+w)z + w\right)$$

satisfy

$$z_4 + z_5 = \frac{2}{3}(1+w),$$
$$z_4 z_5 = \frac{w}{3}.$$

After analysis of the last formulae we again conclude that z_4, z_5 are in the upper half-plane and that the sum of their arguments is equal to the argument of w. As a result, this time we see that the angle between the line $L_{z_4 O}$ and the line L_{Ow} is equal to the angle of the line $L_{z_5 O}$ and the positive real semi-axis. From previous considerations, we know that the ellipse $\mathcal{E}_1$ with z_4, z_5 as focal points touches the real axis at the point $1/2$. One can easily see that the origin O is outside the ellipse $\mathcal{E}_1$. Now, we apply another well-known focal property of ellipses (see Corollary 2.6 of Proposition 2.4) to the ellipse $\mathcal{E}_1$, point O outside $\mathcal{E}_1$ and two tangents t_1 and t_2 to $\mathcal{E}_1$ from the point O. The angle between t_1 and L_{Oz_5} is equal to the angle between t_2 and L_{Oz_4}. One tangent, t_1 is the real axis. Thus, the second tangent coincides with the line L_{Ow}. This proves that the line L_{Ow} is tangent to the ellipse $\mathcal{E}_1$. But, again by focal properties of ellipses, (see Proposition 2.3) among confocal ellipses, there is only one tangent to a given line. This shows that $\mathcal{E}_1 = \mathcal{E}_2$ and $\mathcal{E}_1$ touches the segment $[O, w]$ at the midpoint. This finishes the proof of Siebeck's theorem in the case $n = 3$.

The analysis of the previous proof of Siebeck's theorem in the simplest case, gives us a hint of possible deep connection between this theorem and billiards within ellipse. Namely, in both cases focal properties of ellipses (see Proposition 2.3 and Proposition 2.4) play an important and very similar role. We are going to explore possible connections even in more general situations.

Previous results were extended to the cases with not all roots being simple. Consider a function of the form

$$P(z) = (z - \alpha_1)^{m_1} (z - \alpha_2)^{m_2} \cdots (z - \alpha_n)^{m_n}, \qquad (6.1)$$

where all α_k are distinct, every three noncollinear and $N = m_1 + m_2 + \cdots + m_n$. The zeros of the derivative DP are divided into two groups. In the first group are those α_i such that $m_i > 1$. The second group is formed from zeros of the logarithmic derivative LP of P. Since the positions of the zeros of the first group are known from the beginning, the interesting part is location of the members of the second group.

Thus, consider the function $F(z) = LP(z) = d[\log P(z)]/dz$:

$$F(z) = \frac{m_1}{z - \alpha_1} + \cdots + \frac{m_n}{z - \alpha_n}. \tag{6.2}$$

Historically, the constants m_i in the last expressions were firstly considered as positive integers. Then, step by step, that condition has been relaxed up to the condition that m_i are nonzero real numbers, as we can find in Marden's formulation:

Theorem 6.2. [Mar1966, Theorem 4.2] *The zeros of the function*

$$F(z) = \frac{m_1}{z - \alpha_1} + \cdots + \frac{m_n}{z - \alpha_n}$$

where m_i are real nonzero constants, are the foci of the curve of class $n-1$ which touches each line-segment $[\alpha_i, \alpha_j]$ in a point dividing the line segment in the ratio $m_i : m_j$.

A good account of a century-long path from Theorem 6.1 to Theorem 6.2 can be found in Marden's book [Mar1966], together with references. Thus we are going to omit them here.

Darboux Theorem and Poncelet–Darboux curves

Now we pass to quite a different subject, which appeared in the theory of conics in the context of the Full Poncelet theorem [Pon1822]. We start here with one of Darboux's original formulations of his theorem.

Theorem 6.3 (Darboux [Dar1917, page 248]). *Si une courbe d'ordre $n-1$ contient tous les points d'intersection de n tangents a une conique, elle contient aussi les points d'intersection d'une infinité d'autres systemes de n tangentes a la même conique.*

Darboux had been interested in this matter for about fifty years and he published several variations of the last theorem (see for example [Dar1914]). Slightly changing terminology from [Tra1997], we will say that a curve $\mathcal{S}$ of degree $n - 1$ *is Poncelet–Darboux related to a conic $\mathcal{K}$* if the curve $\mathcal{S}$ and the conic $\mathcal{K}$ satisfy conditions of the previous theorem. The set of all such curves of degree $n-1$ which are Poncelet–Darboux related to a fixed conic $\mathcal{K}$ will be denoted as

$$\text{Pon-Dar}_{n-1}(\mathcal{K}).$$

We will say that the n tangents $t_1, t_2, \ldots, t_n$ of a conic $\mathcal{K}$ from the previous theorem *form a Poncelet n-polygon* $P_n = T_1 T_2 \ldots T_n$, where $T_i = t_i \cap t_{i+1}$ for

$i = 1, \ldots, n - 1$ and $T_n = t_n \cap t_1$ if there exists a conic $\mathcal{C}$ such that $T_i \in \mathcal{C}$ for $i = 1, \ldots, n$. In that case we will say that conics $\mathcal{C}$ and $\mathcal{K}$ are n-*Poncelet related*.

Let us formulate once again the Poncelet Theorem (see Theorems 2.14 and 4.91).

Theorem 6.4 (Poncelet [Pon1822]). *If two conics $\mathcal{C}$ and $\mathcal{K}$ are n-Poncelet related, then there are infinitely many n-polygons circumscribed about $\mathcal{K}$ and inscribed in $\mathcal{C}$. Moreover, an arbitrary point of the conic $\mathcal{C}$ may be chosen for a vertex of such a Poncelet n-polygon.*

If the n tangents from the Darboux Theorem 6.3 form a Poncelet n-polygon inscribed in a conic $\mathcal{C}$, then Darboux proved that the curve $\mathcal{S}$ of degree $n - 1$ completely decomposes. More precisely, Darboux proved the following

Theorem 6.5 (Darboux [Dar1917]). *If a curve $\mathcal{S}$ of degree $n - 1$ is Poncelet–Darboux related to a conic $\mathcal{K}$ and if there is a conic $\mathcal{C}$, component of $\mathcal{S}$ which is n-Poncelet related to the conic $\mathcal{C}$, then for $n = 2k + 1$ the curve $\mathcal{S}$ is completely decomposed on k conics and if $n = 2k$ it is decomposed on $k - 1$ conics and a line.*

We will present the proof of the last theorem in Section 6.3.

In the case of decomposition of a Poncelet–Darboux curve, the conic components are parts of what we call the Poncelet–Darboux grids. Further generalizations of Darboux theorems and Poncelet–Darboux grids have been obtained very recently in [DR2008].

Among other modern investigations in the framework of the Darboux theorems, we should mention here [Tra1988, Tra1997, NT1990] and references therein, where Poncelet–Darboux curves are related to the study of stable bundles, instanton bundles and their decomposition.

We can reformulate the Darboux Theorem 6.3 following [Tra1988]:

Theorem 6.6 (Darboux). *Let $\mathcal{K}$, $\mathcal{S}$ be a non-degenerate conic and a curve of degree $n - 1$ in the projective plane $\mathbf{P}W$ and let $\beta : W \mapsto W^*$ be a non-degenerate bilinear form. Assume that there are n points on $\mathcal{K}$ such that the points of intersection of any two lines associated to these points by β belong to $\mathcal{S}$. Then, $\mathcal{S}$ is Poncelet–Darboux $n - 1$ related to $\mathcal{K}$.*

Some notation and notions

We will use the following notation.

By

$$\{x_1, x_2, \ldots, x_n\}_M \tag{6.3}$$

we will denote a *multiset*, meaning that the number of appearances of an item is important, but not the order; the notion of divisor has synonymous meaning. The standard symmetric functions of n quantities $(a_1, \ldots, a_n)$ will be denoted as

$$\sigma_1(a_1, \ldots, a_n) = \sum_{i=1}^{n} a_i,$$

$$\sigma_2(a_1, \ldots, a_n) = \sum_{i<j}^{n} a_i a_j,$$

$$\ldots$$

$$\sigma_n(a_1, \ldots, a_n) = a_1 a_2 \cdots a_n;$$

when one of the quantities, a_k, is omitted, the symmetric functions of the $n - 1$ rest quantities will be denoted as

$$\sigma_0^k(a_1, \ldots, a_n) = 1 \quad k = 1, \ldots, n;$$

$$\sigma_1^k(a_1, \ldots, a_n) = \sum_{i \neq k}^{n} a_i \quad k = 1, \ldots, n;$$

$$\sigma_2^k(a_1, \ldots, a_n) = \sum_{i<j, i, j \neq k}^{n} a_i a_j \quad k = 1, \ldots, n;$$

$$\ldots$$

$$\sigma_{n-1}^k(a_1, \ldots, a_n) = a_1 a_2 \cdots a_{k-1} a_{k+1} \cdots a_n \quad k = 1, \ldots, n.$$

We will use also vector notation

$$\vec{\sigma}_i(a_1, \ldots, a_n) = (\sigma_i^1(a_1, \ldots, a_n), \sigma_i^2(a_1, \ldots, a_n), \ldots, \sigma_i^n(a_1, \ldots, a_n)), \qquad (6.4)$$

for $i = 0, 1, \ldots, n - 1$;

$$\vec{m} = (m_1, \ldots, m_n),$$

$$\langle \vec{m}, \vec{\sigma}_i \rangle = \sum_{k=1}^{n} m_k \sigma_i^k. \qquad (6.5)$$

Let us recall some traditional notions (see Section 4.2). We assume that a real plane is the real part of the complex plane, where the last one is embedded into a complex projective plane according to the formula $(z_1, z_2) \mapsto (z_1, z_2, 1)$. *The cyclic points* are points on the infinite line with coordinates $\hat{I} = (1, i, 0)$ and $\hat{J} = (1, -i, 0)$. A line is *isotropic* or *minimal* if it is finite and passes through one of the cyclic points. For a given curve, a point is *focal* if it is the intersection of two isotropic tangents to the curve with finite points of contact with the curve. As an example, an ellipse has four focal points: a pair of real foci and a pair of imaginary foci.

6.2 Isofocal deformations

Definition of an integrable dynamical system

Let us start with a function of the form

$$P^0(z) = (z - a_1^0)^{m_1^0}(z - a_2^0)^{m_2^0} \cdots (z - a_n^0)^{m_n^0}, \tag{6.6}$$

where all a_k^0 are distinct, and consider its logarithmic derivative $F^0(z) = LP^0(z) = d[\log P^0(z)]/dz$.

Let us set

$$F^0(z) = \frac{m_1^0}{z - a_1^0} + \cdots + \frac{m_n^0}{z - a_n^0} = \frac{f(z)}{\phi(z)}, \tag{6.7}$$

where

$$\phi(z) = (z - a_1^0)(z - a_2^0) \cdots (z - a_n^0),$$

$$f(z) = \sum_{i=1}^{n} m_i^0 \prod_{j \neq i}(z - a_j^0). \tag{6.8}$$

One can easily see that

$$f(z) = B_n z^{n-1} + \cdots + (-1)^{n-i} B_i z^{i-1} + \cdots + (-1)^{n-1} B_1$$

where, using notation from equations (6.4) and (6.5), we have

$$B_i = \langle \overrightarrow{m}^0, \overrightarrow{\sigma}_{n-i}(a_1^0, \ldots, a_n^0) \rangle. \tag{6.9}$$

We point out two particular cases.

Lemma 6.7.

(a) *The function f is equal to the derivative of ϕ if and only if all m_i^0 are equal to 1.*

(b) *The function f is constant if and only if*

$$\overrightarrow{m}^0 \perp [\overrightarrow{\sigma}_0, \overrightarrow{\sigma}_1, \ldots, \overrightarrow{\sigma}_{n-2}]. \tag{6.10}$$

Before proceeding with introduction of dynamics, we are going to consider the simplest example.

Example 6.8. Let us consider the case $n = 3$. According to the Marden Theorem 6.2 for $n = 3$, there exists a Marden curve $\mathcal{K}$, which is in this case, a *conic*. Its focal points z_1 and z_2 satisfy $f(z_i) = 0$, under the condition $\deg f = 2$. Again, by the Marden Theorem 6.2, the conic $\mathcal{K}$ touches line-segments $[a_i^0, a_j^0]$ in the ratio $m_i^0 : m_j^0$.

Now, since the lines (α_i^0, α_j^0) are tangent to the *conic* $\mathcal{K}$, we may apply the Darboux Theorem 6.3. The triplet of the conic $\mathcal{K}$ and the two polynomials ϕ and f uniquely determine the Poncelet–Darboux curve

$$\mathcal{PD}_\mathcal{K}(\alpha_1^0, \alpha_2^0, \alpha_3^0, m_1^0, m_2^0, m_3^0) = \mathcal{PD}_\mathcal{K}(\phi, f).$$

The curve $\mathcal{PD}_\mathcal{K}(\phi, f)$ is a *conic*.

Thus, the conics $\mathcal{PD}_\mathcal{K}(\phi, f)$ and $\mathcal{K}$ are 3-Poncelet related.

According to the theorems of Poncelet and Darboux, there exists another set of three points $\alpha_1^1, \alpha_2^1, \alpha_3^1$ which belong to the Poncelet–Darboux conic $\mathcal{PD}_\mathcal{K}(\phi, f)$ such that the lines (α_i^1, α_j^1) are tangent to the Marden conic $\mathcal{K}$. The triangle $\alpha_1^1, \alpha_2^1, \alpha_3^1$ is a Poncelet triangle with the caustic $\mathcal{K}$ and the boundary $\mathcal{PD}_\mathcal{K}(\phi, f)$.

Now, we want to apply the Marden Theorem 6.2 with a new polynomial

$$\phi_1(z) = (z - \alpha_1^1)(z - \alpha_2^1)(z - \alpha_3^1)$$

instead of ϕ. In order to do that, we determine *new "masses"* (m_1^1, m_2^1, m_3^1) such that the polynomial f remains unchanged. More precisely, we calculate (m_1^1, m_2^1, m_3^1), up to a scalar factor, from the system of linear equations:

$$
\begin{aligned}
B_3 &= m_1^1 + m_2^1 + m_3^1, \\
B_2 &= m_1^1(\alpha_2^1 + \alpha_3^1) + m_2^1(\alpha_1^1 + \alpha_3^1) + m_3^1(\alpha_1^1 + \alpha_2^1), \\
B_1 &= m_1^1(\alpha_2^1\alpha_3^1) + m_2^1(\alpha_1^1\alpha_3^1) + m_3^1(\alpha_1^1\alpha_2^1),
\end{aligned}
\tag{6.11}
$$

where the constants B_1, B_2, B_3 are determined from equation (6.9)

$$B_i = \langle \overrightarrow{m}^0, \overrightarrow{\sigma}_{n-i}(\alpha_1^0, \alpha_2^0, \alpha_3^0) \rangle.$$

Since f is unchanged, the focal points of the Marden curve in the new case are the same as focal points of $\mathcal{K}$. We deduce now that the Marden curve of the new case is equal to $\mathcal{K}$, because among confocal conics there is at most one inscribed in a triangle.

By a new application of the Marden theorem, we finally get new information concerning Poncelet triangles. We are able to deduce the ratio of tangency of a new triangle by the caustic $\mathcal{K}$:

$$
\begin{aligned}
\alpha_3^1\beta_2^1 : \beta_2^1\alpha_1^1 &= m_3^1 : m_1^1, \\
\alpha_2^1\beta_1^1 : \beta_1^1\alpha_3^1 &= m_2^1 : m_3^1, \\
\alpha_1^1\beta_3^1 : \beta_3^1\alpha_2^1 &= m_1^1 : m_2^1,
\end{aligned}
\tag{6.12}
$$

where β_i^1 denote points of contact of the caustic and the triangle.

If the degree of the polynomial f is equal to 1, then the conic $\mathcal{K}$ is a parabola. Thus we finish the example with a nice classical lemma:

Lemma 6.9 (folklore). *If a parabola touches three lines of a triangle ABC, then the focus belongs to the circumscribed circle of the triangle ABC.*

If the polynomial f is constant, then the conic $\mathcal{K}$ is a circle.

Although Example 6.8 treats the simplest case $n = 3$, it is very instructive. In order to extract a new statement about ratios of tangency of a new Poncelet triangle as it is formulated in equation (6.12), one needs several nontrivial steps of alternative use of the Marden and the Poncelet–Darboux theorems. In case $n > 3$ we cannot follow the same lines, because the Marden curve is not a conic any more and one cannot build the Darboux theorem directly on it. In order to link together the Marden and the Darboux Theorems in general case we need to develop a much more subtle approach.

From Example 6.8, we learned that it was a fruitful idea to pass from one Poncelet triangle to a new one by

a transformation which keeps the polynomial f unchanged.

Following this principle, we introduce a new dynamics, which depends on continuous "time" parameter t and which has quantities

$$(\alpha_1^0, \ldots, \alpha_n^0) \quad \text{and} \quad (m_1^0, \ldots, m_n^0)$$

as *the initial data.* More precisely, we introduce functions:

$$
\begin{aligned}
\alpha_1(t) &= \alpha_1(t, \alpha_1^0, \ldots, \alpha_n^0, m_1^0, \ldots, m_n^0), \\
\alpha_2(t) &= \alpha_2(t, \alpha_1^0, \ldots, \alpha_n^0, m_1^0, \ldots, m_n^0), \\
&\quad \ldots \\
\alpha_n(t) &= \alpha_n(t, \alpha_1^0, \ldots, \alpha_n^0, m_1^0, \ldots, m_n^0), \\
m_1(t) &= m_1(t, \alpha_1^0, \ldots, \alpha_n^0, m_1^0, \ldots, m_n^0), \\
m_2(t) &= m_2(t, \alpha_1^0, \ldots, \alpha_n^0, m_1^0, \ldots, m_n^0), \\
&\quad \ldots \\
m_n(t) &= m_n(t, \alpha_1^0, \ldots, \alpha_n^0, m_1^0, \ldots, m_n^0)
\end{aligned}
\tag{6.13}
$$

in order to satisfy

$$F_t(z) := \frac{m_1(t)}{z - \alpha_1(t)} + \cdots + \frac{m_n(t)}{z - \alpha_n(t)} = \frac{f(z)}{\phi(z) + t f(z)}, \tag{6.14}$$

with the initial conditions

$$
\begin{aligned}
m_i(0) &= m_i^0, \quad i = 1, \ldots, n, \\
\alpha_i(0) &= \alpha_i^0.
\end{aligned}
\tag{6.15}
$$

By condition (6.14), the function f remains unchanged during its evolution. This means that focal points, as zeros of the polynomial f are fixed during the evolution.Thus, we will refer to such dynamics as *isofocal dynamics* or *isofocal deformations*.

The polynomial f plays a role of an isospectral polynomial. From the formula

$$f(z) = \sum_{i=1}^{n} m_i(t) \prod_{j \neq i} (z - \alpha_j(t)) \qquad (6.16)$$

we get the following

Proposition 6.10. *The dynamics (6.13, 6.14, 6.15) has the coefficients of the polynomial f as first integrals:*

$$B_i = \langle \vec{m}(t), \vec{\sigma}_{n-i}(\alpha_1(t), \dots, \alpha_n(t)) \rangle, \quad i = 1, \dots, n. \qquad (6.17)$$

We will use terminology *positions in a moment t* for $\alpha_1(t), \dots, \alpha_n(t))$ and for $(m_1(t), \dots, m_n(t))$ we will use *masses* although these masses will change during the time and might be negative as well.

One of the first integrals is *the law of conservation of masses.*

We will also use the notation

$$\Phi_t(z) := \phi(z) + t f(z) = (z - \alpha_1(t)) \cdots (z - \alpha_n(t)). \qquad (6.18)$$

Moser's trick and the Flaschka coordinates

We consider the function (6.13)

$$F_t(z) = \frac{f(z)}{\Phi_t(z)}$$

and we apply the Moser trick (see [Mos1975a, Mos1975b]) to develop it in a continued fraction of the following form:

$$F_t(z) = \cfrac{1}{z - b_n - \cfrac{a_{n-1}^2}{z - b_{n-1} - \cfrac{a_{n-2}^2}{z - b_{n-2} - \cdots \cfrac{}{\cdots - \cfrac{a_1^2}{z - b_1}}}}} \qquad (6.19)$$

The last formula (6.19) gives us a transformation from our dynamical coordinates $(\alpha_1, \dots, \alpha_n, m_1, \dots, m_n)$ to the new coordinates

$$(a_1, a_2, \dots, a_{n-1}, b_1, b_2, \dots, b_n).$$

The last set of coordinates we will call *the Flaschka coordinates* (see [Fla1974, Mos1975a, Mos1975b]).

To construct the inverse transformation we consider the Flaschka–Lax matrix L_n for the n-point Toda chain (see [Fla1974, Mos1975a, Mos1975b]):

$$
L_k =
\begin{bmatrix}
b_1 & a_1 & & & \\
a_1 & b_2 & & & \\
& & \ddots & & \\
& & & b_{k-1} & a_{k-1} \\
& & & a_{k-1} & b_k
\end{bmatrix}
\tag{6.20}
$$

Let
$$
\delta_k = \det(L_k - zId).
$$

The following well-known difference relations hold:

$$
\delta_k = (z - b_k)\delta_{k-1} - a_{k-1}^2 \delta_{k-2}, \quad k = 3, \dots, n.
\tag{6.21}
$$

The inverse transformation from the Flaschka coordinates to the initial dynamical coordinates is defined by the formula (see [Mos1975a])

$$
F_t(z) = \frac{\delta_{n-1}}{\delta_n}.
\tag{6.22}
$$

From the last formula and from (6.14) we conclude

Lemma 6.11. *The time evolution of δ according to dynamics (6.14) satisfies*

$$
\begin{aligned}
\delta_{n-1}(t) &= \delta_{n-1}(0), \\
\delta_n(t) &= \delta_n(0) + t\delta_{n-1}(0).
\end{aligned}
\tag{6.23}
$$

From Lemma 6.11 one concludes that

1. *only b_n among the Flaschka coordinates depends on t;*
2. *the coordinate b_n depends on t linearly.*

Thus we have the following

Theorem 6.12. *The dynamical system (6.14) becomes trivial in the Flaschka coordinates, where it takes the form*

$$
\begin{aligned}
\dot{a}_1 &= 0 & \dot{b}_1 &= 0, \\
\dot{a}_2 &= 0 & \dot{b}_2 &= 0, \\
& \cdots & & \\
\dot{a}_{n-1} &= 0 & \dot{b}_{n-1} &= 0, \\
& & \dot{b}_n &= -1.
\end{aligned}
\tag{6.24}
$$

From Marden's curve to the Poncelet–Darboux curve

Following Marden, denote by $\mathcal{L}_{\alpha_j^0}$ the line equation of the point $\alpha_j^0 = x_j + iy_j$:

$$\mathcal{L}_{\alpha_j^0} = \lambda x_j + \mu y_j - 1,$$

the equation of all lines passing through the point α_j^0. The equation of the Marden curve (see [Mar1966, equation (4.10)]) is deduced from the condition

$$\sum_{j=1}^{n} \frac{m_j^0}{\mathcal{L}_{\alpha_j^0}} = 0. \tag{6.25}$$

We will denote the last curve by $\mathcal{M}(\alpha_1^0, \alpha_2^0, \ldots, \alpha_n^0, m_1^0, m_2^0, \ldots, m_n^0)$.

Now we pass to the projective plane. We see it as a projective space of quadratic polynomials: to a polynomial $P(z) = a(z - b)(z - c)$ we associate the point $(cb, -(c+b), 1)$. The projection

$$\pi : \mathbf{CP}^1 \times \mathbf{CP}^1 \to \mathbf{CP}^2 \tag{6.26}$$

is branched over the conic $\mathcal{K}$ with the equation

$$z_1^2 = 4z_0 z_2. \tag{6.27}$$

In this presentation, to a linear polynomial $a(z - \alpha_j^0)$ corresponds the line $t_{\mathcal{K}}(\alpha_j^0)$, tangent to the conic $\mathcal{K}$. This line is the set of all quadratic polynomials which have polynomial $z - \alpha_j^0$ as a factor.

To develop further this connection, following Darboux (see [Dar1917]) we introduce a new system of coordinates. Given a plane with standard coordinates (z_0, z_1, z_2), we start from the given conic $\mathcal{K}$. It is given by equation (6.27) and rationally parameterized by $(s^2, 2s, 1)$. The tangent line to the conic $\mathcal{K}$ through the point with the parameter s_0 is given by the equation

$$t_{\mathcal{K}}(s_0) : z_2 s_0^2 - z_1 s_0 + z_0 = 0.$$

On the other hand, for a given point P in the plane with coordinates $P = (\hat{z}_0, \hat{z}_1, \hat{z}_2)$ there correspond two solutions ρ and ρ_1 of the quadratic in s equation

$$\hat{z}_2 s^2 - \hat{z}_1 s + \hat{z}_0 = 0. \tag{6.28}$$

Each solution corresponds to a tangent to the conic $\mathcal{K}$ from the point P. We will call the pair (ρ, ρ_1) *the Darboux coordinates* of the point P. One finds immediately

$$\frac{\hat{z}_0}{\rho \rho_1} = \frac{\hat{z}_1}{\rho + \rho_1} = \hat{z}_2. \tag{6.29}$$

The line $t_\mathcal{K}(\alpha_j^0)$ which corresponds to the point α_j^0 and to the linear polynomial $a(z - \alpha_j^0)$ has the following presentation in the Darboux coordinates:

$$t_\mathcal{K}(\alpha_j^0)(\rho, \rho_1) = (\rho - \alpha_j^0)(\rho_1 - \alpha_j^0).$$

The Poncelet–Darboux curve is expressed by the equation

$$\sum_{j=1}^{n} \frac{m_j^0}{t_\mathcal{K}(\alpha_j^0)} = 0. \tag{6.30}$$

In Darboux coordinates it may be rewritten in the form

$$\frac{f(\rho)}{\phi(\rho)} = \frac{f(\rho_1)}{\phi(\rho_1)}. \tag{6.31}$$

The curve defined by the last equations we will denote as $\mathcal{PD}_\mathcal{K}(\phi, f)$. From equations (6.25) and (6.30) and all previous considerations we get the following theorem.

Theorem 6.13. *There is a birational morphism defined by equations (6.8) and (6.17) between the data of Marden curves*

$$(\alpha_1^0, \alpha_2^0, \dots, \alpha_n^0, m_1^0, m_2^0, \dots, m_n^0),$$

up to permutation of indices $1, 2, \dots, n$, and the data (ϕ, f) of Poncelet–Darboux curves associated with the conic $\mathcal{K}$. The dual of projective closure of a Marden curve $\mathcal{M}(\alpha_1^0, \alpha_2^0, \dots, \alpha_n^0, m_1^0, m_2^0, \dots, m_n^0)$ is birationally equivalent to the Poncelet–Darboux curve $\mathcal{PD}_\mathcal{K}(\phi, f)$ with corresponding data.

Proof. Fix data $(\alpha_1^0, \alpha_2^0, \dots, \alpha_n^0, m_1^0, m_2^0, \dots, m_n^0)$, with all α_i different between each other. Polynomials f and ϕ of degrees $n-1$ and n are constructed according to the formula (6.8). The polynomial ϕ has all zeros simple. Conversely, given a pair of polynomials f, ϕ of degrees $n-1$ and n, respectively, where ϕ has only simple zeros. By factorizing the polynomial ϕ, the data $(\alpha_1^0, \alpha_2^0, \dots, \alpha_n^0)$ is obtained. From the coefficients of the polynomial f we get $B_i, i = 1, \dots, n$. By plugging into the formulae (6.17), the unknown data $(m_1^0, m_2^0, \dots, m_n^0)$ are calculated as solutions of the system of linear equations. The determinant of the system is a function of $(\alpha_1^0, \alpha_2^0, \dots, \alpha_n^0)$ and it vanishes only if $\alpha_i^0 = \alpha_j^0$ for some $i \neq j$. (See the next subsection for a comment on this condition.)

Given a pair of data points in the above correspondence. The relationship between the Poncelet–Darboux curve and the Marden curve which are associated to these data, can be seen as follows. A Poncelet–Darboux curve of degree $n-1$ associated to a smooth conic $\mathcal{K}$ in $\mathbf{P}^2$ is defined by a pencil of divisors of degree n on $\mathcal{K}$. It means that the curve is described by the vertices of a pencil of n-gons circumscribed about $\mathcal{K}$. Let us denote by $l_1 = l_{\alpha_1^0}, \dots, l_n = l_{\alpha_n^0}$ the n lines of a tangent polygon, such that l_i is tangent to $\mathcal{K}$ at the point with rational parameter α_i^0. Then the equation $f = 0$ of the curve is defined by the formula $f = \sum_i m_i^0 l_1 \dots \hat{l}_i \dots l_n$, with $m_i^0 \in \mathbf{C}$. The line l_i corresponds to a polynomial $z - \alpha_i^0 \in \mathbf{C}[z]$. The equation of the Poncelet–Darboux curve may be rewritten in the form $\sum_i m_i^0 / t_K(\alpha_i^0) = 0$. The associated pair of polynomials f, ϕ arises by rewriting the last equation in terms of the Darboux coordinates, according to the formula (6.31). This curve is dual to the Marden curve given by equation (6.25). This ends the proof of the theorem.   □

Discriminant and collisions

The morphism from the previous Theorem 6.13 fails to be one to one for those $\alpha_1^0, \alpha_2^0, \dots, \alpha_n^0$ for which the system of linear equations in $(m_1^0, m_2^0, \dots, m_n^0)$ given by equations (6.17) has determinant $D(\alpha_1^0, \alpha_2^0, \dots, \alpha_n^0)$ equal to zero. The condition

$$D(\alpha_1^0, \alpha_2^0, \dots, \alpha_n^0) = 0$$

is equivalent to $\alpha_i^0 = \alpha_j^0$ for some $i \neq j$.

We will refer to configurations

$$(\alpha_1(t), \alpha_2(t), \dots, \alpha_n(t), m_1(t), m_2(t), \dots, m_n(t))$$

such that $\alpha_i(t) = \alpha_j(t)$ for some $i \neq j$ as *points of collision*. Such a moment t we will call *moment of collision*. If there is no $k \neq i, k \neq j$ such that in addition $\alpha_i(t) = \alpha_j(t) = \alpha_k(t)$ the point of collision is *simple*. The system is *simple* if it has only simple collision points.

We will assume that the system passes smoothly through a collision point in the phase space.

6.3   *n*-volutions, collisions and decomposition of Poncelet–Darboux curves

n-volutions

By an *n-volution* in a set V we will assume a family of multisets $\mathcal{A}_n$, a subset of the nth symmetric product $\mathrm{Sym}_n V$ such that there is a unique function f:

$$f : V \to \mathcal{A}_n,$$

such that $v \in f(v)$. In some classical terminology the notion of cyclic-symmetric correspondences is used.

If $\alpha \in \mathcal{A}_n$ is a multiset such that its cardinality is less than n we will say that it is *a collision point* of the n-volution. In other words $\alpha = \{\alpha_1, \dots, \alpha_n\}_M$ is a collision point if there exist $\alpha_1, \dots, \alpha_k$, $k < n$ and natural numbers $c_1, \dots, c_k$

such that

$$\alpha = \{\alpha_1, \dots, \alpha_n\}_M = c_1\alpha_1 + \dots + c_k\alpha_k.$$

The basic examples of n-volutions are involutions, which correspond to $n = 2$. Then, the notion of a collision point coincides with the notion of a fixed point. A nice case is a set of involutions of a conic. By the Frégier theorem (see Chapter 4.3, Theorem 4.14), we know that for every involution on a conic, there exists a point, *the Frégier point* such that the involution is cut from the conic by lines from the pencil determined by the Frégier point.

We pass now to the case of $\mathbf{CP}^1$ and n-volutions generated by rational functions of order n. Every such n-volution is determined by a pair of polynomials p, q of degree n.

Consider the pencil $p_t(z) = tp(z) + q(z)$. The roots $a_1(t), \dots, a_n(t)$ determine a one-parameter family of n-tuples. From the system

$$tp(a_1(t)) + q(a_1(t)) = 0,$$
$$tp(a_2(t)) + q(a_2(t)) = 0,$$

we get

$$(a_1(t) - a_2(t))r(a_1(t), a_2(t)) = 0.$$

Here r is of degree $n - 1$ in a_1 and symmetric. When t varies, $r(a_1(t), a_2(t)) = 0$ defines a curve.

The question is how to describe all such r. Following the lines of Section 6.2 we pass to $\mathbf{CP}^2$ and correspond to an n-tuple of n linear factors $(z - a_1), \dots, (z - a_n)$ a polygon of n-sides circumscribed about the conic $\mathcal{K}$. Thus, we have the following

Proposition 6.14. *An n-volution defined by polynomials p, q of degree n is associated to a curve from* Pon-Dar$_{n-1}(\mathcal{K})$ *given by the equation*

$$\det A \circ K(z) = 0.$$

The $(n - 1) \times (n + 1)$ matrix A annihilates the pencil generated by p, q and the matrix $K(z)$ is induced by the conic $\mathcal{K}$ and it has the form

$$K(z) = \begin{pmatrix} z_2 & 0 & \dots & \dots & 0 \\ -z_1 & z_2 & \dots & \dots & 0 \\ z_0 & -z_1 & \dots & \dots & 0 \\ 0 & z_0 & \dots & \dots & \\ & & \dots & & z_2 \\ & & \dots & & -z_1 \\ & & \dots & & z_0 \end{pmatrix}.$$

The last equation follows from [Tra1988]. A linear system $L = L(p, q)$ has a *base point* if polynomials p and q have a common root. From [Tra1988] we have that the base points of the linear system correspond to the components of the Poncelet–Darboux curve which are tangent lines to the conic $\mathcal{K}$. From [Tra1988] we also have the following description of the collision points.

Proposition 6.15. *Let C be a Poncelet–Darboux curve with respect to the conic K such that the corresponding linear system $L = L(p, q)$ does not have base points. Then for a given point $P = (s_0, t_0) \in K$ and an integer $k \geq 0$ the following are equivalent:*

(A) *the intersection multiplicity of K and C at P is equal to k;*

(B) *(s_0, t_0) is a zero of order $k + 1$ of a unique polynomial $h \in L(p, q)$.*

Conic component.
Complete decomposition of Poncelet–Darboux curves

From Proposition 6.15 we see that the intersection of a Poncelet–Darboux curve C and the conic K is transversal if and only if the corresponding system is simple in terminology of the Section 6.2. If a Poncelet–Darboux curve C satisfies any of two equivalent conditions of Proposition 6.15 with $k = 1$ at every point of intersection of K and C, we will say that it is *transversal*. Thus, the linear system which corresponds to a transversal Poncelet–Darboux curve contains polynomials with at most double roots.

From now on we will consider transversal Poncelet–Darboux curves. Let us consider first the simplest case of curves of degree 2.

Example 6.16. Let C be a degree 2 Poncelet–Darboux curve with transversal intersection with the conic K. Let four intersection points of C and K be x_1, x_2, x_3, x_4. Let four common bitangents be t_i, $i = 1, \ldots, 4$ and denote points of contact of the tangent t_i with the conic C as y_i and the point of contact with the conic K as a_i.

The conic C is 3-Poncelet related to the conic K. By Poncelet's theorem 6.4, this means that there is a triangle inscribed in C and circumscribed about K with an arbitrary point of C taken as a vertex.

Choose any of the points y_i, say y_1 as a vertex. Then y_1 is a double vertex of a Poncelet triangle. The third vertex of the Poncelet triangle is one of the points x_i and the Poncelet triangle is $y_1 y_1 x_i$. Thus we get

Lemma 6.17. *With the use of previous notation, for any point y_i there is a point x_j such that the line $y_i x_j$ is tangent to the conic K at the point x_j. The triplet $(y_i y_i x_j)$ forms a Poncelet triangle.*

Coming back to the general case of transversal Poncelet–Darboux curves of degree $n - 1$, let us study the case when a curve S has a conic component C which is n-Poncelet related to the conic K. According to the Darboux theorem 6.5, then the curve S is decomposed as a product of k conics if $n = 2k + 1$ or as a product of $k - 1$ conics and a line if $n = 2k$.

Any conic C_i defines a symmetric (2-2)-correspondence of the Euler–Chasles type Φ_i such that a point P belongs to the conic C_i if and only if $\Phi_i(\rho, \rho_1) = 0$ where (ρ, ρ_1) are the Darboux coordinates of the point P. Darboux proved Theorem 6.5 using these correspondences:

Proof of the Darboux Theorem 6.5. Denote by Φ the symmetric (2-2)-correspond-
ence induced by the conic $\mathcal{K}$ as a caustic and the conic $\mathcal{C}$ as a boundary. This is a
correspondence between Darboux coordinates (ρ, ρ_1) of the point of the boundary,
where (ρ, ρ_1) are parameters of the tangents to the conic $\mathcal{K}$ from the point of the
boundary. Vice versa, every such symmetric (2-2)-correspondence determines a
conic from the confocal family as a boundary.

Consider iterations of k reflections along the Poncelet configuration. We get
symmetric (2-2)-correspondences Φ_k, $k \in \mathbf{N}$. They satisfy

$$\Phi_k \circ \Phi_l = \Phi_{k+l} \cdot \Phi_{|k-l|}.$$

(See Section 4.12.)

The condition $\Phi_k(\rho_1, \rho_{k+1}) = 0$ means that there is a sequence ρ_1, ρ_2, ...,
ρ_k, ρ_{k+1} of tangents to the conic $\mathcal{K}$ such that any two consecutive ρ_i, ρ_{i+1} intersect
at the conic $\mathcal{C}$.

Denote by $\mathcal{C}_k$ the conic associated with the correspondence Φ_k. The pairs
of sides of a Poncelet polygon (ρ_i, ρ_{i+k}) intersect at the conic $\mathcal{C}_k$ for fixed k and
arbitrary i. Thus, for fixed k the intersections of sides of a Poncelet polygon
(ρ_l, ρ_{l-k}) belong to the same conic $\mathcal{C}_k$ for any l and for any Poncelet polygon
inscribed in $\mathcal{C}$ and circumscribed about the conic $\mathcal{K}$. In the case of n even, the
opposite sides of a Poncelet polygon intersect on a line L.

The conics $\mathcal{C}_k$ and line L for n even are components of the Darboux curve.
This concludes the proof of the Darboux theorem.   □

Moreover, we have

Lemma 6.18. *Suppose the lines* $t_1, t_2, \ldots, t_n$ *are given such that* $\{P_i\} := t_i \cap t_{i+1} \in$
$\mathcal{C}_i$ *and* $\{P_n\} := t_n \cap t_1 \in \mathcal{C}_n$, *where conics* $\mathcal{C}_i$ *belong to a confocal pencil together*
with the conic $\mathcal{K}$. *Denote related Euler–Chasles correspondences as* Φ_i. *If multisets*
are equal,

$$\{\mathcal{C}_2, \mathcal{C}_3, \ldots, \mathcal{C}_k\}_M = \{\mathcal{C}_{n-1}, \mathcal{C}_{n-2}, \ldots, \mathcal{C}_{n-k+1}\}_M,$$

then there exist a conic $\mathcal{C}_0$ *from the confocal family which contains the point* P_1
and the point $P_{k+1,n-k} := t_{k+1} \cap t_{n-k}$.

Proof. The lemma follows from the fact that in the plane there are two conics
from a confocal family which contain a point P_1. In our notation one conic is $\mathcal{C}_1$,
denote the other one as $\mathcal{C}_0$ with the Euler–Chasles correspondence Φ_0. We have:

$$\Phi_k \circ \cdots \circ \Phi_2 \circ \Phi_1(t_1) = t_{k+1},$$
$$\Phi_k \circ \cdots \circ \Phi_2 \circ \Phi_0(t_1) = t_{k+1}.$$

Using its communicative property, we get from the last equation

$$\Phi_0 \circ \Phi_k \circ \cdots \circ \Phi_2(t_1) = t_{k+1}. \tag{6.32}$$

Consider the intersection of the lines

$$\Phi_{n-k+1} \circ \cdots \circ \Phi_{n-2} \circ \Phi_{n-1}(t_1)$$

and t_{k+1}. Denote by $\hat{\Phi}$ the Euler–Chasles correspondence associated with the conic $\mathcal{C}$ such that

$$\Phi_{n-k+1} \circ \cdots \circ \Phi_{n-2} \circ \Phi_{n-1}(t_1) \cap t_{k+1} \in \mathcal{C}.$$

Thus we have

$$\hat{\Phi} \circ \Phi_{n-k+1} \circ \cdots \circ \Phi_{n-2} \circ \Phi_{n-1}(t_1) = t_{k+1}.$$

From the assumption of the lemma, the last equation and from equation (6.32) it follows that $\mathcal{C} = \mathcal{C}_0$, giving the proof of the lemma.   □

There is an important special case of Lemma 6.18, when all the conics are equal: $\mathcal{C}_2 = \mathcal{C}_3 = \cdots = \mathcal{C}_k = \mathcal{C}_{n-1} = \mathcal{C}_{n-2} = \cdots = \mathcal{C}_{n-k+1}$. In the case of Poncelet n-tangles even more is true: $\mathcal{C}_i = \mathcal{C}_j$ for any $i, j = 1, \ldots, n$.

Exercise 6.19. All vertices of a triangle lie on a conic $\mathcal{C}_1$, one of the sides touches a conic $\mathcal{C}_2$ and the other two sides touch the conics $t\mathcal{C}_1 + \mathcal{C}_2 = 0$ and $s\mathcal{C}_1 + \mathcal{C}_2 = 0$ respectively. If the three points of contact are not collinear, prove

$$(I_3 - I_1ts)^2 - 4I_4(I_2 + I_1(s + t)) = 0,$$

where I_1, I_2, I_3, I_4 denote the coefficients of the characteristic polynomial $P(t) = \det(t\mathcal{C}_1 + \mathcal{C}_2) = I_1t^3 + I_2t^2 + I_3t + I_4$ of the pair of conics $(\mathcal{C}_1, \mathcal{C}_2)$ (see Section 4.7).

Exercise 6.20. Let $P_1, P_2, \ldots, P_n, \ldots$ be points on a conic $\mathcal{C}_1$ such that there exists a conic $\mathcal{C}_2$ to which all sides P_iP_{i+1} are tangent. Assume $P_{k+2} \neq P_k$. Prove that lines $P_1P_{k+1}, P_2P_{k+2}, \ldots P_iP_{i+k}$ touch the conic $t_k\mathcal{C}_1 + \mathcal{C}_2 = 0$ where

$$t_2 = 0, \quad t_3 = \frac{I_3^2 - 4I_2I_4}{4I_1I_4}, \quad t_4 = \frac{2I_3t_3 - 4I_4}{I_1t_3^2}$$

and

$$t_{k+1} = \frac{(I_3^2 - 4I_2I_3) - 4I_1I_4t_k}{I_1^2t_k^2t_{k-1}}, \quad k > 4.$$

Here I_1, I_2, I_3, I_4 are the coefficients of the characteristic polynomial, as in the previous exercise.

Definition 6.21. The union of k conics if $n = 2k + 1$ or $k - 1$ conic and the line if $n = 2k$ together with conics from the previous Lemma 6.18 form *the complete projective Poncelet–Darboux grid*.

Notice that transversal conics from Lemma 6.18 do not form a decomposition of a Poncelet–Darboux curve. We continue to study Poncelet–Darboux curves in the Euclidean plane in Section 8.5. There we will give also higher-dimensional generalizations of the Darboux theorem.

The condition *for two conics* $\mathcal{K}$ *and* $\mathcal{C}$ to be n-Poncelet related was derived, as we saw, by Cayley.

Here, we want to present a condition of different type: the condition *for two polynomials* ϕ, f of degree n in order that corresponding Poncelet–Darboux curve $\mathcal{S} = \mathcal{PD}_{\mathcal{K}}(\phi, f) \in \text{Pon-Dar}_{n-1}(\mathcal{K})$ has a conic component n-Poncelet related to the conic $\mathcal{K}$.

Theorem 6.22. *Let a transversal Poncelet–Darboux curve,*

$$\mathcal{S} = \mathcal{PD}_{\mathcal{K}}(\phi, f) \in \text{Pon-Dar}_{n-1}(\mathcal{K}),$$

be given, where polynomials f and ϕ are without common zeroes.

(i) *For $n = 2k + 1$, curve $\mathcal{S}$ is completely decomposed to k conics only if there exist four values t_1, t_2, t_3, t_4 such that*

$$\deg \text{GCD} \left(\Phi_{t_i}(z), \frac{d}{dz} \Phi_{t_i}(z) \right) = k, \tag{6.33}$$

for $i = 1, 2, 3, 4$;

(ii) *for $n = 2k$, curve $\mathcal{S}$ is completely decomposed to $k - 1$ conics and a line only if there exist four values t_1, t_2, t_3, t_4 such that*

$$\deg \text{GCD} \left(\Phi_{t_i}(z), \frac{d}{dz} \Phi_{t_i}(z) \right) = k, \tag{6.34}$$

for $i = 1, 2$, and

$$\deg \text{GCD} \left(\Phi_{t_i}(z), \frac{d}{dz} \Phi_{t_i}(z) \right) = k - 1, \tag{6.35}$$

for $i = 3, 4$.

Here, we set $\Phi_{t_i}(z) := \phi(z) + t_i f(z)$, while GCD *stands for the greatest common divisor of polynomials.*

Proof. Suppose that a Poncelet–Darboux curve $\mathcal{S}$ of degree $n - 1$ completely decomposes. Then, denote by $\mathcal{C}$ its conic component which is n Poncelet related to the conic $\mathcal{K}$. By assumption of transversality, conics $\mathcal{C}$ and $\mathcal{K}$ intersect in four points. As in Example 6.16 let four intersection points $\mathcal{C} \cap \mathcal{K} = \{x_1, x_2, x_3, x_4\}$. Denote four common bitangents t_i, $i = 1, \ldots, 4$ and denote points of contact of the tangent t_i with the conic $\mathcal{C}$ as y_i and the point of contact with the conic $\mathcal{K}$ as a_i.

The conic $\mathcal{C}$ is n-Poncelet related to the conic $\mathcal{K}$ and according to Poncelet theorem 6.4 there is a polygon of n sides inscribed in $\mathcal{C}$ and circumscribed about $\mathcal{K}$ with an arbitrary point of $\mathcal{C}$ taken as a vertex.

Suppose n is odd: $n = 2k + 1$. Choose any of the points y_i, say y_1 as a vertex. Then y_1 is a double vertex of a Poncelet $2k + 1$-polygon. Moreover, the next $k - 1$ vertices $c_1, \ldots, c_{k-1}$ are also double vertices. The last vertex of the Poncelet n-tangle is one of the points x_i and the Poncelet n-tangle is $c_{k-1} \ldots c_1 y_1 y_1 c_1 \ldots c_{k-1} x_i$.

Suppose now that n is even: $n = 2k$. In this case there are two pairs of distinguished Poncelet n-tangles. Two of them are of the form

$$c_{k-1} \ldots c_1 y_j y_j c_1 \ldots c_{k-1} y_i y_i,$$

for example for $(i, j) = (1, 2)$ and for $(i, j) = (3, 4)$. Each of them connect a pair of common tangents and it has all other vertices as double. Another pair of distinguished Poncelet n-tangles connects pairs of intersection points. For example the first one connects x_1 and x_2 while the second connects x_3 with x_4. All their other vertices are double. Thus these two Poncelet n-tangles are of the form $d_{k-1} \ldots d_1 x_1 d_1 \ldots d_{k-1} x_2$ and $e_{k-1} \ldots e_1 x_3 e_1 \ldots e_{k-1} x_4$.

Lemma 6.23. *If a transversal Poncelet–Darboux curve*

$$\mathcal{S} = \mathcal{P}\mathcal{D}_{\mathcal{K}}(\phi, f) \in \text{Pon-Dar}_{n-1}(\mathcal{K})$$

has a conic component $\mathcal{C}$ which is n-Poncelet related to the conic $\mathcal{K}$ then:

(i) *for $n = 2k + 1$, there exist four values t_1, t_2, t_3, t_4 and four polynomials Q_1, Q_2, Q_3, Q_4 such that*

$$\begin{aligned}
\Phi_{t_1}(z) &:= t_1 \phi(z) + f(z) = (z - a_1) Q_1^2(z), \\
\Phi_{t_2}(z) &:= t_2 \phi(z) + f(z) = (z - a_2) Q_2^2(z), \\
\Phi_{t_3}(z) &:= t_3 \phi(z) + f(z) = (z - a_3) Q_3^2(z), \\
\Phi_{t_4}(z) &:= t_4 \phi(z) + f(z) = (z - a_4) Q_4^2(z);
\end{aligned} \qquad (6.36)$$

(ii) *for $n = 2k$, there exist four values t_1, t_2, t_3, t_4 and four polynomials Q_1, Q_2, Q_3, Q_4 such that*

$$\begin{aligned}
\Phi_{t_1}(z) &:= t_1 \phi(z) + f(z) = (z - a_1)(z - a_2) Q_1^2(z), \\
\Phi_{t_2}(z) &:= t_2 \phi(z) + f(z) = (z - a_3)(z - a_4) Q_2^2(z), \\
\Phi_{t_3}(z) &:= t_3 \phi(z) + f(z) = Q_3^2(z), \\
\Phi_{t_4}(z) &:= t_4 \phi(z) + f(z) = Q_4^2(z).
\end{aligned} \qquad (6.37)$$

From conditions (6.36) and (6.37) respectively, conditions (6.33) and (6.34) together with (6.35) immediately follow. This proves the theorem. □

For the opposite direction, observe that from conditions (6.33) and (6.34) together with (6.35) of Theorem 6.22, conditions (6.36) and (6.37) of Lemma 6.23 follow by use of the transversality condition. From the transversality it follows that all multiple zeros of the polynomials in the pencil generated by f and ϕ are of the second degree. Now, from conditions (6.36) and (6.37) one can easily prove the following

Lemma 6.24. *Suppose conditions* (6.36) *and* (6.37) *are satisfied. Denote by* Γ_1 *and* Γ_2 *the following elliptic curves:*

$$\Gamma_1 : y^2 = (z - a_1)(z - a_2)(z - a_3)(z - a_4),$$
$$\Gamma_2 : Y^2 = (X + t_1)(X + t_2)(X + t_3)(X + t_4).$$

Then, there is an $n : 1$ *morphism,*

$$h : \Gamma_1 \to \Gamma_2, \quad h(z, y) = (X, Y),$$

where

$$X = \frac{f(z)}{\phi(z)}, \quad Y = y \frac{Q_1(z)Q_2(z)Q_3(z)Q_4(z)}{\phi^2(z)}.$$

From the last lemma we see that the question of decomposition of the Poncelet–Darboux curve defined by polynomials f and ϕ corresponds to a study of an unramified covering of degree n of the elliptic curve Γ_1 over the elliptic curve Γ_2.

We can say more about such coverings.

Lemma 6.25.

(a) *There is a constant* N *such that*

$$Q_1(z)Q_2(z)Q_3(z)Q_4(z) = N \left(\phi(z) \frac{d}{dz} f(z) - \frac{d}{dz} \phi(z) f(z) \right).$$

(b) *There is a relation between holomorphic differentials of elliptic curves* Γ_1 *and* Γ_2:

$$\frac{dz}{y} = N \frac{dX}{Y}.$$

Proof. It follows by straightforward calculations.   □

A systematic study of the elliptic coverings was established by Jacobi in [Jac1829]. We pass now to Jacobi's notation. By applying rational-linear transformations, we come to the canonical form of elliptic curves Γ_2 and Γ_1 and to the relation between differentials

$$\frac{N \, dy}{\sqrt{(1 - y^2)(1 - \lambda y^2)}} = \frac{dx}{\sqrt{(1 - x^2)(1 - k \cdot x^2)}}.$$

The constants k and λ are the modules of the elliptic curves Γ_1 and Γ_2.

Then, if all above conditions are satisfied, for a general point $R \in \Gamma_2$ there is a value u such that $y(R) = \text{sn}\,(u/N, \lambda)$ and there are n points P_j, $j = 0, \ldots, n-1$ in $h^{-1}(R) \subset \Gamma_1$ with parameters u_j, $j = 0, \ldots, n-1$ such that $x_j(P_j) = \text{sn}\,(u_j, k)$.

Definition 6.26. Using the above notation, we will say that a covering h is *cyclic* if there exists η such that $n\eta \equiv 0$ and $u_j = u_0 + j\eta$, $j = 0, \ldots, n-1$.

Lemma 6.27. *The existence of an n-Poncelet conic component corresponds to the condition that the covering is cyclic.*

Proof. A Poncelet polygon is given by choosing $P \in \Gamma_1$ such that $nP \equiv 0$ in Γ_1. The vertices of the polygon correspond to parameters $u_j = u_0 + j\eta$, $j = 0, \ldots, n-1$, where η corresponds to the point P. Denote by $\mathcal{C}_P$ a conic which corresponds to P in the pencil of conics generated with the conic $\mathcal{K}$ and its four tangents at the points a_1, a_2, a_3, a_4. Then the conic $\mathcal{C}_P$ is n-Poncelet related to the conic $\mathcal{K}$. $\square$

Following Jacobi, for a given n odd and given module k, we have explicit formulae for the transformations. Let

$$K(k) = \int_0^{\frac{\pi}{2}} \frac{d\theta}{\sqrt{1 - k^2 \sin^2 \theta}}, \qquad K'(k) = \int_0^{\frac{\pi}{2}} \frac{d\theta}{\sqrt{1 - k'^2 \sin^2 \theta}},$$

where $k'^2 = 1 - k^2$.

Theorem 6.28. *For a given n odd and for*

$$\omega = \frac{mK + m'iK'}{n},$$

where integers m and m' have no common divisors which divide n, the transformation is defined by

$$\phi(z) = \frac{x}{N} \prod_{r=1}^{(n-1)/2} \left(1 - \frac{x^2}{\operatorname{sn}^2 4r\omega}\right),$$

$$f(z) = \prod_{r=1}^{(n-1)/2} (1 - x^2 \cdot \operatorname{sn}^2 4r\omega),$$

$$N = (-1)^{(n-1)/2} \prod_{r=1}^{(n-1)/2} \left(1 - \frac{\operatorname{sn}(K - 4r\omega)}{\operatorname{sn}^2(4r\omega)}\right),$$

$$\lambda = \prod_{r=1}^{(n-1)/2} \left(\operatorname{sn}^4(K - 4r\omega)\right).$$

The transformation corresponds to an n-Poncelet trajectory, where

$$x_i = \operatorname{sn}(u + 4(i+1)\omega, k) \quad i = 0, \ldots, n-1, \quad y = \operatorname{sn}(u/N, \lambda).$$

Proof. It follows from [Jac1829, Sections 20 and 21] and Lemma 6.27. All the coverings are cyclic. $\square$

Let us consider three arithmetic functions. Suppose a natural number $n \in \mathbf{N}$ is given by its prime decomposition,

$$n = p_1^{k_1} p_2^{k_2} \cdots p_r^{k_r},$$

where p_i are different prime numbers. Following [BM1993], we define a function $t(n)$, *the number of primitive n-torsion points on an elliptic curve*:

$$t(n) := (p_1^2 - 1)p_1^{2(k_1-1)}(p_2^2 - 1)p_2^{2(k_2-1)} \cdots (p_r^2 - 1)p_r^{2(k_r-1)}.$$

As an example, for $n = p$ a prime number, $t(p) = p^2 - 1$. The function t is a multiplicative arithmetic function.

As the second arithmetic function, we introduce a function $\sigma'(n)$ as a multiplicative function which is for n odd equal to

$$\sigma'(n) = n\left(1 + \frac{1}{p_1}\right)\left(1 + \frac{1}{p_2}\right)\cdots\left(1 + \frac{1}{p_r}\right), \quad n \text{ odd}.$$

For $n = 2^k$ we define

$$\sigma'(2^k) := 2^{k+1} - 1.$$

For example, for $n = p$ a prime number, we have $\sigma'(p) = p + 1 = \sigma(p)$. The σ' function in this case is equal to the σ function, where the last function represents the sum of divisors of p. The number of transformations listed in Theorem 6.28 for a given odd number n is equal to $\sigma'(n)$.

The third arithmetic function we are going to consider is a well-known Euler function $\varphi(n)$ counting the numbers smaller than n relatively prime to n:

$$\varphi(n) = n\left(1 - \frac{1}{p_1}\right)\left(1 - \frac{1}{p_2}\right)\cdots\left(1 - \frac{1}{p_r}\right).$$

Theorem 6.29. *Let m be an arbitrary odd number. For $n = 2^k \cdot m$ where $k = 0, 1$ all above elliptic coverings are cyclic.*

For $n = 2^k \cdot m$ and every $k > 1$, there are elliptic coverings of the above form which are not cyclic.

Proof. It has been proven by classics that the function $\sigma'(n)$ counts the number of degree n elliptic coverings of the above form assuming the module k being fixed.

From [BM1993] we know that the function t represents the number of Poncelet polygons in total up to the porism, with fixed caustic and confocal pencil of conics.

The proof is concluded by the next two Lemmata 6.30 and 6.31.   □

Lemma 6.30. *For n odd the following identity holds:*

$$t(n) = \sigma'(n) \cdot \varphi(n).$$

Lemma 6.31. *For $n = 2^k$, $k \geq 1$, the following inequality holds:*

$$t(n) \leq \sigma'(n) \cdot \varphi(n).$$

The equality holds only for $k = 1$.

Both lemmata are proved by direct calculation.

Thus, we come to the converse of Theorem 6.22.

Theorem 6.32. *Let m be an arbitrary odd number. For $n = 2^k \cdot m$ where $k = 0, 1$ the conditions of Theorem 6.22 are sufficient as well.*

Proof. Let a transversal Poncelet–Darboux curve $\mathcal{S} = \mathcal{PD}_{\mathcal{K}}(\phi, f) \in \text{Pon-Dar}_{n-1}(\mathcal{K})$ be given, where polynomials f and ϕ are without common zeros and of degree $n-1$ and n. We now assume that conditions (6.33) and (6.34) together with (6.35) are satisfied. Conditions (6.36) and (6.37) can be deduced using the transversality assumption. From the transversality, it follows that all multiple zeros of the polynomials in the pencil generated by f and ϕ are of the second degree. From conditions (6.36) and (6.37) follows the existence of an elliptic covering of degree n, as it was described in Lemma 6.24. Since $n = 2^k \cdot m$ where $k = 0, 1$, from Theorem 6.29 it follows that the covering is cyclic. The existence of an n-Poncelet component now follows from Lemma 6.27. Complete decomposition follows from the Darboux Theorem 6.5. This concludes the proof. □

Conic choreography of the "$\binom{n}{2}$-body" problem

In order to get effective description of polynomials f and ϕ which satisfy Theorem 6.22, we add three more statements.

Exercise 6.33. A necessary and sufficient condition that polynomials

$$F(x) = a_0 x^m + \cdots + a_{m-1} x + a_m,$$
$$G(x) = b_0 x^n + \cdots + b_{n-1} x + b_n,$$

of degree m and n respectively have a common factor of degree at least r is that all the determinants of order $m + n - 2r + 2$ of the following $(m + n - 2r + 2) \times (m + n - r + 1)$ matrix $R_r(F, G)$ are equal to zero:

$$R_r(F, G) = \begin{pmatrix} a_0 & a_1 & & \cdots & a_m & \cdots & \cdots & 0 \\ 0 & a_0 & a_1 & \cdots & & a_m & \cdots & 0 \\ & \cdots & & & \cdots & & & 0 \\ 0 & \cdots & \cdots & & a_0 & a_1 & \cdots & a_m \\ b_0 & b_1 & & \cdots & b_n & \cdots & \cdots & 0 \\ 0 & b_0 & b_1 & \cdots & & b_n & \cdots & 0 \\ & \cdots & & & \cdots & & & 0 \\ 0 & \cdots & \cdots & & b_0 & b_1 & \cdots & b_n \end{pmatrix}. \tag{6.38}$$

For $r = 1$, the formula (6.38) gives the $(m + n) \times (m + n)$ matrix $R_1(F, G)$, which is the standard resultant of the two polynomials F and G of degrees m and n. Denote by $\Delta_0(F, G)$ the $(m + n) \times (m + n)$-matrix obtained from $R_1(F, G)$ by rearranging the b rows in the opposite order:

$$\Delta_0(F,G) = \begin{pmatrix} a_0 & a_1 & & \cdots & a_m & \cdots & \cdots & 0 \\ 0 & a_0 & a_1 & \cdots & & a_m & \cdots & 0 \\ & \cdots & & & \cdots & & \cdots & 0 \\ 0 & \cdots & \cdots & & a_0 & a_1 & \cdots & a_m \\ 0 & \cdots & \cdots & & b_0 & b_1 & \cdots & b_n \\ & \cdots & & & & \cdots & & 0 \\ 0 & b_0 & b_1 & \cdots & & b_n & \cdots & 0 \\ b_0 & b_1 & & \cdots & b_n & \cdots & \cdots & 0 \end{pmatrix}. \tag{6.39}$$

Denote now by $\Delta_i(F,G)$ the diagonal square submatrix of dimension $(m+n-2i) \times (m+n-2i)$ of the matrix $\Delta_0(F,G)$. Having this notation at hand, we can easily state another version of the previous Exercise 6.33.

Exercise 6.34. A necessary and sufficient condition that polynomials

$$F(x) = a_0 x^m + \cdots + a_{m-1} x + a_m,$$
$$G(x) = b_0 x^n + \cdots + b_{n-1} x + b_n$$

of degree m and n respectively have a common factor of degree exactly r is that all the determinants of the matrices $\Delta_i(F,G)$ are equal to zero, for $i = 0, \ldots, r-1$ and that the determinant of the matrix $\Delta_r(F,G)$ is nonzero:

$$\det \Delta_0(F,G) = 0,$$
$$\det \Delta_1(F,G) = 0,$$
$$\cdots$$
$$\det \Delta_{r-1}(F,G) = 0,$$
$$\det \Delta_r(F,G) \neq 0.$$

The half of the conditions of the second part of Theorem 6.22 and of the second half of Lemma 6.30 can be restated as follows.

Exercise 6.35. For polynomials ϕ and f of degree $n = 2k$ and $n-1$ respectively, there exist constants t_1 and t_2 such that polynomials $F_{t_i} = \phi + t_i f$ are full squares if and only if polynomials ϕ and f can be represented in the form

$$\phi(z) = \frac{\phi_1(z)\phi_2(z)}{t_2 - t_1},$$
$$f(z) = \frac{t_1 - t_2}{t_1 t_2}(\phi_1^2(z) + \phi_2^2(z)) - 2\frac{t_1 + t_2}{t_1 t_2}\phi(z).$$

However, a direct application of the previous three statements is not very effective. Thus, instead, we use parametrization of transformations, for small odd numbers, obtained by classics algebraically.

Assume n is an odd number. Then all transformations are given by the formulae

$$\phi(x) = P^2 + (2PQ + Q^2)x^2,$$
$$f(x) = x(P^2 + 2PQ + Q^2x^2),$$

where P and Q are polynomials in x^2:

$$P(x) = \alpha + \gamma x^2 + \epsilon x^4 + \cdots$$
$$Q(x) = \beta + \delta x^2 + \zeta x^4 + \cdots,$$

of degree p in x^2 both if $n = 4p + 3$ and of degree p and $p - 1$ respectively if $n = 4p + 1$. Then

$$\frac{1}{N} = 1 + \frac{2\beta}{\alpha},$$

where N has the same meaning as in Lemma 6.25.

Case $n = 3$

For $n = 3$ we have

$$P = \alpha, \quad Q = \beta.$$

Thus

$$f(x) = (\alpha^2 + 2\alpha \cdot \beta)x + \beta^2 x^3,$$
$$\phi(x) = \alpha^2 + (2\alpha \cdot \beta + \beta^2)x^2.$$

Now, one can easily get the initial data of the isofocal deformations, which correspond to the completely decomposable situation for $n = 3$.

Proposition 6.36. *The initial data of completely decomposable situation are given up to projective-linear transformations, by the formulae*

$$\alpha_1 = 0,$$

$$\alpha_{23} = \pm\sqrt{-\frac{\alpha^2 + 2\alpha\beta}{\beta^2}},$$

$$m_1 = -\frac{\beta^2(2\alpha\beta + \beta^2)}{\alpha^2 + 2\alpha\beta},$$

$$m_2 = m_3 = \frac{\alpha^2(\alpha^2 + 2\alpha\beta) + \beta^2(2\alpha\beta + \beta^2)}{2(\alpha^2 + 2\alpha\beta)}.$$

Case $n = 5$

For $n = 5$ we have

$$P(x) = \alpha + \gamma x^2, \quad Q(x) = \beta.$$

Thus,

$$\phi(x) = x((\alpha^2 + 2\alpha \cdot \beta) + (\beta^2 + 2\alpha\gamma + 2\gamma\beta)x^2 + \gamma x^4),$$
$$f(x) = \alpha^2 + (2\alpha\beta + 2\alpha\gamma + \beta^2)x^2 + (\gamma^2 + 2\beta\gamma)x^4.$$

Let

$$A = \frac{-(2\alpha\gamma + 2\gamma\beta + \beta^2) + \sqrt{D}}{2\gamma^2},$$

$$B = \frac{-(2\alpha\gamma + 2\gamma\beta + \beta^2) - \sqrt{D}}{2\gamma^2},$$

$$D = 2\alpha\gamma + 2\gamma\beta + \beta^2 - 4\gamma^2(\alpha^2 + 2\alpha\beta).$$

Now we have

Proposition 6.37. *The initial data of a completely decomposable situation are given up to projective-linear transformations, by the formulae*

$$\alpha_1 = 0,$$
$$\alpha_{23} = \pm\sqrt{A},$$
$$\alpha_{45} = \pm\sqrt{B},$$
$$m_1 = \frac{\gamma^2 + 2\alpha\gamma}{AB},$$
$$m_2 = m_3$$
$$= \frac{1}{2(B-A)}\left(2\alpha\gamma + 2\beta\alpha + \beta^2 + \frac{A+B}{AB}(\gamma^2 + 2\alpha\gamma) - A\alpha^2 + \frac{\gamma^2 + 2\alpha\gamma}{B}\right),$$
$$m_4 = m_5$$
$$= \frac{1}{2(B-A)}\left(2\alpha\gamma + 2\beta\alpha + \beta^2 + \frac{A+B}{AB}(\gamma^2 + 2\alpha\gamma) - B\alpha^2 + \frac{\gamma^2 + 2\alpha\gamma}{A}\right).$$

Case $n = 7$

For $n = 7$ we have

$$P = \alpha + \gamma x^2, \quad Q = \beta + \delta x^2,$$

and

$$f = x(\alpha^2 + 2\alpha\beta + (2(\alpha\gamma + \gamma\beta + \alpha\delta)x^2 + (\gamma^2 + 2(\gamma\delta + \beta\delta)x^4 + \delta^2 x^6),$$
$$\phi = \alpha^2 + (2(\alpha\gamma + \alpha\beta) + \beta^2)x^2 + (\gamma^2 + 2(\gamma\beta + \alpha\delta + \beta\delta))x^4 + (2\gamma\delta + \delta^2)x^6.$$

Denote by A_1, A_2, A_3 the three roots of the polynomial

$$(\alpha^2 + 2\alpha\beta) + 2(\alpha\gamma + \gamma\beta + \alpha\delta)x + (\gamma^2 + 2(\gamma\delta + \beta\delta))x^2 + \delta^2 x^3,$$

and the seven zeroes of the polynomial f are

$$\alpha_0 = 0, \qquad \alpha_{12} = \pm\sqrt{A_1},$$
$$\alpha_{34} = \pm\sqrt{A_2}, \quad \alpha_{56} = \pm\sqrt{A_3}.$$

To calculate the initial weights, introduce the new coordinates

$$Z_1 = m_2 + m_3, \quad \hat{Z}_1 = m_2 - m_3,$$
$$Z_2 = m_4 + m_5, \quad \hat{Z}_2 = m_4 - m_5,$$
$$Z_3 = m_6 + m_7, \quad \hat{Z}_3 = m_6 - m_7 .$$

Now, one gets the initial weights from the system of linear equations

$$m_1 = \frac{2\gamma\delta + \delta^2}{\alpha^2};$$

$$Z_1 + Z_2 + Z_3 = d_1,$$
$$\alpha_2^2 Z_1 + \alpha_4^2 Z_2 + \alpha_6^2 Z_3 = d_2,$$
$$\alpha_2^2(\alpha_4^2 + \alpha_6^2)Z_1 + \alpha_4^2(\alpha_2^2 + \alpha_6^2)Z_2 + \alpha_6^2(\alpha_4^2 + \alpha_2^2)Z_3 = d_3,$$
$$\alpha_2 \hat{Z}_1 + \alpha_4 \hat{Z}_2 + \alpha_6 \hat{Z}_3 = 0,$$
$$\alpha_2(\alpha_4^2 + \alpha_6^2)\hat{Z}_1 + \alpha_4(\alpha_2^2 + \alpha_6^2)\hat{Z}_2 + \alpha_6(\alpha_4^2 + \alpha_2^2)\hat{Z}_3 = 0,$$
$$\alpha_2\alpha_4^2\alpha_6^2\hat{Z}_1 + \alpha_4\alpha_2^2\alpha_6^2\hat{Z}_2 + \alpha_6\alpha_4^2\alpha_2^2\hat{Z}_3 = 0,$$

where

$$d_1 = \alpha^2 - \frac{2\gamma\delta + \delta^2}{\alpha^2},$$
$$d_2 = (\gamma^2 + 2(\gamma\beta + \alpha\delta) + 2\beta\delta) - \alpha^2(\gamma^2 + 2\gamma\delta + 2\beta\delta),$$
$$d_3 = (\gamma^2 + 2(\gamma\beta + \alpha\delta) - \frac{2\gamma\delta + \delta^2}{\alpha^2}(2(\alpha\gamma + \gamma\beta + \alpha\delta) + \beta^2).$$

Chapter 7

Ellipsoidal Billiards and Their Periodical Trajectories

Let $\mathcal{M}$ be an n-dimensional Riemann manifold, and $\Omega \subset \mathcal{M}$ a domain with the boundary composed of several smooth hypersurfaces. The billiard [KT1991] inside Ω is a dynamical system where a material point of the unit mass is freely moving inside the domain and obeying the reflection law at the boundary, i.e., having congruent impact and reflection angles with the space tangent to the boundary at any bouncing point. It is also assumed that the reflection is absolutely elastic, i.e., that the velocity of the material point does not change before and after impacts.

In Section 7.1 we are going to consider the case when $\mathcal{M}$ is in Euclidean space, and the billiard boundary is composed of a few confocal quadrics. There, we present our results on analytical description of periodical trajectories of such billiards [DR2004, DR2005, DR2006a]. In Section 7.2, using algebro-geometric integration of discrete billiard form [MV1991], we obtain the corresponding generalization of Cayley's condition [DR1998a, DR1998b]. Poncelet's theorem and Cayley's conditions for billiards within quadrics in the Lobachevsky space, obtained in [DJR2003], are presented in Section 7.3. After this, in Section 7.4, we review topological properties of elliptical billiards [DR2009], and we conclude the chapter with results on integrable potential perturbations of such billiards [Dra1998b].

Let us mention that billiards in other geometries were also studied, see, for example [GT2002, Rad2003, Tab2002].

7.1 Periodical trajectories inside k confocal quadrics in Euclidean space

Darboux was the first who considered a higher-dimensional generalization of the Poncelet theorem. Namely, he investigated light-rays in the three-dimensional case

$(d = 3)$ and announced the corresponding Full Poncelet theorem in [Dar1870] in 1870.

Higher-dimensional generalizations of the Full Poncelet Theorem $(d \geq 3)$ were obtained quite recently in [CCS1993], and the related Cayley-type conditions were derived by the authors [DR2004].

The main goal of this section is to present a detailed proof of a Cayley-type condition for the generalized Full Poncelet Theorem, together with discussions and examples.

Consider an ellipsoid in $\mathbf{R}^d$:

$$\frac{x_1^2}{a_1} + \cdots + \frac{x_d^2}{a_d} = 1, \quad a_1 > \cdots > a_d > 0,$$

and the related system of Jacobian elliptic coordinates $(\lambda_1, \ldots, \lambda_d)$ ordered by the condition

$$\lambda_1 > \lambda_2 > \cdots > \lambda_d.$$

If we write

$$Q_\lambda(x) = \frac{x_1^2}{a_1 - \lambda} + \cdots + \frac{x_d^2}{a_d - \lambda},$$

then any quadric from the corresponding confocal family is given by an equation of the form

$$\mathcal{Q}_\lambda : \ Q_\lambda(x) = 1. \tag{7.1}$$

The famous Chasles theorem (see Corollary 4.12) states that any line in the space $\mathbf{R}^d$ is tangent to exactly $d - 1$ quadrics from a given confocal family. The next lemma gives an important condition on these quadrics.

Lemma 7.1. *Suppose a line ℓ is tangent to quadrics $\mathcal{Q}_{\alpha_1}, \ldots, \mathcal{Q}_{\alpha_{d-1}}$ from the family (7.1). Then Jacobian coordinates $(\lambda_1, \ldots, \lambda_d)$ of any point on ℓ satisfy the inequalities $\mathcal{P}(\lambda_s) \geq 0$, $s = 1, \ldots, d$, where*

$$\mathcal{P}(x) = (a_1 - x) \ldots (a_d - x)(\alpha_1 - x) \ldots (\alpha_{d-1} - x).$$

Proof. Let x be a point of ℓ, $(\lambda_1, \ldots, \lambda_d)$ its Jacobian coordinates, and y a vector parallel to ℓ. The equation $Q_\lambda(x + ty) = 1$ is quadratic with respect to t. Its discriminant is

$$\Phi_\lambda(x, y) = Q_\lambda(x, y)^2 - Q_\lambda(y)(Q_\lambda(x) - 1),$$

where

$$Q_\lambda(x, y) = \frac{x_1 y_1}{a_1 - \lambda} + \cdots + \frac{x_d y_d}{a_d - \lambda}.$$

By [Mos1980],

$$\Phi_\lambda(x, y) = \frac{(\alpha_1 - \lambda) \ldots (\alpha_{d-1} - \lambda)}{(a_1 - \lambda) \ldots (a_d - \lambda)}.$$

For each of the coordinates $\lambda = \lambda_s$, $(1 \leq s \leq d)$, the quadratic equation has a solution $t = 0$; thus, the corresponding discriminants are non-negative. This is obviously equivalent to $\mathcal{P}(\lambda_s) \geq 0$. $\qquad\square$

Billiard inside a Domain Bounded by Confocal Quadrics

Suppose that a bounded domain $\Omega \subset \mathbf{R}^d$ is given such that its boundary $\partial\Omega$ lies in the union of several quadrics from the family (7.1). Then, in elliptic coordinates, Ω is given by

$$\beta_1' \leq \lambda_1 \leq \beta_1'', \quad \ldots, \quad \beta_d' \leq \lambda_d \leq \beta_d'',$$

where $a_{s+1} \leq \beta_s' < \beta_s'' \leq a_s$ for $1 \leq s \leq d-1$ and $-\infty < \beta_d' < \beta_d'' \leq a_d$.

Consider a billiard system within Ω and let $\mathcal{Q}_{\alpha_1}, \ldots, \mathcal{Q}_{\alpha_{d-1}}$ be caustics of one of its trajectories. For any $s = 1, \ldots, d$, the set Λ_s of all values taken by the coordinate λ_s on the trajectory is, according to Lemma 7.1, included in $\Lambda_s' = \{\lambda \in [\beta_s', \beta_s''] : \mathcal{P}(\lambda) \geq 0\}$. By [Knö1980], each of the intervals (a_{s+1}, a_s), $(2 \leq s \leq d)$ contains at most two of the values $\alpha_1, \ldots, \alpha_{d-1}$, the interval $(-\infty, a_d)$ contains at most one of them, while none is included in $(a_1, +\infty)$. Thus, for each s, the following three cases are possible:

First case: $\alpha_i, \alpha_j \in [\beta_s', \beta_s'']$, $\alpha_i < \alpha_j$. Since any line which contains a segment of the trajectory touches $\mathcal{Q}_{\alpha_i}$ and $\mathcal{Q}_{\alpha_j}$, the whole trajectory is placed between these two quadrics. The elliptic coordinate λ_s has critical values at points where the trajectory touches one of them, and remains monotonic elsewhere. Hence, meeting points with with $\mathcal{Q}_{\alpha_i}$ and $\mathcal{Q}_{\alpha_j}$ are placed alternately along the trajectory and $\Lambda_s = \Lambda_s' = [\alpha_i, \alpha_j]$.

Second case: Among $\alpha_1, \ldots, \alpha_{d-1}$, only α_i is in $[\beta_s', \beta_s'']$. $\mathcal{P}$ is non-negative in exactly one of the intervals: $[a_{s+1}, \alpha_i]$, $[\alpha_i, a_s]$, let us take in the first one. Then the trajectory has bounces only on $\mathcal{Q}_{\beta_s'}$. If $\alpha_i \neq \beta_s''$, the billiard particle never reaches the boundary $\mathcal{Q}_{\beta_s''}$. The coordinate λ_s has critical values at meeting points with $\mathcal{Q}_{\beta_s'}$ and the caustic $\mathcal{Q}_{\alpha_i}$, and remains monotonic elsewhere. Hence, $\Lambda_s = \Lambda_s' = [\beta_s', \alpha_i]$. If $\mathcal{P}$ is non-negative in $[\alpha_i, a_s]$, then we obtain $\Lambda_s = \Lambda_s' = [\alpha_i, \beta_s'']$.

Third case: The segment $[\beta_s', \beta_s'']$ does not contain any of values $\alpha_1, \ldots, \alpha_{d-1}$. Then $\mathcal{P}$ is non-negative in $[\beta_s', \beta_s'']$. The coordinate λ_s has critical values only at meeting points with boundary quadrics $\mathcal{Q}_{\beta_s'}$ and $\mathcal{Q}_{\beta_s''}$, and changes monotonically between them. This implies that the billiard particle bounces off them alternately. Obviously, $\Lambda_s = \Lambda_s' = [\beta_s', \beta_s'']$.

Let $[\gamma_s', \gamma_s''] := \Lambda_s = \Lambda_s'$. Notice that the trajectory meets quadrics of any pair $\mathcal{Q}_{\gamma_s'}$, $\mathcal{Q}_{\gamma_s''}$ alternately. Thus, any periodic trajectory has the same number of intersection points with each of them.

Let us make a few remarks on the case when $\gamma_s' = a_{s+1}$ or $\gamma_s'' = a_s$. This means that either a part of $\partial\Omega$ is a degenerate quadric from the confocal family or Ω is not bounded, from one side at least, by a quadric of the corresponding type. Discussion of the case when Ω is bounded by a coordinate hyperplane does not differ from the one we have just made. On the other hand, non-existence of a part of the boundary means that the coordinate λ_s will have extreme values at the points of intersection of the trajectory with the corresponding hyperplane.

Since a closed trajectory intersects any hyperplane an even number of times, it follows that the coordinate λ_s is taking each of its extreme values an even number of times during the period.

Theorem 7.2. *A trajectory of the billiard system within Ω with caustics $\mathcal{Q}_{\alpha_1}, \ldots,$ $\mathcal{Q}_{\alpha_{d-1}}$ is periodic with exactly n_s points at $\mathcal{Q}_{\gamma'_s}$ and n_s points at $\mathcal{Q}_{\gamma''_s}$ ($1 \leq s \leq d$) if and only if*

$$\sum_{s=1}^{d} n_s \left(\mathcal{A}(P_{\gamma'_s}) - \mathcal{A}(P_{\gamma''_s}) \right) = 0$$

on the Jacobian of the curve

$$\Gamma \; : \; y^2 = \mathcal{P}(x) := (a_1 - x) \cdots (a_d - x)(\alpha_1 - x) \cdots (\alpha_{d-1} - x).$$

Here, $\mathcal{A}$ denotes the Abel-Jacobi map, where $P_{\gamma'_s}$, $P_{\gamma''_s}$ are points on Γ with coordinates $P_{\gamma'_s} = \left(\gamma'_s, (-1)^s \sqrt{\mathcal{P}(\gamma'_s)} \right)$, $P_{\gamma''_s} = \left(\gamma''_s, (-1)^s \sqrt{\mathcal{P}(\gamma''_s)} \right)$.

Proof. Following Jacobi [Jac1884a] and Darboux [Dar1914], let us consider the equations

$$\sum_{s=1}^{d} \frac{d\lambda_s}{\sqrt{\mathcal{P}(\lambda_s)}} = 0, \quad \sum_{s=1}^{d} \frac{\lambda_s d\lambda_s}{\sqrt{\mathcal{P}(\lambda_s)}} = 0, \quad \cdots, \quad \sum_{s=1}^{d} \frac{\lambda_s^{d-2} d\lambda_s}{\sqrt{\mathcal{P}(\lambda_s)}} = 0, \qquad (7.2)$$

where, for any fixed s, the square root $\sqrt{\mathcal{P}(\lambda_s)}$ is taken with the same sign in all of the expressions. Then (7.2) represents a system of differential equations of a line tangent to $\mathcal{Q}_{\alpha_1}, \ldots, \mathcal{Q}_{\alpha_{d-1}}$. Besides that,

$$\sum_{s=1}^{d} \frac{\lambda_s^{d-1} d\lambda_s}{\sqrt{\mathcal{P}(\lambda_s)}} = 2d\ell, \qquad (7.3)$$

where $d\ell$ is the element of the line length.

Attributing all possible combinations of signs to $\sqrt{\mathcal{P}(\lambda_1)}, \ldots, \sqrt{\mathcal{P}(\lambda_d)}$, we can obtain 2^{d-1} non-equivalent systems (7.2), which correspond to 2^{d-1} different tangent lines to $\mathcal{Q}_{\alpha_1}, \ldots, \mathcal{Q}_{\alpha_{d-1}}$ from a generic point of the space. Moreover, the systems corresponding to a line and its reflection to a given hypersurface $\lambda_s = \mathrm{const}$ differ from each other only in signs of the roots $\sqrt{\mathcal{P}(\lambda_s)}$.

Solving (7.2) and (7.3) as a system of linear equations with respect to

$$\frac{d\lambda_s}{\sqrt{\mathcal{P}(\lambda_s)}},$$

we obtain

$$\frac{d\lambda_s}{\sqrt{\mathcal{P}(\lambda_s)}} = \frac{2d\ell}{\prod_{i \neq s}(\lambda_s - \lambda_i)}.$$

Thus, along a billiard trajectory, the differentials $(-1)^{s-1}\dfrac{d\lambda_s}{\sqrt{\mathcal{P}(\lambda_s)}}$ stay always positive, if we assume that the signs of the square roots are chosen appropriately on each segment.

From these remarks and the discussion preceding this theorem, it follows that the value of the integral $\int\dfrac{\lambda_s^i d\lambda_s}{\sqrt{\mathcal{P}(\lambda_s)}}$ between two consecutive common points of the trajectory and the quadric $\mathcal{Q}_{\gamma_s'}$ (or $\mathcal{Q}_{\gamma_s''}$) is equal to

$$2(-1)^{s-1}\int_{\gamma_s'}^{\gamma_s''}\frac{\lambda_s^i d\lambda_s}{+\sqrt{\mathcal{P}(\lambda_s)}}.$$

Now, if $\mathbf{p}$ is a finite polygon representing a billiard trajectory and having exactly n_s points at $\mathcal{Q}_{\gamma_s'}$ and n_s at $\mathcal{Q}_{\gamma_s''}$ $(1 \le s \le d)$, then

$$\sum\int^{\mathbf{P}}\frac{\lambda_s^i d\lambda_s}{\sqrt{\mathcal{P}(\lambda_s)}} = 2\sum(-1)^{s-1}n_s\int_{\gamma_s'}^{\gamma_s''}\frac{\lambda_s^i d\lambda_s}{+\sqrt{\mathcal{P}(\lambda_s)}}, \quad (1 \le i \le d).$$

Finally, the polygonal line is closed if and only if

$$\sum(-1)^s n_s\int_{\gamma_s'}^{\gamma_s''}\frac{\lambda_s^i d\lambda_s}{\sqrt{\mathcal{P}(\lambda_s)}} = 0, \quad (1 \le i \le d-1),$$

which was needed. $\qquad\qquad\qquad\qquad\qquad\qquad\qquad\qquad\qquad\qquad\qquad\qquad\square$

Example 7.3. Consider two domains Ω' and Ω'' in $\mathbf{R}^3$. Let Ω' be bounded by the ellipsoid $\mathcal{Q}_0$ and the two-folded hyperboloid $\mathcal{Q}_\beta$, $a_2 < \beta < a_1$, in such a way that Ω' is placed between the branches of $\mathcal{Q}_\beta$. On the other hand, suppose Ω'' is bounded by $\mathcal{Q}_0$, the right-hand branch of $\mathcal{Q}_\beta$ (this one which is placed in the half-space $x_1 > 0$) and the plane $x_3 = 0$. Elliptic coordinates of points inside both Ω' and Ω'' satisfy

$$0 \le \lambda_3 \le a_3, \quad \beta \le \lambda_1 \le a_1.$$

Consider billiard trajectories within these two domains, with caustics $\mathcal{Q}_{\mu_1}$ and $\mathcal{Q}_{\mu_2}$, $a_3 < \mu_1 < a_2$, $a_2 < \mu_2 < a_1$. Since $\mu_2 \le \beta$, the segments Λ_s $(s \in \{1,2,3\})$ of all possible values of elliptic coordinates along a trajectory, for both domains, are

$$\Lambda_1 = [\beta, a_1], \quad \Lambda_2 = [\mu_1, a_2], \quad \Lambda_3 = [0, a_3].$$

In Ω'', existence of a periodic trajectory with caustics $\mathcal{Q}_{\mu_1}$ and $\mathcal{Q}_{\mu_2}$, which becomes closed after n bounces at $\mathcal{Q}_0$ and $2m$ bounces at $\mathcal{Q}_\beta$, is equivalent to the equality

$$n\big(\mathcal{A}(P_0) - \mathcal{A}(P_{a_3})\big) + 2m\big(\mathcal{A}(P_\beta) - \mathcal{A}(P_{\mu_1})\big) = 0,$$

on the Jacobian of the corresponding hyperelliptic curve. In Ω', existence of a trajectory with the same properties is equivalent to

$$n\big(\mathcal{A}(P_0) - \mathcal{A}(P_{a_3})\big) + 2m\big(\mathcal{A}(P_\beta) - \mathcal{A}(P_{\mu_1})\big) = 0 \quad \text{and} \quad n \ \text{is even}.$$

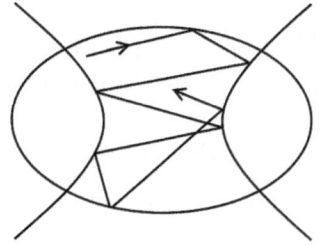 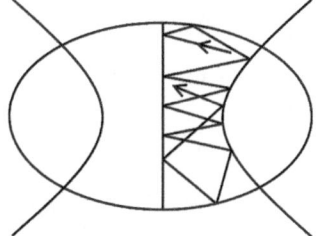

Figure 7.1: A trajectory inside Ω' Figure 7.2: A trajectory inside Ω''

The fact that the second equality implies the first one is due to the following geometrical fact: any billiard trajectory in Ω' can be transformed to a trajectory in Ω'' applying the symmetry with respect to the x_2x_3-plane to all its points placed under this plane. Notice that this correspondence of trajectories is 2 to 1 – in such a way, a generic billiard trajectory in Ω'' corresponds to exactly two trajectories in Ω'. An example of such corresponding billiard trajectories is shown on Figures 7.1 and 7.2.

Billiard Ordered Game

Our next step is to introduce a notion of bounces "from outside" and "from inside". More precisely, let us consider an ellipsoid $\mathcal{Q}_\lambda$ from the confocal family (7.1) such that $\lambda \in (a_{s+1}, a_s)$ for some $s \in \{1, \ldots, d\}$, where $a_{d+1} = -\infty$.

Observe that along a billiard ray which reflects at $\mathcal{Q}_\lambda$, the elliptic coordinate λ_i has a local extremum at the point of reflection.

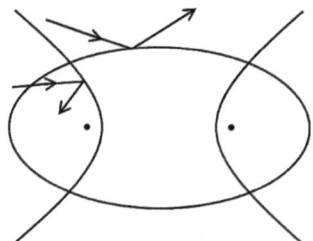 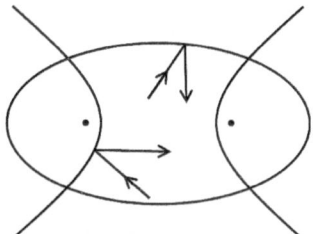

Figure 7.3: Reflection from outside   Figure 7.4: Reflection from inside

Definition 7.4. A ray reflects *from outside* at the quadric $\mathcal{Q}_\lambda$ if the reflection point is a local maximum of the Jacobian coordinate λ_s, and it reflects *from inside* if the reflection point is a local minimum of the coordinate λ_s.

On Figures 7.3 and 7.4 reflections from inside and outside an ellipse and a hyperbola are sketched.

Let us remark that in the case when $\mathcal{Q}_\lambda$ is an ellipsoid, the notions introduced in Definition 7.4 coincide with the usual ones.

Assume now a k-tuple of confocal quadrics $\mathcal{Q}_{\beta_1}, \ldots, \mathcal{Q}_{\beta_k}$ from the confocal pencil (7.1) is given, and $(i_1, \ldots, i_k) \in \{-1, 1\}^k$.

Definition 7.5. *The billiard ordered game joined to quadrics* $\mathcal{Q}_{\beta_1}, \ldots, \mathcal{Q}_{\beta_k}$, with the *signature* $(i_1, \ldots, i_k)$ is the billiard system with trajectories having bounces at $\mathcal{Q}_{\beta_1}, \ldots, \mathcal{Q}_{\beta_k}$ respectively, such that

the reflection at $\mathcal{Q}_{\beta_s}$ is from inside if $i_s = +1$,

the reflection at $\mathcal{Q}_{\beta_s}$ is from outside if $i_s = -1$.

Note that any trajectory of a billiard ordered game has $d - 1$ caustics from the same family (7.1).

Suppose $\mathcal{Q}_{\beta_1}, \ldots, \mathcal{Q}_{\beta_k}$ are ellipsoids and consider a billiard ordered game with the signature $(i_1, \ldots, i_k)$. In order that trajectories of such a game stay bounded, the following condition has to be satisfied:

$$i_s = -1 \ \Rightarrow \ i_{s+1} = i_{s-1} = 1 \text{ and } \beta_{s+1} < \beta_s, \ \beta_{s-1} < \beta_s.$$

(Here, we identify indices 0 and $k + 1$ with k and 1 respectively.)

Example 7.6. On Figure 7.5, a trajectory corresponding to the 7-tuple

$$(\mathcal{Q}_1, \mathcal{Q}_2, \mathcal{Q}_1, \mathcal{Q}_3, \mathcal{Q}_2, \mathcal{Q}_3, \mathcal{Q}_1),$$

with the signature $(1, -1, 1, -1, 1, 1, 1)$, is shown.

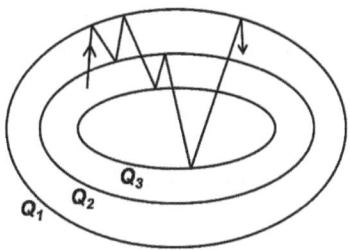

Figure 7.5: Billiard ordered game

Theorem 7.7. *Given a billiard ordered game within k ellipsoids* $\mathcal{Q}_{\beta_1}, \ldots, \mathcal{Q}_{\beta_k}$ *with the signature* $(i_1, \ldots, i_k)$. *Its trajectory with caustics* $\mathcal{Q}_{\alpha_1}, \ldots, \mathcal{Q}_{\alpha_{d-1}}$ *is k-periodic if and only if*

$$\sum_{s=1}^{k} i_s \left(\mathcal{A}(P_{\beta_s}) - \mathcal{A}(P_\alpha) \right)$$

is equal to a sum of several expressions of the form: $\left(\mathcal{A}(P_{\alpha_p}) - \mathcal{A}(P_{\alpha_{p'}})\right)$ *on the Jacobian of the curve* $\Gamma : y^2 = \mathcal{P}(x)$, *where* $P_{\beta_s} = \left(\beta_s, +\sqrt{\mathcal{P}(\beta_s)}\right)$, $\alpha = \min\{a_d, \alpha_1, \ldots, \alpha_{d-1}\}$ *and* $\mathcal{Q}_{\alpha_p}$, $\mathcal{Q}_{\alpha_{p'}}$ *are pairs of caustics of the same type.*

When $\mathcal{Q}_{\beta_1} = \cdots = \mathcal{Q}_{\beta_k}$ and $i_1 = \cdots = i_k = 1$ we obtain the Cayley-type condition for the billiard motion inside an ellipsoid in $\mathbf{R}^d$.

We are going to treat in more detail the case of the billiard motion between two ellipsoids.

Proposition 7.8. *The condition that there exists a closed billiard trajectory between two ellipsoids* $\mathcal{Q}_{\beta_1}$ *and* $\mathcal{Q}_{\beta_2}$, *which bounces exactly* m *times to each of them, with caustics* $\mathcal{Q}_{\alpha_1}, \ldots, \mathcal{Q}_{\alpha_{d-1}}$, *is*

$$
\text{rank} \begin{pmatrix}
f_1'(P_{\beta_2}) & f_2'(P_{\beta_2}) & \cdots & f_{m-d+1}'(P_{\beta_2}) \\
f_1''(P_{\beta_2}) & f_2''(P_{\beta_2}) & \cdots & f_{m-d+1}''(P_{\beta_2}) \\
& & \cdots & \\
& & \cdots & \\
f_1^{(m-1)}(P_{\beta_2}) & f_2^{(m-1)}(P_{\beta_2}) & \cdots & f_{m-d+1}^{(m-1)}(P_{\beta_2})
\end{pmatrix} < m - d + 1.
$$

Here

$$
f_j = \frac{y - B_0 - B_1(x - \beta_1) - \cdots - B_{d+j-2}(x - \beta_1)^{d+j-2}}{x^{d+j-1}}, \quad 1 \le j \le m - d + 1,
$$

and $y = B_0 + B_1(x - \beta_1) + \cdots$ *is the Taylor expansion around the point symmetric to* P_{β_1} *with respect to the hyperelliptic involution of the curve* Γ. (*All notations are as in Theorem 7.7.*)

Example 7.9. Consider a billiard motion in the three-dimensional space, with ellipsoids $\mathcal{Q}_0$ and $\mathcal{Q}_\gamma$ as boundaries ($0 < \gamma < a_3$) and caustics $\mathcal{Q}_{\alpha_1}$ and $\mathcal{Q}_{\alpha_2}$. Such a motion closes after four bounces from inside to $\mathcal{Q}_0$ and four bounces from outside to $\mathcal{Q}_\gamma$ if and only if

$$\text{rank } X < 2.$$

The matrix X is given by

$$
\begin{aligned}
X_{11} &= -3C_0 + C_1\gamma + 3B_0 + 2B_1\gamma + B_2\gamma^2, \\
X_{12} &= -4C_0 + C_1\gamma + 4B_0 + 3B_1\gamma + 2B_2\gamma^2 + B_3\gamma^3, \\
X_{21} &= 6C_0 - 3C_1\gamma + C_2\gamma^2 - 6B_0 - 3B_1\gamma - B_2\gamma^2, \\
X_{22} &= 10C_0 - 4C_1\gamma - 10B_0 - 6B_1\gamma - 3B_2\gamma^2 - B_3\gamma^3, \\
X_{31} &= -10C_0 + 6C_1\gamma - 3C_2\gamma^2 + C_3\gamma^3 + 10B_0 + 4B_1\gamma + B_2\gamma^2, \\
X_{32} &= -20C_0 - 10C_1\gamma - 4C_2\gamma^2 + C_3\gamma^3 + 20B_0 + 10B_1\gamma + 4B_2\gamma^2 + B_3\gamma^3,
\end{aligned}
$$

and the expressions

$$-\sqrt{(a_1 - x)(a_2 - x)(a_3 - x)(\alpha_1 - x)(\alpha_2 - x)} = B_0 + B_1 x + B_2 x^2 + \cdots,$$
$$+\sqrt{(a_1 - x)(a_2 - x)(a_3 - x)(\alpha_1 - x)(\alpha_2 - x)} = C_0 + C_1(x - \gamma) + \cdots$$

are Taylor expansions around points $x = 0$ and $x = \gamma$ respectively.

Example 7.10. Using the same notation as in the previous example, let us consider trajectories with four bounces from inside to each of Q_0 and Q_γ, as shown on Figure 7.6.

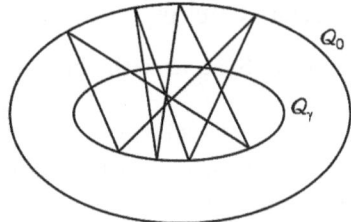

Figure 7.6: A closed trajectory of the billiard ordered game with eight alternate bounces from inside of two ellipses

The explicit condition for periodicity of such a trajectory is

$$\text{rank } X < 2,$$

with

$$X_{11} = -4C_0 + C_1\gamma + 3B_1\gamma + 2B_2\gamma^2 + B_3\gamma^3,$$
$$X_{12} = -3C_0 + C_1\gamma + 3B_0 + 2B_1\gamma + B_2\gamma^2,$$
$$X_{21} = -6C_0 + C_2\gamma^2 + 6B_0 + 6B_1\gamma + 5B_2\gamma^2 + 3B_3\gamma^3,$$
$$X_{22} = -6C_0 + C_1\gamma + C_2\gamma^2 + 6B_0 + 5B_1\gamma + 3B_2\gamma^2,$$
$$X_{31} = -4C_0 + C_3\gamma^3 + 4B_0 + 4B_1\gamma + 4B_2\gamma^2 + 3B_3\gamma^3,$$
$$X_{32} = -4C_0 + C_2\gamma^2 + C_3\gamma^3 + 4B_0 + 4B_1\gamma + 3B_2\gamma^2.$$

7.2 Ellipsoidal billiard as a discrete time system

Another approach to the description of periodic billiard trajectories is based on the technique of discrete Lax representation.

In this section, first, we are going to list the main steps of algebro-geometric integration of the elliptic billiard, following [MV1991]. Then, the connection between periodic billiard trajectories and points of finite order on the corresponding hyperelliptic curve will be established, and, using results from Section 3.8, the Cayley-type conditions will be derived, as they were obtained by the authors in [DR1998a, DR1998b]. In addition, here we provide a more detailed discussion concerning trajectories with the period not greater than d and the cases of the singular isospectral curve.

Following [MV1991], the billiard system will be considered as a system with discrete time. Using its integration procedure, the connection between periodic billiard trajectories and points of finite order on the corresponding hyperelliptic curve will be established.

XYZ Model and Isospectral Curves

Elliptical Billiard as a Mechanical System with the Discrete Time. Let the ellipsoid in $\mathbf{R}^d$ be given by

$$(Ax, x) = 1.$$

We can assume that A is a diagonal matrix, with different eigenvalues. The billiard motion within the ellipsoid is determined by the following equations:

$$x_{k+1} - x_k = \mu_k y_{k+1},$$
$$y_{k+1} - y_k = \nu_k A x_k,$$

where

$$\mu_k = -\frac{2(Ay_{k+1}, x_k)}{(Ay_{k+1}, y_{k+1})}, \qquad \nu_k = -\frac{2(Ax_k, y_k)}{(Ax_k, Ax_k)}.$$

Here, x_k is a sequence of points of billiard bounces, while

$$y_k = \frac{x_k - x_{k-1}}{|x_k - x_{k-1}|}$$

are the momenta.

Connection between Billiard and XYZ Model. To the billiard system with discrete time, Heisenberg's XYZ model can be joined, in the way described by Veselov and Moser in [MV1991] and which is going to be presented here.

Consider the mapping $\varphi : (x, y) \mapsto (x', y')$ given by

$$x'_k = Jy_{k+1} = J(y_k + \nu_k A x_k), \quad y'_k = -J^{-1}x_k, \quad J = A^{-\frac{1}{2}}.$$

Notice that the dynamics of φ contains the billiard dynamics

$$x''_k = Jy'_{k+1} = -x_{k+1}, \quad y''_k = -J^{-1}x'_k = -y_{k+1},$$

and define the sequence $(\bar{x}_k, \bar{y}_k)$ as

$$(\bar{x}_0, \bar{y}_0) := (x_0, y_0), \quad (\bar{x}_{k+1}, \bar{y}_{k+1}) := \varphi(\bar{x}_k, \bar{y}_k),$$

which obeys the relations

$$\bar{x}_{k+1} = J\bar{y}_k + \nu_k J^{-1}\bar{x}_k, \quad \bar{y}_{k+1} = -J^{-1}\bar{x}_k,$$

where the parameter ν_k is such that $|\bar{y}_k| = 1$, $(A\bar{x}_k, \bar{x}_x) = 1$. This can be rewritten as

$$\bar{x}_{k+1} + \bar{x}_{k-1} = \nu_k J^{-1} \bar{x}_k.$$

Now, for the sequence $q_k := J^{-1}\bar{x}_k$, we have

$$q_{k+1} + q_{k-1} = \nu_k J^{-1} q_k, \quad |q_k| = 1.$$

These equations represent the equations of the discrete Heisenberg XYZ system.

Theorem 7.11. [MV1991] *Let $(\bar{x}_k, \bar{y}_k)$ be the sequence connected with an elliptical billiard in the described way. Then $q_k = J^{-1}\bar{x}_k$ is a solution of the discrete Heisenberg system.*

Conversely, if q_k is a solution to the Heisenberg system, then the sequence $x_k = (-1)^k J q_{2k}$ is a trajectory of the discrete billiard within an ellipsoid.

Integration of the Discrete Heisenberg XYZ System. The usual scheme of algebro-geometric integration contains the following [MV1991]. First, the sequence $L_k(\lambda)$ of matrix polynomials has to be determined, together with a factorization

$$L(\lambda) = B(\lambda)C(\lambda) \mapsto C(\lambda)B(\lambda) = B'(\lambda)C'(\lambda) = L'(\lambda),$$

such that the dynamics $L \mapsto L'$ corresponds to the dynamics of the system q_k. For each problem, finding this sequence of matrices requires a separate search and a mathematician with excellent intuition. All matrices L_k are mutually similar, and they determine the same *isospectral curve*

$$\Gamma \ : \ \det(L(\lambda) - \mu I) = 0.$$

The factorization $L_k = B_k C_k$ gives splitting of the spectrum of L_k. Denote by ψ_k the corresponding eigenvectors, $\psi_k \in \mathbf{P}\mathcal{M}(\Gamma)^d$, where $\mathcal{M}(\Gamma)$ is the set of all meromorphic functions on Γ. Denote by D_k the pole divisor of ψ_k.

The sequence of divisors is linear on the Jacobian of the isospectral curve, and this enables us to find, conversely, eigenfunctions ψ_k, then matrices L_k, and, finally, the sequence (q_k).

Now, integration of the discrete XYZ system by this method will be shortly presented. Details of the procedure can be found in [MV1991].

The equations of discrete XYZ model are equivalent to the isospectral deformation

$$L_{k+1}(\lambda) = A_k(\lambda)L_k(\lambda)A_k^{-1}(\lambda),$$

where

$$L_k(\lambda) = J^2 + \lambda q_{k-1} \wedge J q_k - \lambda^2 q_{k-1} \otimes q_{k-1},$$
$$A_k(\lambda) = J - \lambda q_k \otimes q_{k-1}.$$

The equation of the isospectral curve Γ : $\det(L(\lambda) - \mu I) = 0$ can be written in the form

$$\nu^2 = \prod_{i=1}^{d-1}(\mu - \mu_i) \prod_{j=1}^{d}(\mu - J_j^2), \tag{7.4}$$

where $\nu = \lambda \prod_{i=1}^{d-1}(\mu - \mu_i)$ and $\mu_1, \ldots, \mu_{d-1}$ are zeroes of the function

$$\phi_\mu(x, Jy) = \sum_{i=1}^{d} \frac{F_i(x, y)}{\mu - J_i^2},$$

$$F_i = x_i^2 + \sum_{j \neq i} \frac{(x \wedge Jy)_{ij}^2}{J_i^2 - J_j^2}, \quad x = q_{k-1}, \quad y = q_k.$$

It can be proved that $\mu_1, \ldots, \mu_{d-1}$ are parameters of the caustics corresponding to the billiard trajectory [Mos1980]. Another way for obtaining the same conclusion is to calculate them directly by taking the first segment of the billiard trajectory to be parallel to a coordinate axis.

If eigen-vectors ψ_k of matrices $L_k(\lambda)$ are known, it is possible to determine uniquely members of the sequence q_k. Let D_k be the divisor of poles of function ψ_k on curve Γ. Then [MV1991]

$$D_{k+1} = D_k + P_\infty - P_0,$$

where P_∞ is the point corresponding to the value $\mu = \infty$ and P_0 to $\mu = 0$, $\lambda = (q_k, J^{-1}q_{k+1})^{-1}$.

Remark that, if transformation T is applied to the system, such that $[T, J] = 0$, $T \in O(d)$, then we obtain $L'_k = TL_kT^{-1}$, $\psi'_k = T\psi_k$, $D'_k = D_k$, and $T = \mathrm{diag}\,(\pm 1, \ldots, \pm 1)$.

Characterization of Periodical Billiard Trajectories

In the next lemmata, we establish a connection between periodic billiard sequences q_k and periodic divisors D_k.

Lemma 7.12. [DR1998a] *A sequence ψ_k is n-periodic if and only if the sequence q_k is also periodic with the period n or $q_{k+n} = -q_k$ for all k.*

Proof. If, for all k, $q_{k+n} = q_k$, or $q_{k+n} = -q_k$ for all k, then, obviously, $L_{k+n} = L_k$. Thus, the sequence of eigenvectors ψ_k is periodic with the period n. It follows that the sequence of divisors is also periodic.

Now suppose that $\psi_{k+n} = c_k \psi_k$. We have

$$\psi_{k+1} = A_k(\lambda)\psi_k.$$

Let μ_1 and μ_2 be values of parameter μ which correspond to the value $\lambda = 1$ on the curve Γ, and
$$\Psi_k = (\psi_k(1, \mu_1), \psi_k(1, \mu_2)).$$
From $\psi_{k+1} = A_k(\lambda)\psi_k$, we obtain $A_k(1) = \Psi_{k+1}\Psi_k^{-1}$. It follows that
$$A_k(1) = \frac{c_{k+1}}{c_k} A_{k+n}(1).$$
From the condition $\det A_k = \det A_{k+1}$ for all k, we have $c_k = c_{k+1}$. Thus, the sequence
$$A_k(1) = J - q_k \otimes q_{k-1}$$
is n-periodic. From there,
$$q_{k+n} = \alpha_k q_k, \quad q_{k+n-1} = \frac{1}{\alpha_k} q_{k-1}.$$
Since $|q_k| = 1$, we have $\alpha_k = 1$ or $\alpha_k = -1$, where all α_k are equal to each other, which proves the assertion. $\qquad\square$

Lemma 7.13. [DR1998a] *A billiard trajectory is, up to the central symmetry, periodic with period n if and only if the sequence of functions functions ψ_k joined to the corresponding Heisenberg XYZ system is also periodic, with period $2n$.*

Proof. Let $x_{k+n} = \alpha x_k$ for all k, $\alpha \in \{-1, 1\}$. Join to a billiard trajectory (x_k, y_k) the corresponding flow $(\bar{x}_k, \bar{y}_k)$. Since
$$(\bar{x}_{2k}, \bar{y}_{2k}) = (-1)^k(x_k, y_k), \quad (\bar{x}_{2k+1}, \bar{y}_{2k+1}) = \phi(\bar{x}_{2k}, \bar{y}_{2k}),$$
where the mapping ϕ is linear, we obtain
$$\bar{x}_{k+2n} = \alpha(-1)^n \bar{x}_k.$$

From there, it immediately follows that $q_{k+2n} = \alpha(-1)^n q_k$. According to Lemma 7.12, the divisor sequence D_k is $2n$-periodic. $\qquad\square$

Applying the previous lemma, we obtain the main statement of this section:

Theorem 7.14. [DR1998b] *A condition on a billiard trajectory inside ellipsoid $\mathcal{Q}_0$ in $\mathbf{R}^d$, with non-degenerate caustics $\mathcal{Q}_{\mu_1}, \ldots, \mathcal{Q}_{\mu_{d-1}}$, to be periodic, up to the central symmetry (and the symmetries of the $L - A$ pair), with the period $n \geq d$ is*

$$\mathrm{rank} \begin{pmatrix} B_{n+1} & B_n & \cdots & B_{d+1} \\ B_{n+2} & B_{n+1} & \cdots & B_{d+2} \\ \cdots\cdots\cdots\cdots\cdots\cdots\cdots \\ B_{2n-1} & B_{2n-2} & \cdots & B_{n+d-1} \end{pmatrix} < n - d + 1,$$

where

$$\sqrt{(x - \mu_1) \cdots (x - \mu_{d-1})(x - a_1) \cdots (x - a_d)} = B_0 + B_1 x + B_2 x^2 + \cdots.$$

Proof. The trajectory is periodic with period n if, by Lemma 7.13, the corresponding divisor sequence on the curve Γ has the period $2n$, i.e., $2n(P_\infty - P_0) = 0$ on $\mathcal{J}(\Gamma)$. Curve Γ is hyperelliptic with genus $g = d - 1$. Taking $\mathcal{A}(P_\infty)$ to be the neutral on $\mathcal{J}(\Gamma)$ we get the desired result by applying Lemma 3.140.   □

Cases of Singular Isospectral Curve.   When all $a_1, \ldots, a_d, \mu_1, \ldots, \mu_{d-1}$ are mutually different, then the isospectral curve has no singularities in the affine part. However, singularities appear in the following three cases and their combinations:

(i) $a_i = \mu_j$ for some i, j. The isospectral curve (7.4) decomposes into a rational and a hyperelliptic curve. Geometrically, this means that the caustic corresponding to μ_i degenerates into the hyperplane $x_i = 0$. The billiard trajectory can be asymptotically tending to that hyperplane (and therefore cannot be periodic), or completely placed in this hyperplane. Therefore, closed trajectories appear when they are placed in a coordinate hyperplane. Such a motion can be discussed as in the case of dimension $d - 1$.

(ii) $a_i = a_j$ for some $i \neq j$. The boundary $\mathcal{Q}_0$ is symmetric.

(iii) $\mu_i = \mu_j$ for some $i \neq j$. The billiard trajectory is placed on the corresponding confocal quadric hypersurface.

In the cases (ii) and (iii) the isospectral curve Γ is a hyperelliptic curve with singularities. In spite of their different geometrical nature, they both need the same analysis of the condition $2nP_0 \sim 2nE$ for the singular curve (7.4).

An immediate consequence of Lemma 3.141 is that Theorem 7.14 can be applied not only for the case of the regular isospectral curve, but in the cases (ii) and (iii), too. Therefore, the following interesting property holds.

Theorem 7.15. *If the billiard trajectory within an ellipsoid in d-dimensional Eucledean space is periodic, up to the central symmetry, with period $n < d$, then it is placed in one of the n-dimensional planes of symmetry of the ellipsoid.*

Proof. This follows immediately from Theorem 7.14 and the fact that the section of a confocal family of quadrics with a coordinate hyperplane is again a confocal family.   □

Note that all trajectories periodic with period n up to the central symmetry, are closed after $2n$ bounces. A statement sharper than Theorem 7.15 for the trajectories closed after n bounces, where $n \leq d$, can be obtained in the elementary fashion:

Proposition 7.16. *If the billiard trajectory within an ellipsoid in d-dimensional Euclidean space is periodic with period $n \leq d$, then it is placed in one of the $(n-1)$-dimensional planes of symmetry of the ellipsoid.*

Proof. First, consider the case $n = d$. Let $x_1 \cdots x_d$ be a periodic trajectory, and $(N, x) = \alpha$ the equation of the hyperplane spanned by its vertices. Here, N is a vector normal to the hyperplane and α is a constant. Since all lines normal to the surface of the ellipsoid at the points of reflection belong to this hyperplane, it follows that $(AN, x_i) = (N, Ax_i) = 0$. Thus, $(AN, x) = 0$ is also an equation of the hyperplane, so $\alpha = 0$ and the vectors N, AN are collinear. From here, the claim follows immediately.

The case $n < d$ can be proved similarly, or applying Theorem 7.15 and the previous case of this proposition. $\qquad\square$

This property can be seen easily for $d = 3$.

Example 7.17. Consider the billiard motion in an ellipsoid in the three-dimensional space, with $\mu_1 = \mu_2$, when the segments of the trajectory are placed on generatrices of the corresponding one-folded hyperboloid, confocal to the ellipsoid. If there existed a periodic trajectory with period $n = d = 3$, the three bounces would have been coplanar, and the intersection of that plane and the quadric would have consisted of three lines, which is impossible. It is obvious that any periodic trajectory with period $n = 2$ is placed along one of the axes of the ellipsoid. So, there is no periodic trajectories contained in a confocal quadric surface, with period less than or equal to 3.

7.3 Poncelet's theorem and Cayley's condition in the Lobachevsky space

Veselov proved the integrability of the billiard system within an ellipsoid in the Lobachevsky space in [Ves1990]. He showed that its motion corresponds to certain translations of the Jacobian variety of some hyperelliptic curve and gave explicit formulae of the motion in terms of theta-functions. The oldest reference to Poncelet's theorem in Lobachevsky space we found was [MB1951].

The aim of this section is to present an analogue of the Poncelet and Cayley theorems for the billiard motion within an ellipsoid in the Lobachevsky space [DJR2003].

Integration of billiard motion in the Lobachevsky space

For a brief account of Veselov's results on a billiard in the Lobachevsky space [Ves1990], let us consider the $(d+1)$-dimensional Minkowski space $V = \mathbf{R}^{d,1}$ with the symmetric bilinear form

$$\langle \xi, \eta \rangle = -\xi_0 \eta_0 + \xi_1 \eta_1 + \cdots + \xi_d \eta_d.$$

One sheet of the hyperboloid $\langle \xi, \xi \rangle = -1$ with the induced metric is a model of the d-dimensional Lobachevsky space $\mathbf{H}^d$. An ellipsoid Γ in this space is determined by the equation

$$\mathcal{E} = \left\{ \xi \in \mathbf{H}^d, \ -\frac{\xi_0^2}{a_0} + \frac{\xi_1^2}{a_1} + \cdots + \frac{\xi_d^2}{a_d} = 0 \right\}, \tag{7.5}$$

with $a_0 > a_1 \geq a_2 \geq \cdots \geq a_d > 0$.

All segments of the billiard trajectory within this ellipsoid are tangent to $d-1$ confocal quadric surfaces (including multiplicity), fixed for a given trajectory (Theorem 3 in [Ves1990]). Denote by μ_i, $i = 1, \ldots, d-1$ the numbers such that equations of these caustics are

$$-\frac{x_0^2}{a_0 - \mu_i} + \frac{x_1^2}{a_1 - \mu_i} + \cdots + \frac{x_d^2}{a_d - \mu_i} = 0, \quad (1 \leq i \leq d-1). \tag{7.6}$$

Then the points of reflection from the boundary $\mathcal{E}$ correspond to the shift $D_{k+1} = D_k + Q_- - Q_+$ on the Jacobi variety of the isospectral curve Γ,

$$\Gamma : \quad (\mu - a_0) \ldots (\mu - a_d) = c \cdot \lambda^2 (\mu - \mu_1) \ldots (\mu - \mu_{d-1}), \tag{7.7}$$

where c is a constant, and Q_+, Q_- are the points on the curve Γ over $\mu = 0$. (See Theorem 2 of [Ves1990]. The curve $\mathcal{C}$ is the isospectral curve of the $L - A$ pair considered there.)

Let us note that Veselov considered only the case of the regular (hyperelliptic) curve $\mathcal{C}$ [Ves1990]. However, his consideration holds for the singular case, too.

Poncelet's Theorem and Cayley-type Conditions

Suppose a periodical billiard trajectory inside the ellipsoid $\mathcal{E}$ in the Lobachevsky space is given. All trajectories with the same caustics have the same isospectral curve. If the period of the given trajectory is n, then $n(Q_+ - Q_-) = 0$ on $\mathcal{J}(\mathcal{C})$, and vice-versa. Thus, all these trajectories close after n bounces. Therefore, a Poncelet-type theorem for the billiard in the Lobachevsky space is derived from Veselov's results:

Proposition 7.18. [DJR2003] *Suppose a periodical billiard trajectory inside an ellipsoid in the Lobachevsky space is given. Then any billiard trajectory which shares the same caustic quadrics is also periodical, with the same period.*

The analytical condition for periodical billiard trajectories, for both singular and non-singular isospectral curve, is stated as follows:

Theorem 7.19. [DJR2003] *The condition of a billiard trajectory inside the ellipsoid (7.5) in the d-dimensional Lobachevsky space, with non-degenerate caustics (7.6),* *to be periodic, up to central symmetry, with the period $n \geq d$ is*

$$\text{rank} \begin{bmatrix} B_{n+1} & B_n & \cdots & B_{d+1} \\ B_{n+2} & B_{n+1} & \cdots & B_{d+2} \\ \cdots & \cdots & \cdots & \cdots \\ B_{2n-1} & B_{2n-2} & \cdots & B_{n+d-1} \end{bmatrix} < n - d + 1,$$

where $\sqrt{(x - a_0) \ldots (x - a_d)(x - \mu_1) \ldots (x - \mu_{d-1})} = B_0 + B_1 x + B_2 x^2 + \cdots$. *There are no such trajectories with period less than d.*

Proof. Introducing new coordinates $x = \mu$, $y = \sqrt{c}\lambda(\mu - \mu_1) \ldots (\mu - \mu_{d-1})$, the isospectral curve (7.7) is transformed to

$$y^2 = (x - a_0) \ldots (x - a_d)(x - \mu_1) \ldots (x - \mu_{d-1}),$$

and we obtain the result by Lemma 3.142.   □

Cases of the singular isospectral curve can be discussed in the same way as in the Euclidean case (see the previous section). So, the complete analogue of Theorem 7.15 also holds in the Lobachevsky space:

Theorem 7.20. *[DJR2003] If the billiard trajectory within an ellipsoid in d-dimensional Lobachevsky space is periodic, up to central symmetry, with period $n < d$, then it is placed in one of the n-dimensional planes of symmetry of the ellipsoid.*

At the end of this section, it is interesting to note that the Cayley-type conditions obtained in the Lobachevsky case are of the completely same form as those for the billiard in the Euclidean space. This intriguing coincidence is explained by the trajectorial equivalence of elliptical billiards in these two geometries [DJR2003].

7.4 Topological properties of an elliptical billiard

The object of this section is to give a topological description of elliptical billiards using Fomenko graphs [DR2009] (see also [DR2010]). A detailed description of this kind of topological classification of integrable systems can be found in [BF2004, BMF1990, BO2006] and references therein, while a concise summary for a reader not acquainted with it is given here. Although plane elliptical billiards are quite well known, we may see that such a description will give a new and exciting insight into its properties – note, for example, appearance of a non-orientable periodic trajectory along one of the axes of the billiard boundary in Proposition 7.23, see Figure 7.10.

In the book [BF2004], one may find a large list of Fomenko graphs for known integrable systems, such as integrable cases of rigid body motion and integrable geodesic flows on surfaces. This paper provides an additional set of such examples.

Some other examples, connected with near-integrable dynamics can be found in [SRK2005] while bifurcations of Liouville foliations in a certain class of integrable systems with two degrees of freedom, were classified using Fomenko graphs in [RRK2008].

Let us note that topological properties of elliptical billiards, viewed as discrete dynamical systems, have been recently studied in [WD2002]. However, here we consider billiards as continuous systems, which enables us to use tools developed by Fomenko and his school.

Fomenko graphs

Now, we are going to give a concise description of the representation of isoenergy surfaces and some of their topological invariants by Fomenko graphs. For all details, see [BF1994, BO2006] and references therein.

Let $(\mathcal{M}, \omega, H, K)$ be an integrable system given on a four-dimensional surface with a symplectic form ω. Here H, K denote independent functions on $\mathcal{M}$, commuting with respect to the symplectic structure on $\mathcal{M}$.

Level sets are subsets of $\mathcal{M}$ given by equations $H = $ const, $K = $ const. The *Liouville foliation* of $\mathcal{M}$ is its decomposition into connected components of the level sets. A leaf of a foliation is *regular* if dH, dK are independent at each point, otherwise it is called *singular*. Singular leaves satisfying some additional conditions will be called *non-degenerate*. For these non-degeneracy conditions, see [BF1994].

We suppose that *isoenergy surface* $H = $ const are compact, thus each foliation leaf will also be compact. Thus, by the Arnol'd–Liouville theorem [Arn1978], each regular leaf is diffeomorphic to the two-dimensional torus $\mathbf{T}^2$ and the motion in its neighborhood is completely described by, for example, the action-angle coordinates. We say that isoenergy surfaces are *Liouville equivalent* if they are topologically conjugate and the homeomorphism preserves their Liouville foliations.

The set of topological invariants that describe completely isoenergy surfaces containing regular and non-degenerate leaves not containing fixed points of H, up to their Liouville equivalence, consists of:

- *The oriented graph G*, whose vertices correspond to the singular connected components of the level sets of K, and edges to one-parameter families of Liouville tori;

- *The collection of Fomenko atoms*, such that each atom marks exactly one vertex of the graph G;

- *The collection of pairs of numbers (r_i, ε_i)*, with $r_i \in ([0, 1) \cap \mathbf{Q}) \cup \{\infty\}$, $\varepsilon_i \in \{-1, 1\}$, $1 \leq i \leq n$. Here, n is the number of edges of the graph G and each pair (r_i, ε_i) marks an edge of the graph;

- *The collection of integers $n_1, n_2, \ldots, n_s$.* The numbers n_k correspond to certain connected components of the subgraph G^0 of G. G^0 consists of all vertices of G and the edges marked with $r_i = \infty$. The connected components marked with integers n_k are those that do not contain a vertex corresponding to an isolated critical circle (an **A** atom) on the manifold Q.

Let us clarify the meaning of these invariants.

First, we are going to describe the construction of the graph G from the manifold Q. Each singular leaf of the Liouville foliation corresponds to exactly one vertex of the graph. If we cut Q along such leaves, the manifold will fall apart into connected sets, each one consisting of a one-parameter family of Liouville tori. Each of these families is represented by an edge of the graph G. The vertex of G which corresponds to a singular leaf $\mathcal{L}$ is incident to the edge corresponding to the family $\mathcal{T}$ of tori if and only if $\partial \mathcal{T} \cap \mathcal{L}$ is nonempty.

Note that $\partial \mathcal{T}$ has two connected components, each corresponding to a singular leaf. If the two singular leaves coincide, then the edge creates a loop connecting one vertex to itself.

Now, when the graph is constructed, one needs to add the orientation to each edge. This may be done arbitrarily, but, once determined, the orientation must stay fixed because the values of numerical Fomenko invariants depend on it.

The Fomenko atom which corresponds to a singular leaf $\mathcal{L}$ of a singular level set is determined by the topological type of the set $\mathcal{L}_\varepsilon$. The set $\mathcal{L}_\varepsilon \supset \mathcal{L}$ is the connected component of $\{\, p \in Q \mid c - \varepsilon < K(p) < c + \varepsilon \,\}$, where $c = K(\mathcal{L})$, and $\varepsilon > 0$ is such that c is the only critical value of the function K on Q in interval $(c - \varepsilon, c + \varepsilon)$.

Let $\mathcal{L}_\varepsilon^+ = \{\, p \in \mathcal{L}_\varepsilon \mid K(p) > c \,\}$, $\mathcal{L}_\varepsilon^- = \{\, p \in \mathcal{L}_\varepsilon \mid K(p) < c \,\}$. Each of the sets $\mathcal{L}_\varepsilon^+$, $\mathcal{L}_\varepsilon^-$ is a union of several connected components, each component being a one-parameter family of Liouville tori. Each of these families corresponds to the beginning of an edge of the graph G incident to the vertex corresponding to $\mathcal{L}$.

Let us say a few words on the topological structure of the set $\mathcal{L}$. This set consists of at least one fixed point or closed one-dimensional orbit of the Poisson action Φ on $\mathcal{M}$ and several (possibly none) two-dimensional orbits of the action, which are called *separatrices*.

The trajectories on each of these two-dimensional separatrices is homoclinically or heteroclinically tending to the lower-dimensional orbits. The Liouville tori of each of the families in $\mathcal{L}_\varepsilon \setminus \mathcal{L} = \mathcal{L}_\varepsilon^+ \cup \mathcal{L}_\varepsilon^-$ tend, as the integral K approaches c, to a closure of a subset of the separatrix set.

Fomenko Atoms. Fomenko and his school completely described and classified non-degenerate leaves that do not contain fixed points of the Hamiltonian H.

If $\mathcal{L}$ is not an isolated critical circle, then a sufficiently small neighborhood of each one-dimensional orbit in $\mathcal{L}$ is isomorphic to either two cylinders intersecting along the base circle, and then the orbit is *orientable*, or to two Möbius bands intersecting each other along the joint base circle, then the one-dimensional orbit is *non-orientable*.

The number of closed one-dimensional orbits in $\mathcal{L}$ is called *the complexity* of the corresponding atom.

Fomenko atoms of complexity 1.   There are exactly three such atoms.

The atom A.   This atom corresponds to a normally elliptic singular circle, which is isolated on the isoenergy surface $\mathcal{Q}$. A small neighborhood of such a circle in $\mathcal{Q}$ is diffeomorphic to a solid torus. One of the sets $\mathcal{L}_\varepsilon^+$, $\mathcal{L}_\varepsilon^-$ is empty, the other one is connected. Thus only one edge of the graph G is incident with the vertex marked with the letter atom **A**.

The atom B.   In this case, $\mathcal{L}$ consists of one orientable normally hyperbolic circle and two two-dimensional separatrices – it is diffeomorphic to a direct product of the circle $\mathbf{S}^1$ and the plane curve given by the equation $y^2 = x^2 - x^4$. Because of its shape, we will refer to this curve as the *'figure eight'*. The set $\mathcal{L}_\varepsilon \setminus \mathcal{L}$ has three connected components, two of them being placed in $\mathcal{L}_\varepsilon^+$ and one in $\mathcal{L}_\varepsilon^-$, or vice versa. Let us fix that two of them are in $\mathcal{L}_\varepsilon^+$. Each of these two families of Liouville tori limits as K approaches c to only one of the separatrices. The tori in $\mathcal{L}_\varepsilon^-$ tend to the union of the separatrices.

The atom A^*. $\mathcal{L}$ consists of one non-orientable hyperbolic circle and one two-dimensional separatrix. It is homeomorphic to the smooth bundle over $\mathbf{S}^1$ with the 'figure eight' as fiber and the structural group consisting of the identity mapping and the central symmetry of the 'figure eight'. Both $\mathcal{L}_\varepsilon^+$ and $\mathcal{L}_\varepsilon^-$ are 1-parameter families of Liouville tori, one limiting to the separatrix from outside the 'figure eight' and the other from the interior part of the 'figure eight'.

Fomenko atoms of complexity 2.   There are six such atoms. However, we give here the description only of those appearing in the examples of this paper, see Figures 7.9, 7.14, 7.15.

The atom C_2. $\mathcal{L}$ consists of two orientable circles γ_1, γ_2 and four heteroclinic two-dimensional separatrices $\mathcal{S}_1$, $\mathcal{S}_2$, $\mathcal{S}_3$, $\mathcal{S}_4$. Trajectories on $\mathcal{S}_1$, $\mathcal{S}_3$ are approaching γ_1 as time tends to ∞, and γ_2 as time tends to $-\infty$, while those placed on $\mathcal{S}_2$, $\mathcal{S}_4$ approach γ_2 as time tends to ∞, and γ_1 as time tends to $-\infty$. Each of the sets $\mathcal{L}_\varepsilon^+$, $\mathcal{L}_\varepsilon^-$ contains two families of Liouville tori. As K approaches c, the tori from one family in $\mathcal{L}_\varepsilon^-$ deform to $\mathcal{S}_1 \cup \mathcal{S}_2$, and from the other one to $\mathcal{S}_3 \cup \mathcal{S}_4$. The tori from one family in $\mathcal{L}_\varepsilon^+$ is deformed to $\mathcal{S}_1 \cup \mathcal{S}_4$, and from the other to $\mathcal{S}_2 \cup \mathcal{S}_3$.

The atom D_1. $\mathcal{L}$ consists of two orientable circles γ_1, γ_2 and four two-dimensional separatrices $\mathcal{S}_1$, $\mathcal{S}_2$, $\mathcal{S}_3$, $\mathcal{S}_4$. Trajectories of $\mathcal{S}_1$, $\mathcal{S}_2$ homoclinically tend to γ_1, γ_2 respectively. Trajectories on $\mathcal{S}_3$ are approaching γ_1 as time tends to ∞, and γ_2 as time tends to $-\infty$, while those placed on $\mathcal{S}_4$, approach γ_2 as time tends to ∞, and

γ_1 as time tends to $-\infty$. One of the sets $\mathcal{L}_\varepsilon^+$, $\mathcal{L}_\varepsilon^-$ contains three, and the other one family of Liouville tori. Lets say that $\mathcal{L}_\varepsilon^+$ contains three families. As K approaches to c, these families deform to $\mathcal{S}_1$, $\mathcal{S}_2$ and $\mathcal{S}_3 \cap \mathcal{S}_4$ respectively, while the family contained in $\mathcal{L}_\varepsilon^-$ deform to the whole separatrix set.

This concludes the complete list of all atoms appearing in this section.

Numerical Fomenko Invariants. Each edge of the Fomenko graph corresponds to a one-parameter family of Liouville tori. Let us cut each of these families along one Liouville torus. The manifold Q will disintegrate into pieces, each corresponding to the singular level set, i.e., to a part of the Fomenko graph containing only one vertex and the initial segments of the edges incident to this vertex. To reconstruct Q from these pieces, we need to identify the corresponding boundary tori. This can be done in different ways. Thus, the basis of cycles is fixed in a certain canonical way on each boundary torus. Denote by b_i^-, b_i^+ the bases corresponding to the beginning and the end of an edge respectively. Denote by

$$\begin{pmatrix} \alpha_i & \beta_i \\ \gamma_i & \delta_i \end{pmatrix}$$

the transformation matrix from b_i^- to b_i^+.

Marks r_i. They are defined as

$$r_i = \begin{cases} \left\{ \dfrac{\alpha_i}{\beta_i} \right\}, & \text{if } \beta_i \neq 0, \\ \infty, & \text{if } \beta_i = 0. \end{cases}$$

Marks ε_i. They are

$$\varepsilon_i = \begin{cases} \operatorname{sign} \beta_i, & \text{if } \beta_i \neq 0, \\ \operatorname{sign} \alpha_i, & \text{if } \beta_i = 0. \end{cases}$$

Marks n_k. To determine marks n_k, we cut all the edges of the graphs having finite marks r_i. In this way the graph will fall apart into a few connected parts. Consider only parts not containing **A**-atoms. To each edge of the graph having a vertex in a given connected part, we join the following integer:

$$\Theta_i = \begin{cases} \left[\dfrac{\alpha_i}{\beta_i} \right], & \text{if } e_i \text{ is has the initial vertex in the connected part,} \\[2ex] \left[\dfrac{-\delta_i}{\beta_i} \right], & \text{if } e_i \text{ is has the ending vertex in the connected part,} \\[2ex] \left[\dfrac{-\gamma_i}{\alpha_i} \right], & \text{if both vertices of } e_i \text{ are in the connected part.} \end{cases}$$

Then to the connected part of the graph, we join the mark

$$n_k = \sum \Theta_i.$$

Rotation Functions.   Consider an arbitrary edge of a given Fomenko graph and recall that it represents a one-parameter family of Liouville tori. Let us choose one such a torus and fix on it a basis (λ, μ) of cycles. By the Arnol'd–Liouville theorem, the motion on the torus is linear, thus there are coordinates $(\varphi_1, \varphi_2) \in \mathbf{S}^1 \times \mathbf{S}^1$ such that the Hamiltonian vector field can be written in the form

$$a \frac{\partial}{\partial \varphi_1} + b \frac{\partial}{\partial \varphi_2},$$

while the coordinate lines $\varphi_2 = \mathrm{const}$, $\varphi_1 = \mathrm{const}$ are equivalent to basic cycles λ, μ. *Rotation number* on the torus is $\rho = \dfrac{a}{b}$. If we continuously extend the basis (λ, μ) to the other tori of the family, the collection of obtained rotation numbers will represent a *rotation function*.

Isoenergy Surfaces of Billiard Systems

Let $\mathcal{M}^n$ be an n-dimensional Riemann manifold, and $\Omega \subset \mathcal{M}$ a domain with the boundary composed of several smooth hypersurfaces. The billiard [KT1991] inside Ω is a dynamical system where a material point of the unit mass is freely moving inside the domain and obeying the reflection law at the boundary, i.e., having congruent impact and reflection angles with the space tangent to the boundary at any bouncing point. It is also assumed that the reflection is absolutely elastic, i.e., that the speed of the material point does not change before and after impact.

It is important to remark the change of the total energy of the system; only the intensity of the velocity vector of the point will be changed while the trajectories will be the same at each energy level. That is why, for the complete analysis of the billiard system, we may fix the speed and investigate only one energy level.

The isoenergy space for the billiard system inside Ω is

$$\mathcal{B} = \{\ (x, v)\ \mid\ x \in \Omega,\ v \in \mathrm{T}_x \mathcal{M},\ |v| = 1\ \} / \sim,$$
$$(x, u) \sim (x, v) \quad \Leftrightarrow \quad x \in \partial\Omega\ \text{ and }\ u - v \perp \mathrm{T}_x \partial\Omega.$$

Although the smoothness of motion of the billiard particle is violated at the boundary, let us note that, unexpectedly, the isoenergy space will be smooth, assuming that the billiard boundary is smooth. Even more, this manifold is smooth also when the boundary is only composed of a few smooth parts, if the billiard reflection can be continuously defined at the angle points. This will always be the case in our examples, when the boundary is placed on a few confocal conics, since they are orthogonal to each other. Thus, the reflection at the intersection points is defined as follows: the velocity vector v after the reflection at the non-smooth point of the boundary, changes to $-v$.

Plane Elliptical Billiards

An important example of an integrable billiard system is the billiard within an ellipse in $\mathbf{E}^2$. The integrability of such a billiard motion is due to a nice and elementary fact: each segment of a trajectory is tangent to the same conic confocal with the boundary [Arn1978, Ber1987]. Moreover, the boundary of any integrable billiard in a plane domain is composed of segments of several confocal conics [Bol1990].

In this section, we will describe the topology of the plane billiards with the elliptic boundary. We will use the Fomenko graphs [BMF1990] to represent the isoenergy surfaces.

Suppose that the ellipse in the plane is given by

$$\mathcal{E} \; : \; \frac{x^2}{a} + \frac{y^2}{b} = 1, \quad a > b > 0.$$

The family of conics confocal with $\mathcal{E}$ is

$$\mathcal{C}_\mu \; : \; \frac{x^2}{a - \mu} + \frac{y^2}{b - \mu} = 1, \quad \mu \in \mathbf{R}.$$

We are going to consider billiard systems inside a bounded domain Ω in $\mathbf{E}^2$, whose boundary is a union of arcs of several confocal conics. As in our introduction, we denote its isoenergy surface by $\mathcal{B}$.

Proposition 7.21. *The isoenergy surface corresponding to the billiard system within an ellipse in $\mathbf{E}^2$ is represented by the Fomenko graph on Figure 7.7.*

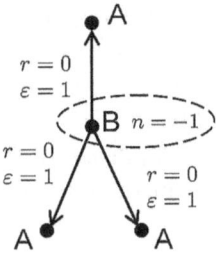

Figure 7.7: Fomenko graph corresponding to
the billiard within an ellipse

The rotation functions corresponding to the edges of the graph are monotonic, and the limits of these functions on the lower edges are equal to ∞ approaching to $\mathbf{A}$-atom and 2 approaching to $\mathbf{B}$-atom; on the upper edge the limit is ∞ approaching to $\mathbf{A}$-atom and 1 approaching to $\mathbf{B}$-atom.

Proof. The isoenergy surface $\mathcal{B}$ is the solid torus with the identification on $\sim$ on the boundary. This identification glues pairs of points, while points contained on two curves – representing flows in positive and negative directions along the boundary of the billiard table, are not glued with any other. The torus with the two curves on the boundary is shown on Figure 7.8. These two curves correspond

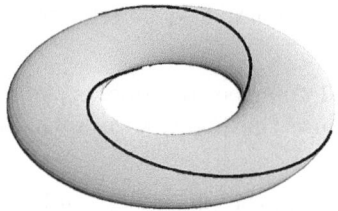

Figure 7.8: The isoenergy surface for the billiard system within an ellipse

to the limit flow with the caustic $\mathcal{C}_0 = \mathcal{E}$, and they are represented with the lower **A**-atoms on the graph. For the parameter $0 < \mu < b$, i.e., when the caustic is an ellipse, we have two Liouville tori over each value of μ – one for each direction of rotation.

Let us describe the level set $\mu = b$. This level set contains exactly those billiard trajectories that pass through foci of the ellipse. Among them, there is one periodic trajectory – the motion along the x-axis, while others have the well-known property that their segments alternately pass through the left and right foci of the ellipse. These trajectories can be naturally divided into two classes – one is composed of the trajectories where the particle is moving upward through the left focus and downward through the right one, and the second contains the trajectories with the reverse property. Denote these two classes by $\mathcal{S}_1$ and $\mathcal{S}_2$ respectively. All their trajectories are homoclinically tending to the x-axis, thus $\mathcal{S}_1$ and $\mathcal{S}_2$ are separatrices. Note that the periodic trajectory is orientable, so this level set corresponds to the **B**-atom.

Note that $\mathcal{S}_1$ (resp. $\mathcal{S}_2$) is the limit set of the family of Liouville tori corresponding to the flow with elliptic caustic in the clockwise (resp. counterclockwise) direction.

When $b < \mu < a$, i.e., when the caustic is a hyperbola, there is only one torus for each value. Finally, to the $\mu = a$, a periodic motion along the y-axis takes place, and it is represented by the upper **A**-atom.   □

In a similar way, we can obtain the following:

Proposition 7.22. *The isoenergy surface corresponding to the billiard system in the domain limited with two confocal ellipses in* $\mathbf{E}^2$ *is represented by the Fomenko graph on Figure 7.9.*

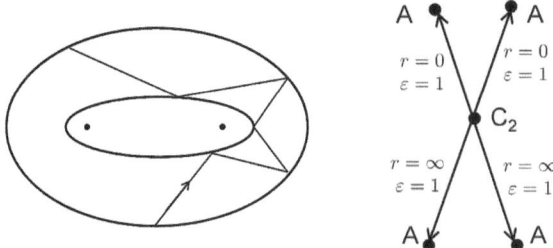

Figure 7.9: Billiard between two confocal ellipses and its Fomenko graph

Proof. As in Proposition 7.21, the lower **A**-atoms correspond to the limit flows on the boundary of the outer ellipse.

The level set for the caustic $\mu = b$ contains two periodic trajectories placed on the x-axis and all trajectories such that the continuations of their segments contain the foci. Let us note that such a trajectory is heteroclinic and placed only in one of the half-planes with the x-axis as the edge. Thus, this level set has four separatrices: two in the upper half-plane, two in the lower one. In each half-plane, one separatrix contains trajectories tending to the left periodic trajectory on the x-axis when time tends to ∞ and to the right one in $-\infty$, and opposite for orbits on the other separatrix. Liouville tori corresponding to elliptic caustics tend when μ tends to b, to two separatrices with the same movement direction – clockwise or counterclockwise.

For hyperbolic caustics, the billiard table becomes split into two domains, symmetric with respect to the x-axis. Each family of Liouville tori tend to the two separatrices from the level $\mu = b$ that are contained in the same half-plane.

This analysis shows that the atom that describes level set $\mu = b$ is $\mathbf{C_2}$.

The upper **A**-atoms correspond to the periodic orbits along the y-axis. □

Now we will consider the domain as being determined by an ellipse and a confocal hyperbola. Let us note that the phase space in the cases, where the boundary is not smooth at every point, must be considered with a special attention. Since confocal conics are always orthogonal in the intersection points, the periodic flows along smooth arcs may appear as limits of the billiard motion, when the caustic tends to the conic containing the arc. Thus, although the tangent space to the boundary in the intersection points is not defined, we will take that the "allowed" velocity vectors in such points are those tangent to the curves containing these points, with the opposite vectors identified with each other.

Proposition 7.23. *Consider the billiard domain with the border composed of an ellipse and a confocal hyperbola. Then:*

1. *If the domain is outside the hyperbola, then the isoenergy surface is represented by the graph on Figure 7.10.*

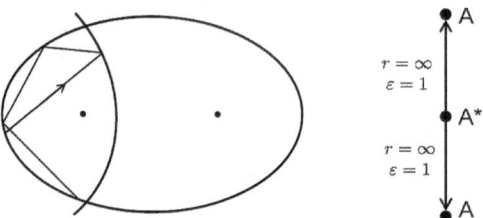

Figure 7.10: Billiard between ellipse and one branch of
a hyperbola and its Fomenko graph

2. *If the domain is inside the hyperbola, then the isoenergy surface is represented*
   by the graph on Figure 7.11.

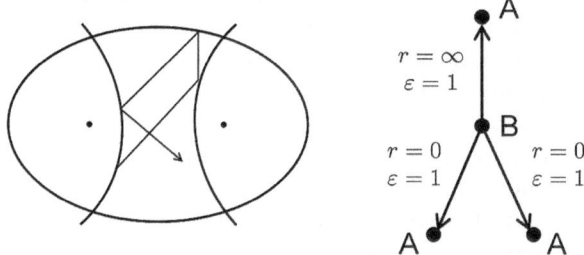

Figure 7.11: Billiard inside ellipse and hyperbola and its Fomenko graph

Proof. In the first case, each of the **A**-atoms corresponds to the limit periodic orbit
along one of the smooth arcs constituting the boundary of the billiard domain.
The **A***-atom represents the periodic orbit along the x-axis and its homoclinic
trajectories.

   In the second case, the lower **A**-atoms correspond to the limit motion on two
arcs of the ellipse, the **B**-atom to a periodic orbit on the x-axis and its homoclinic
orbits, and the upper **A** atom to a periodic orbit along the y-axis. It is interesting
to note that this system is Liouville equivalent to the billiard within the ellipse
(compare Figures 7.7 and 7.11). □

   We conclude this section by analyzing an example when the border of the
billiard table is continuously changed, to see how the bifurcations in the isoenergy
surfaces appear.

Proposition 7.24. *Consider a billiard domain limited by two confocal ellipses and*
two arcs contained in different branches of a confocal hyperbola.

1. *If the domain is inside the hyperbola, then the isoenergy surface is represented*
   by the graph on Figure 7.12.

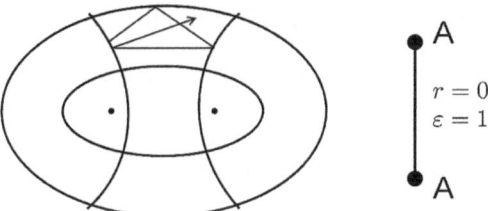

Figure 7.12: Billiard motion between two ellipses and hyperbola and the corresponding Fomenko graph

2. *If the x-axis is a part of the boundary, then the isoenergy surface is represented by the graph on Figure 7.13. Note that **V** corresponds to a degenerated singular leaf containing two orientable periodic orbits and two heteroclinic separatrices.*

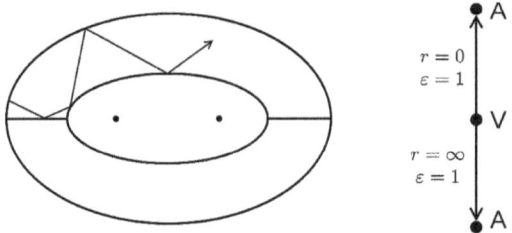

Figure 7.13: Billiard motion between two ellipses and a degenerate hyperbola with the Fomenko graph

3. *If the domain is as shown on the left side of Figure 7.14, then the isoenergy surface is represented by the graph on the right side of Figure 7.14.*

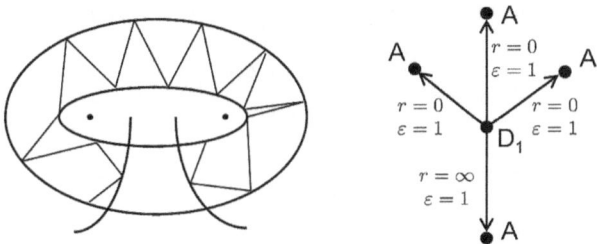

Figure 7.14: Billiard motion between two ellipses and hyperbola and the corresponding Fomenko graph

Remark 7.25. Near-integrable bifurcations of billiards from Propositions 7.23 and 7.24 can be studied by tools developed in [TRK2003].

Billiards on Ellipsoids and Liouville surfaces

In this section, we are going to analyze the topology of the billiard motion on the ellipsoid in $\mathbf{E}^3$, with the boundary cut by a confocal quadric surface. Next, we will consider the billiards within generalized ellipses on Liouville surfaces. Obtained results are going to be compared with those from the previous section.

Topology of Geodesic Motion on Ellipsoid in $\mathbf{E}^3$. Since the segments of billiard trajectories on the ellipsoid are placed on geodesic lines, it is essential to consider topology of the isoenergy surfaces for the geodesic motion on the ellipsoid. This topology is completely described in [BF1994].

We suppose that the ellipsoid in $\mathbf{E}^3$ is given by the equation

$$\mathcal{E} \; : \; \frac{x^2}{a} + \frac{y^2}{b} + \frac{z^2}{c} = 1, \qquad 0 < c < b < a. \tag{7.8}$$

Theorem 7.26. [BF1994, BF2004] *The Fomenko graph for the Jacobi problem of geodesic lines on an ellipsoid is presented in Figure 7.15. The rotation function*

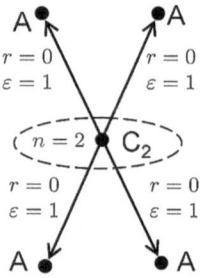

Figure 7.15: Fomenko graph for the Jacobi problem

that corresponds to the lower and upper edges of the graph are

$$\rho_{\text{lower}}(\alpha) = \frac{\int_c^\alpha \Phi(\lambda, \alpha)d\lambda}{\int_b^a \Phi(\lambda, \alpha)d\lambda} \quad (\alpha \in (c, b)),$$

$$\rho_{\text{upper}}(\alpha) = \frac{\int_c^b \Phi(\lambda, \alpha)d\lambda}{\int_\alpha^a \Phi(\lambda, \alpha)d\lambda} \quad (\alpha \in (b, a)),$$

where

$$\Phi(\lambda, \alpha) = \frac{\lambda}{\sqrt{-\lambda(a - \lambda)(b - \lambda)(c - \lambda)(\alpha - \lambda)}}.$$

It is interesting to remark that this theorem is used to prove that the Jacobi problem of geodesic lines on an ellipsoid and the Euler case of the rigid body motion are orbitally equivalent (see [BF1994, BF2004]).

Billiards on Ellipsoid in E^3. A generalized ellipse on the surface of the ellipsoid (7.8) is the intersection of this ellipsoid with a confocal hyperboloid. Let us note that such an intersection consists of two disjoint closed curves, which are curvature lines on the ellipsoid. This intersection divides the surface of the ellipsoid into three domains: two of them are simply connected and congruent to each other, while the third one, placed between them is 1-connected.

More precisely, any surface confocal with the ellipsoid (7.8) is given by an equation of the form (see Figure 7.16)

$$\mathcal{Q}_\lambda \; : \; Q_\lambda = \frac{x^2}{a-\lambda} + \frac{y^2}{b-\lambda} + \frac{z^2}{c-\lambda} = 1. \tag{7.9}$$

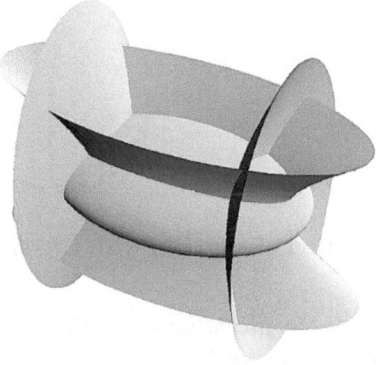

Figure 7.16: Confocal quadrics

Let the Jacobi coordinates $(\lambda_1, \lambda_2, \lambda_3)$ related to this system of confocal quadric surfaces be ordered by the condition $\lambda_1 > \lambda_2 > \lambda_3$.

Then the one-sheeted hyperboloid $\mathcal{Q}_\beta$, $(c < \beta < b)$, cuts out the following three domains on the surface of $\mathcal{E}$: $\Omega_1^\beta = \{\lambda_2 > \beta, \; z > 0\}$, $\Omega_2^\beta = \{\lambda_2 > \beta, \; z < 0\}$ and $\Omega_3^\beta = \{\lambda_2 < \beta\}$. The first two domains are symmetric with respect to the xy-plane, and the third one is the ring placed between them on $\mathcal{E}$, as shown on Figure 7.17.

Similarly, the two-sheeted hyperboloid $\mathcal{Q}_\gamma$, $(b < \gamma < a)$, determines the domains $\Omega_1^\gamma = \{\lambda_1 < \gamma, \; x > 0\}$, $\Omega_2^\gamma = \{\lambda_1 < \gamma, \; x < 0\}$ and $\Omega_3^\gamma = \{\lambda_1 > \gamma\}$, see Figure 7.18.

Proposition 7.27. *The isoenergy surfaces corresponding to the billiard systems within the domains Ω_1^β and Ω_1^γ on $\mathcal{E}$ is represented by the Fomenko graph on Figure 7.7.*

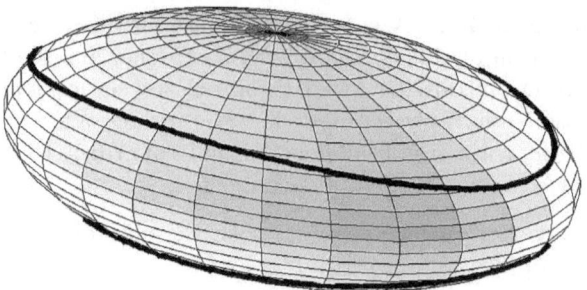

Figure 7.17: Intersection of ellipsoid with one-sheeted hyperboloid

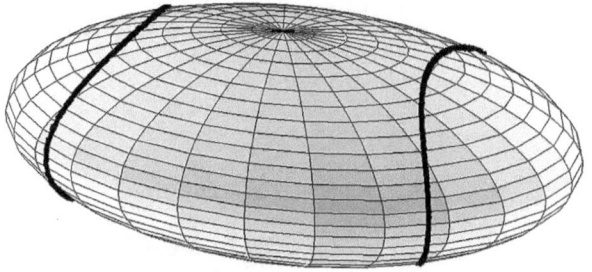

Figure 7.18: Intersection of ellipsoid with two-sheeted hyperboloid

The rotation functions for the billiard within Ω_1^β, for the lower and upper edges of the Fomenko graph respectively, are

$$\rho_{\text{lower}}^\beta(\alpha) = \frac{\int_\beta^\alpha \Phi(\lambda, \alpha)d\lambda}{\int_b^a \Phi(\lambda, \alpha)d\lambda} \quad (\alpha \in (\beta, b)),$$

$$\rho_{\text{upper}}^\beta(\alpha) = \frac{\int_\beta^b \Phi(\lambda, \alpha)d\lambda}{\int_\alpha^a \Phi(\lambda, \alpha)d\lambda} \quad (\alpha \in (b, a)).$$

The rotation functions for the billiard within Ω_1^γ, for the lower and upper edges of the Fomenko graph respectively, are

$$\rho_{\text{lower}}^\gamma(\alpha) = \frac{\int_\alpha^\gamma \Phi(\lambda, \alpha)d\lambda}{\int_c^b \Phi(\lambda, \alpha)d\lambda} \quad (\alpha \in (b, \gamma)),$$

$$\rho_{\text{upper}}^\gamma(\alpha) = \frac{\int_b^\gamma \Phi(\lambda, \alpha)d\lambda}{\int_c^\alpha \Phi(\lambda, \alpha)d\lambda} \quad (\alpha \in (c, b)).$$

Φ is defined as in Theorem 7.26.

A billiard within an Ellipse on a Liouville surface. Now, let us state the main result of this section.

Theorem 7.28. *All billiard systems within an ellipse on an arbitrary Liouville surface are Liouville equivalent. The Fomenko graph corresponding to an isoenergy surface of such a system is represented on Figure 7.7.*

Proof. Families of confocal ellipses and hyperbolas, together with corresponding billiard systems, were defined and described by Darboux [Dar1914] (for a recent account on this topic, see [DR2006b, DR2008]). Since such billiards have all topological properties analogous to billiards within an ellipse in the Euclidean plane, the theorem is proved.   □

Billiards inside Ellipsoid in $\mathbf{E}^3$

Here, we are going to analyze the topology of the billiard motion within an ellipsoid in $\mathbf{E}^3$. We suppose that the equation of the ellipsoid is (7.8).

Each trajectory of the billiard motion within $\mathcal{E}$ has exactly two caustics from the confocal family $\mathcal{Q}_\lambda$. Suppose these two surfaces are $\mathcal{Q}_{\lambda_1}$ and $\mathcal{Q}_{\lambda_2}$, $\lambda_2 \le \lambda_1$. Then the bifurcation set of an isoenergy surface of the system is given on Figure 7.19.

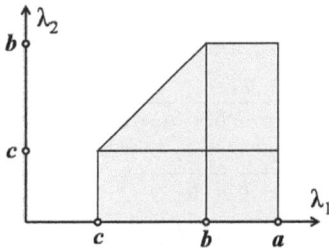

Figure 7.19: The bifurcation set for the billiard within
an ellipsoid in $\mathbf{E}^3$

The bifurcations of the Liouville tori for this system were investigated in [Knö1980, Aud1994, DFRR2001].

Now, let us explain in detail how the behaviour of the billiard motion is connected to the diagram on Figure 7.19.

Over each point in the gray area on the diagram, there are one or two three-dimensional Liouville tori. The marked edges will correspond to the degenerated level sets.

Now, let us describe in detail the level sets lying over different points in the bifurcation set on Figure 7.19.

Regular Leaves. Over each point inside rectangles $[c, b] \times [0, c]$, $[b, a] \times [0, c]$ and the triangle with vertices (c, c), (b, c), (b, b), there are two Liouville tori $\mathbf{T}^3$. Over each point inside rectangle $[b, a] \times [c, b]$, there is only one $\mathbf{T}^3$.

- Points inside $[c, b] \times [0, c]$ correspond to the motion with an ellipsoid and a one-sheeted hyperboloid as caustics. One Liouville torus is formed by the trajectories winding in the positive direction around the z-axis, and the other one by those winding in the negative direction.
- Points inside $[b, a] \times [0, c]$ correspond to the case when one caustic is ellipsoid and the other a two-sheeted hyperboloid. Each Liouville torus is formed by the trajectories winding in one direction around the x-axis.
- Points inside triangle (c, c)-(b, c)-(b, b) correspond to the motion with both caustics being different one-sheeted hyperboloids. Each of the Liouville tori is formed by the trajectories winding in one direction around the x-axis.
- Points inside rectangle $[b, a] \times [c, b]$ correspond to the case when both caustics are hyperboloids, but of different type. Then, all corresponding billiard trajectories form one Liouville torus.

Singular Leaves.   Singular leaves are lying over edges of the diagram represented on Figure 7.19. They appear when one or both caustics are degenerated.

First, let us clarify the notion of a degenerated quadric from confocal family (7.9), as well as the geometrical meaning of the billiard motion with such a caustic.

The degenerated quadrics $\mathcal{Q}_\lambda$, $\lambda \in \{a, b, c\}$ are the following curve lying in the coordinate planes

$$\mathcal{Q}_a \; : \; -\frac{y^2}{a - b} - \frac{z^2}{a - c} = 1, \quad x = 0,$$

$$\mathcal{Q}_b \; : \; \frac{x^2}{a - b} - \frac{z^2}{b - c} = 1, \qquad y = 0,$$

$$\mathcal{Q}_c \; : \; \frac{x^2}{a - c} + \frac{y^2}{b - c} = 1, \qquad z = 0.$$

Notice that $\mathcal{Q}_c$ is an ellipse in the xy-plane, $\mathcal{Q}_b$ is a hyperbola in the xz-plane, while $\mathcal{Q}_c$ is, in the real case, the empty set. Nevertheless, we will consider $\mathcal{Q}_a$ as an abstract curve in the yz-plane. The degenerated quadrics are depicted on Figure 7.20.

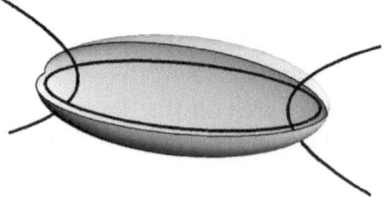

Figure 7.20: Ellipsoid $\mathcal{E}$ and degenerated quadrics $\mathcal{Q}_b$, $\mathcal{Q}_c$

The billiard motion with some of these three degenerated caustics $\mathcal{Q}_\lambda$ is either such that each segment of the trajectory intersects the curve $\mathcal{Q}_\lambda$ or the whole trajectory is lying in the coordinate plane containing this curve. In the latter case, the motion reduces to the plane billiard one of the ellipses:

$$\mathcal{E}_a \;:\; \frac{y^2}{b} + \frac{z^2}{c} = 1, \quad x = 0,$$

$$\mathcal{E}_b \;:\; \frac{x^2}{a} + \frac{z^2}{c} = 1, \quad y = 0,$$

$$\mathcal{E}_c \;:\; \frac{x^2}{a} + \frac{y^2}{b} = 1, \quad z = 0.$$

Notice that, in the case $\lambda = a$, only plane trajectories exist.

Each of these three ellipses is the intersection of $\mathcal{E}$ with one of the coordinate hyperplanes (see Figure 7.21).

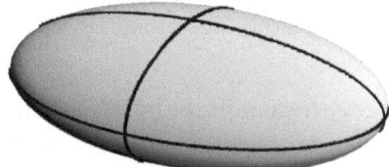

Figure 7.21: Ellipsoid $\mathcal{E}$ and ellipses $\mathcal{E}_a$, $\mathcal{E}_b$, $\mathcal{E}_c$

Besides these cases with a degenerated caustic, there is one more special case – when one of the caustics is the boundary ellipsoid $\mathcal{E}$. This case can be considered as the limit case with the caustic $\mathcal{Q}_\lambda$, $\lambda \to 0_+$, and its trajectories are geodesic lines on $\mathcal{E}$.

Now, let us analyze the edges of the bifurcation diagram on Figure 7.19.

Edge $[c, a] \times \{0\}$. As already mentioned, the motion corresponding to this edge is the geodesic motion on $\mathcal{E}$. Thus, the topology of the corresponding subset in the isoenergy surface is completely described by the Fomenko graph on Figure 7.15.

Let us summarize this case:

- Point $(c, 0)$ corresponds to the motion along the curve $\mathcal{E}_c$. There are two one-dimensional tori $\mathbf{T}^1$ lying over the point $(c, 0)$ of the bifurcation diagram – each one corresponding to the flow in one direction.
- Inner points of the segment $[c, b]$ on λ_1-axis correspond to the geodesic motion on $\mathcal{E}$ with a one-sheeted hyperboloid as caustic. Such geodesic lines fill the ring between two intersection curves of $\mathcal{E}$ and the hyperboloid. Over each point inside the segment, there are two tori $\mathbf{T}^2$ – each one corresponding to winding in one direction around the z-axis.
- The level set corresponding to $(b, 0)$ contains two periodic trajectories and four two-dimensional separatrices. The periodic trajectories are placed along

the curve $\mathcal{E}_b$. Trajectories on the separatrices are passing through the umbilical points of the ellipsoid.

- Inner points of the segment $[b, a]$ on λ_1-axis correspond to the geodesic motion on $\mathcal{E}$ with a two-sheeted hyperboloid as caustic. Such geodesic lines fill the ring between two intersection curves of $\mathcal{E}$ and the hyperboloid. Over each point inside the segment, there are two tori $\mathbf{T}^2$ – each one corresponding to winding in one direction around the x-axis.

- Point $(a, 0)$ corresponds to the motion along $\mathcal{E}_a$. There are two one-dimensional tori $\mathbf{T}^1$ lying over this point – each one corresponding to the motion in one direction.

Edge $\{a\} \times [0, b]$.   This edge corresponds to the billiard motion with the degenerated caustic $\mathcal{Q}_a$. All trajectories of such a motion are placed in the yz-plane, thus this edge in fact corresponds to the billiard within an ellipse. The topology of the set lying over this edge is completely described by the Fomenko atom on Figure 7.7.

Edge $[b, a] \times \{b\}$.   This edge corresponds to the billiard motion with one degenerate caustic $\mathcal{Q}_b$, and the other caustic being a two-sheeted hyperboloid.

7.5   Integrable potential perturbations of an elliptical billiard

In this section, we are going to present results on integrable potential perturbations of the billiard system within an ellipse. The analysis of an ellipsoidal billiard with quadratic potential is given in [Fed2001].

The equation [Koz1995]

$$\lambda V_{xy} + 3\left(yV_x - xV_y\right) + (y^2 - x^2)V_{xy} + xy\left(V_{xx} - V_{yy}\right) = 0 \tag{7.10}$$

is a special case of the Bertrand-Darboux equation [Ber1852, Dar1901, Whi1927], which represents the necessary and sufficient condition for a natural mechanical system with two degrees of freedom,

$$H = \frac{1}{2}(p_x^2 + p_y^2) + V(x, y),$$

to be separable in elliptical coordinates or some of their degenerations.

Solutions of equation (7.10) in the form of Laurent polynomials in x, y were described in [Dra1996]. Such solutions are in [Dra2002] naturally related to the well-known hypergeometric functions of Appell. This relation automatically gives a wider class of solutions of equation (7.10) – new potentials are obtained for non-integer parameters, giving a huge family of integrable billiards within an ellipse with potentials. Similar formulae for potential perturbations for the Jacobi problem for geodesics on an ellipsoid from [Dra1996, Dra1998b] are given. They show the existence of a connection between separability of classical systems on one hand, and the theory of hypergeometric functions on the other which is still not completely understood. Basic references for the Appell functions are [App1880, AKdF1926, VK1995].

The function F_4 is one of the four hypergeometric functions in two variables, which are introduced by Appell [App1880, AKdF1926]:

$$F_4(a, b, c, d; x, y) = \sum \frac{(a)_{m+n}(b)_{m+n}}{(c)_m (d)_n} \frac{x^m y^n}{m! \, n!},$$

where $(a)_n$ is the standard Pochhammer symbol:

$$(a)_n = \frac{\Gamma(a+n)}{\Gamma(a)} = a(a+1)\ldots(a+n-1),$$

$$(a)_0 = 1.$$

(For example $m! = (1)_m$.)

The series is convergent for $\sqrt{x} + \sqrt{y} \leq 1$. The functions F_4 can be analytically continued to the solutions of the equations

$$x(1-x)\frac{\partial^2 F}{\partial x^2} - y^2 \frac{\partial^2 F}{\partial y^2} - 2xy\frac{\partial^2 F}{\partial x \partial y}$$
$$+ [c - (a+b+1)x]\frac{\partial F}{\partial x} - (a+b+1)y\frac{\partial F}{\partial y} - abF = 0,$$

$$y(1-y)\frac{\partial^2 F}{\partial y^2} - x^2 \frac{\partial^2 F}{\partial x^2} - 2xy\frac{\partial^2 F}{\partial x \partial y}$$
$$+ [c' - (a+b+1)y]\frac{\partial F}{\partial y} - (a+b+1)x\frac{\partial F}{\partial x} - abF = 0.$$

Potential perturbations of a billiard inside an ellipse

A billiard system which describes a particle moving freely within the ellipse

$$\frac{x^2}{A} + \frac{y^2}{B} = 1$$

is completely integrable and it has an additional integral

$$K_1 = \frac{\dot{x}^2}{A} + \frac{\dot{y}^2}{B} - \frac{(\dot{x}y - \dot{y}x)^2}{AB}.$$

We are interested now in potential perturbations $V = V(x, y)$ such that the perturbed system has an integral $\tilde{K}_1$ of the form

$$\tilde{K}_1 = K_1 + k_1(x, y),$$

where $k_1 = k_1(x, y)$ depends only on coordinates. This specific condition leads to Equation (7.10) on V with $\lambda = A - B$.

In [Dra1996, Dra1998b] the Laurent polynomial solutions of equation (7.10) were given. Writing

$$V_\gamma = \tilde{y}^{-\gamma}\left((1 - \gamma)\,\tilde{x}\,F_4(1, 2 - \gamma, 2, 1 - \gamma, \tilde{x}, \tilde{y}) + 1\right), \tag{7.11}$$

where $\tilde{x} = x^2/\lambda$, $\tilde{y} = -y^2/\lambda$, the more general result was obtained in [Dra2002]:

Theorem 7.29. *Every function V_γ given with (7.11) and $\gamma \in \mathbf{C}$ is a solution of equation* (7.10).

This theorem gives new potentials for non-integer γ. For integer γ, one obtains the Laurent solutions.

Mechanical Interpretation. With $\gamma \in \mathbf{R}^-$ and the coefficient multiplying V_γ positive, we have a potential barrier along the x-axis. We can consider billiard motion in the upper half-plane. Then we can assume that a cut is done along the negative part of the y-axis, in order to get a unique-valued real function as a potential.

Solutions of Equation (7.10) are also connected with interesting geometric subjects.

Ellipsoidal billiard with quadratic potential

In [Fed2001], a billiard system within an ellipsoid

$$\mathcal{E} \ : \ (x, a^{-1}x) = 1, \quad a = (a_1, \ldots, a_n), \quad 0 < a_1 < \cdots < a_n$$

in the n-dimensional space was considered, with the quadratic potential of the form

$$\sigma(x_1^2 + \cdots + x_n^2)/2, \quad \sigma = \text{const.}$$

This potential is also called *Hook potential*.

Let $\mathcal{B} \ : \ (x, v) \mapsto (\tilde{x}, \tilde{v})$ be the billiard mapping, where $x, \tilde{x} \in \mathcal{E}$ are consecutive points of reflection on the boundary, and $v, \tilde{v}$ the velocity vectors at $x, \tilde{x}$ after the reflection.

The next theorem gives explicit formulae for $\mathcal{B}$.

Theorem 7.30 ([Fed2001]). *The billiard mapping is equivalent to*

$$\tilde{x} = -\frac{1}{\nu}\left[(\sigma - (v, a^{-1}v))x + 2(x, a^{-1}v)v\right],$$

$$\tilde{v} = -\frac{1}{\nu}\left[(\sigma - (v, a^{-1}v))v - 2\sigma(x, a^{-1}v)x\right] + \mu a^{-1}\tilde{x},$$

$$\nu = \sqrt{4\sigma(x, a^{-1}v)^2 + (\sigma - (v, a^{-1}v))^2}, \quad \mu = \frac{2(\tilde{v}, a^{-1}\tilde{x})}{(\tilde{x}, a^{-2}\tilde{x})}.$$

It turns out that the billiard mapping $\mathcal{B}$ has also the Lax pair representation.

Proposition 7.31 ([Fed2001]). *The billiard mapping $\mathcal{B}$ is, up to the symmetry $(x, v) \mapsto (-x, -v)$, equivalent to the equation $\tilde{L}(\lambda) = M(\lambda)L(\lambda)M^{-1}(\lambda)$, where*

$$L(\lambda) = \begin{pmatrix} q_\lambda(x, v) & q_\lambda(v, v) - \sigma \\ -q_\lambda(x, x) + 1 & -q_\lambda(x, v) \end{pmatrix}, \quad q_\lambda(x, y) = \sum_{i=1}^{n} \frac{x_i y_i}{\lambda - a_i},$$

$$M(\lambda) = \begin{pmatrix} [\sigma - (v, a^{-1}v)]\lambda + 2(x, a^{-1}v)\mu & 2\sigma(x, a^{-1}v)\lambda - [\sigma - (v, a^{-1}v)]\mu \\ -2(x, a^{-1}v)\lambda & [\sigma - (v, a^{-1}v)]\lambda \end{pmatrix},$$

and $\tilde{L}(\lambda)$ is depending on $\tilde{x}, \tilde{v}$ in the same way as $L(\lambda)$ on x, v.

A billiard with the Hook potential has the following geometric properties:

Proposition 7.32 ([Fed2001]). *Segments of billiard trajectories within ellipsoid $\mathcal{E}$ in $\mathbf{E}^n$ with Hook potential with $\sigma > 0$ are arcs of ellipses. All ellipses containing segments of a given billiard trajectory are touching the same n quadrics, confocal with $\mathcal{E}$. Parameters of the quadrics are roots of the characteristic polynomial*

$$(\lambda - a_1) \dots (\lambda - a_n) \det L(\lambda).$$

Observe that the number of caustics for motion with the Hook potential is n, while for an unperturbed system this number is $n - 1$, according to Chasles Theorem 4.78. We learned from Fedorov even more: the number of caustics is equal to $n + k - 1$ if the degree of separable perturbation is of degree $2k$.

Jacobi problem for geodesics on an ellipsoid

The Jacobi problem for geodesics (see Example 4.84) on an ellipsoid,

$$\frac{x^2}{A} + \frac{y^2}{B} + \frac{z^2}{C} = 1,$$

has an additional integral

$$K_1 = \left(\frac{x^2}{A^2} + \frac{y^2}{B^2} + \frac{z^2}{C^2} \right) \left(\frac{\dot{x}^2}{A} + \frac{\dot{y}^2}{B} + \frac{\dot{z}^2}{C} \right).$$

Potential perturbations $V = V(x, y, z)$, such that perturbed systems have integrals of the form

$$\tilde{K}_1 = K_1 + k(x, y, z)$$

satisfy the following system:

$$\left(\frac{x^2}{A^2} + \frac{y^2}{B^2} + \frac{z^2}{C^2}\right) V_{xy} \frac{A-B}{AB} - 3\frac{yV_x}{B^2A} + 3\frac{xV_y}{A^2B}$$

$$+\left(\frac{x^2}{A^3} - \frac{y^2}{B^3}\right) V_{xy} + \frac{xy}{AB}\left(\frac{V_{yy}}{A} - \frac{V_{xx}}{B}\right) + \frac{zxV_{zy}}{CA^2} - \frac{zyV_{zx}}{CB^2} = 0,$$

$$\left(\frac{x^2}{A^2} + \frac{y^2}{B^2} + \frac{z^2}{C^2}\right) V_{yz} \frac{B-C}{BC} - 3\frac{zV_y}{C^2B} + 3\frac{yV_z}{B^2C}$$

$$+\left(\frac{y^2}{B^3} - \frac{z^2}{C^3}\right) V_{yz} + \frac{yz}{BC}\left(\frac{V_{zz}}{B} - \frac{V_{yy}}{C}\right) + \frac{xyV_{xz}}{AB^2} - \frac{xzV_{xy}}{AC^2} = 0, \tag{7.12}$$

$$\left(\frac{x^2}{A^2} + \frac{y^2}{B^2} + \frac{z^2}{C^2}\right) V_{zx} \frac{C-A}{AC} - 3\frac{xV_z}{A^2C} + 3\frac{zV_x}{C^2A}$$

$$+\left(\frac{z^2}{C^3} - \frac{x^2}{A^3}\right) V_{zx} + \frac{xz}{AC}\left(\frac{V_{xx}}{C} - \frac{V_{zz}}{A}\right) + \frac{zyV_{xy}}{BC^2} - \frac{yxV_{yz}}{BA^2} = 0.$$

This system is an analogue of the Bertrand–Darboux equation (7.10) (see [Dra2002]).

Let

$$\frac{x^2C(A-C)}{z^2(B-A)A} = \hat{x}, \qquad \frac{y^2C(C-B)}{z^2(B-A)B} = \hat{y}.$$

The following statement was also proved in [Dra2002]:

Theorem 7.33. *For every $\gamma \in \mathbf{C}$, the function*

$$V_\gamma = (-\gamma + 1)\left(\frac{z^2}{x^2}\right)^\gamma F_4(1, -\gamma + 2, 2, -\gamma + 1, \hat{x}, \hat{y})$$

is a solution of the system (7.12).

Thus, by solving the Bertrand–Darboux equation and its generalizations, as it is done in Theorems 7.29 and 7.33, one gets large families of separable mechanical systems with two degrees of freedom. It is well known that separable systems of two degrees of freedom are necessarily of the Liouville type, see [Whi1927].

Now the natural question of a Poncelet-type theorem describing periodic solutions of such perturbed billiard systems arises. It appears that again Darboux studied such a question, since in [Dar1914], he analyzed generalizations of the Poncelet theorem in the case of the Liouville surfaces (see Section 5.5).

Chapter 8

Billiard Law and Hyperelliptic Curves

8.1 Generalized Cayley's curve

We continue our investigation from Sections 5.7 and 7.1. Suppose a confocal family of quadrics is given in d-dimensional space. Then every line in the space determines one remarkable Riemann surface.

Definition 8.1. Let ℓ be a line not contained in any quadric of the given confocal family in the projective space $\mathbf{P}^d$. The *generalized Cayley curve* C_ℓ is the variety of hyperplanes tangent to quadrics of the confocal family at the points of ℓ.

This curve is naturally embedded in the dual space $\mathbf{P}^{d*}$.

On Figure 8.1, we see the planes which correspond to one point of the line ℓ in the three-dimensional space.

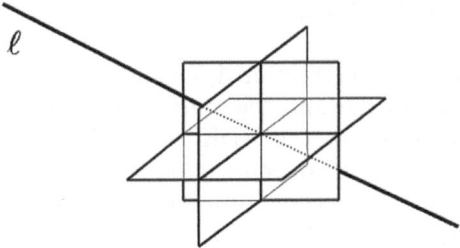

Figure 8.1: Three points of the generalized Cayley curve
in dimension 3

Proposition 8.2. *The generalized Cayley curve is a hyperelliptic curve of genus $g = d - 1$, for $d \geq 3$. Its natural realization in $\mathbf{P}^{d*}$ is of degree $2d - 1$.*

The natural involution τ_ℓ on the generalized Cayley's curve $\mathcal{C}_\ell$ maps to each other the tangent planes at the points of intersection of ℓ with any quadric of the confocal family. It is easy to see that the fixed points of this involution are hyperplanes corresponding to the $d-1$ quadrics that are touching ℓ and to $d+1$ degenerate quadrics of the confocal family.

The equation of the generalized Cayley curve corresponding to a confocal family of the form (4.19) can be written as

$$y^2 = \mathcal{P}(x),$$

with

$$\mathcal{P}(x) = (x - a_1) \cdots (x - a_d)(x - \alpha_1) \cdots (x - \alpha_{d-1}). \tag{8.1}$$

It is important to note that the constants $\alpha_1, \ldots, \alpha_{d-1}$, corresponding to the quadrics that are touching ℓ cannot take arbitrary values. More precisely, following [Aud1994, Knö1980], we can state

Proposition 8.3. *There exists a line in $\mathbf{E}^d$ that is tangent to $d-1$ distinct non-degenerate quadrics $\mathcal{Q}_{\alpha_1}, \ldots, \mathcal{Q}_{\alpha_{d-1}}$ from the family (4.19) if and only if the set $\{a_1, \ldots, a_d, \alpha_1, \ldots, \alpha_{d-1}\}$ can be ordered as $b_1 < b_2 < \cdots < b_{2d-1}$, such that $\alpha_j \in \{b_{2j-1}, b_{2j}\}$ $(1 \le j \le d-1)$.*

It was observed in [DR2006b] that the generalized Cayley curve is isomorphic to the Veselov-Moser isospectral curve.

This curve is also naturally isomorphic to the curves studied by Knörrer, Donagi and Reid (RDK curves). We begin the next section with the construction of this isomorphism, in order to establish the relationship between billiard law and algebraic structure on the Jacobian Jac $(\mathcal{C}_\ell)$.

8.2   Billiard law and algebraic structure on variety $\mathcal{A}_\ell$

The aim of this section is to construct in $\mathcal{A}_\ell$ an algebraic structure that is naturally connected with the billiard motion. We start with showing that the generalized Cayley's curve is isomorphic to the RDK curve. Then, in order to get better understanding and intuition, we are going to describe in detail the billiard algebra, and prove some of its nice geometrical properties, for the case of dimension 3, i.e., when the corresponding curve is of genus 2. The general construction is given right after.

Morphism between Generalized Cayley's Curve and RDK Curve

Now we are going to establish the connection between a generalized Cayley's curve defined above and the curves studied by Knörrer, Donagi, Reid and to trace out the relationship between billiard constructions and the algebraic structure of the corresponding Abelian varieties.

Suppose a line ℓ in $\mathbf{E}^d$ is tangent to quadrics $\mathcal{Q}_{\alpha_1}, \ldots, \mathcal{Q}_{\alpha_{d-1}}$ from the confocal family (4.19). Denote by $\mathcal{A}_\ell$ the family of all lines which are tangent to the same $d-1$ quadrics. Note that according to the corollary of the One Reflection Theorem (see Theorem 5.25), the set $\mathcal{A}_\ell$ is invariant to the billiard reflection on any of the confocal quadrics.

We begin with the next simple observation.

Lemma 8.4. *Let the lines ℓ and ℓ' obey the reflection law at the point z of a quadric $\mathcal{Q}$ and suppose they are tangent to a confocal quadric $\mathcal{Q}_1$ at the points z_1 and z_2. Then the intersection of the tangent spaces $T_{z_1}\mathcal{Q}_1 \cap T_{z_2}\mathcal{Q}_1$ is contained in the tangent space $T_z\mathcal{Q}$.*

Proof. It follows from the One Reflection Theorem: since the poles z_1, z_2 and w of the planes $T_{z_1}\mathcal{Q}_1$, $T_{z_2}\mathcal{Q}_1$ and $T_z\mathcal{Q}$ with respect to the quadric $\mathcal{Q}_1$ are colinear, the planes belong to a pencil. $\square$

Following [Knö1980], together with $d-1$ affine confocal quadrics $\mathcal{Q}_{\alpha_1}, \ldots,$ $\mathcal{Q}_{\alpha_{d-1}}$, one can consider their projective closures $\mathcal{Q}_{\alpha_1}^p, \ldots, \mathcal{Q}_{\alpha_{d-1}}^p$ and the intersection V of two quadrics in $\mathbf{P}^{2d-1}$:

$$x_1^2 + \cdots + x_d^2 - y_1^2 - \cdots - y_{d-1}^2 = 0, \tag{8.2}$$

$$a_1 x_1^2 + \cdots + a_d x_d^2 - \alpha_1 y_1^2 - \cdots - \alpha_{d-1} y_{d-1}^2 = x_0^2. \tag{8.3}$$

Denote by $F = F(V)$ the set of all $(d-2)$-dimensional linear subspaces of V. For a given $L \in F$, denote by F_L the closure in F of the set $\{\, L' \in F \mid \dim L \cap L' = d-3 \,\}$. It was shown in [Rei1972] that F_L is a nonsingular hyperelliptic curve of genus $d-1$. Note that for $d = 3$, i.e., when the curve $\mathcal{C}_\ell$ is of the genus 2, an isomorphism between $F(V)$ and the Jacobian of the hyperelliptic curve was established in [NR1969].

The projection

$$\pi' : \mathbf{P}^{2d-1} \setminus \{(x,y)|x=0\} \to \mathbf{P}^d, \quad \pi'(x,y) = x,$$

maps $L \in F(V)$ to a subspace $\pi'(L) \subset \mathbf{P}^d$ of the codimension 2. $\pi'(L)$ is tangent to the quadrics $\mathcal{Q}_{\alpha_1}^{p*}, \ldots, \mathcal{Q}_{\alpha_{d-1}}^{p*}$ that are dual to $\mathcal{Q}_{\alpha_1}^p, \ldots, \mathcal{Q}_{\alpha_{d-1}}^p$.

Thus, the space dual to $\pi'(L)$, denoted by $\pi^*(L)$, is a line tangent to the quadrics $\mathcal{Q}_{\alpha_1}^p, \ldots, \mathcal{Q}_{\alpha_{d-1}}^p$.

We can reinterpret the generalized Cayley's curve $\mathcal{C}_\ell$, which is a family of tangent hyperplanes, as a set of lines from $\mathcal{A}_\ell$ which intersect ℓ. Namely, for almost every tangent hyperplane there is a unique line ℓ', obtained from ℓ by the billiard reflection. Having this identification in mind, it is easy to prove the following

Corollary 8.5. *There is a birational morphism between the generalized Cayley's curve $\mathcal{C}_\ell$ and Reid–Donagi–Knörrer's curve F_L, with $L = \pi^{*-1}(\ell)$, defined by*

$$j : \ell' \mapsto L', \quad L' = \pi^{*-1}(\ell'),$$

where ℓ' is a line obtained from ℓ by the billiard reflection on a confocal quadric.

Proof. It follows from the previous lemmata and Lemma 4.1 and Corollary 4.2 from [Knö1980]. $\qquad\qquad\qquad\qquad\qquad\qquad\qquad\qquad\qquad\qquad\qquad\qquad\square$

Thus, Lemma 8.4 gives a link between the dynamics of ellipsoidal billiards and algebraic structure of certain Abelian varieties. This link provides a two way interaction: to apply algebraic methods in the study of the billiard motion, but also vice versa, to use billiard constructions in order to get more effective, more constructive and more observable understanding of the algebraic structure.

In the following section, we are going to use this link in constructing *the billiard algebra*, which is a group structure in $\mathcal{A}_\ell$.

Genus 2 Case

Before we proceed in the general case, we want to emphasize the billiard constructions involved in the first nontrivial case, of genus 2.

Leading Principle, Definition and First Properties of the Operation. We formulate *The Leading Principle*:

The sum of the lines in any virtual reflection configuration is equal to zero if the four tangent planes at the points of reflection belong to a pencil.

Recall that, by Definition 5.29, such a configuration of four lines in a VRC, with tangent planes in a pencil, we call *a Double Reflection Configuration (DRC)*.

Neutral element

Let us fix a line $\mathcal{O} \in \mathcal{A}_\ell$.

Opposite element

First, we define $-\mathcal{O} := \mathcal{O}$.

For a given line $x \in \mathcal{C}_\mathcal{O}$, define

$$-x := \tau_\mathcal{O}(x),$$

where $\tau_\mathcal{O}$ is the hyperelliptic involution of the curve $\mathcal{C}_\mathcal{O}$.

Proposition 8.6. *For any $x \in \mathcal{C}_\mathcal{O}$, both lines x and $-x$ are obtained from $\mathcal{O}$ by the reflection on the same quadric $\mathcal{Q}_x$ from the confocal family.*

Moreover, let π be the unique plane orthogonal to $\mathcal{O}$ from the pencil determined by the tangent planes to $\mathcal{Q}_x$ at the intersection points with $\mathcal{O}$, and $\mathcal{Q}_\mathcal{O}$ the unique quadric from the confocal family such that π is tangent to it. Then the intersection point of π with $\mathcal{O}$ belongs to $\mathcal{Q}_\mathcal{O}$.

Proof. Follows from the degenerate case of the Double Reflection Theorem applied on quadrics $\mathcal{Q}_x$ and $\mathcal{Q}_\mathcal{O}$ with $\mathcal{O} = l_1 = l_2$. See Figure 8.2.   □

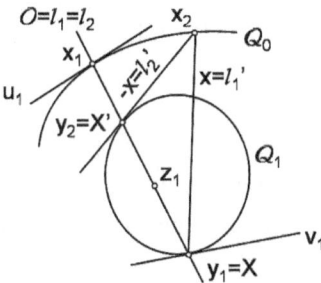

Figure 8.2: Proposition 8.6

Corollary 8.7. *Plane* π *and quadric* $\mathcal{Q}_\mathcal{O}$ *do not depend on* x. *Line* $\mathcal{O}$ *reflects to itself on* $\mathcal{Q}_\mathcal{O}$.

Suppose now that $x \in \mathcal{A}_\ell$, but x does not belong to $\mathcal{C}_\mathcal{O}$. Thus x does not intersect $\mathcal{O}$ and the two lines generate a linear projective space. This space intersects $\mathcal{A}_\ell$ along the divisor

$$\mathcal{O} + x + p + q.$$

(See, for example, [Tyu1975], [GH1978b].)

According to the Double Reflection Theorem, one can see that the lines $\mathcal{O}$, x, p, q form a double reflection configuration. We define $-x$ such that it forms a double reflection configuration with $\mathcal{O}$, $-p$, $-q$.

From the definition, it is immediately seen that $-(-x) = x$ for every $x \in \mathcal{A}_\ell$.

The following property is a consequence of Proposition 8.6 and the Double Reflection Theorem (Theorem 5.27).

Proposition 8.8. *Lines* x *and* $-x$ *intersect each other and they satisfy the reflection law on* $\mathcal{Q}_\mathcal{O}$.

The following example is an illustration for the construction we have just made.

Example 8.9. Take the line $\mathcal{O}$ to be orthogonal to one of the coordinate hyperplanes. Then $\mathcal{Q}_\mathcal{O} = \pi$ coincides with this hyperplane and, additionally, among caustics $\mathcal{Q}_1, \ldots, \mathcal{Q}_{d-1}$ there cannot be two quadrics of the same type.

According to [Aud1994], exactly the case of all quadrics of different type corresponds to the case where $\mathcal{A}_\ell$ consists of only one real connected component, isomorphic to the connected component of zero of $\mathrm{Jac}\,(\mathcal{C}_\mathcal{O})(\mathbf{R})$.

In this case, for any line $x \in \mathcal{A}_\ell$, the opposite element $-x$ may be defined as the line symmetric to x with respect to plane $\mathcal{Q}_\mathcal{O}$.

Addition

We are going to define operation

$$+ \; : \; \mathcal{A}_\ell \times \mathcal{A}_\ell \to \mathcal{A}_\ell.$$

Define $\mathcal{O} + x = x + \mathcal{O} = x$, for all $x \in \mathcal{A}_\ell$.

For $s_1, s_2 \in \mathcal{C}_\mathcal{O}$, define $s_1 + s_2$ as the line that forms a double reflection configuration with $-s_1, -s_2, \mathcal{O}$. Obviously, $s_1 + s_2 = s_2 + s_1$.

Notice that $-s_1, -s_2$ are unique lines from $\mathcal{A}_\ell$ that intersect both $s_1 + s_2$ and $\mathcal{O}$ (see [Tyu1975]). Thus, we have:

Lemma 8.10. *Each line $x \in \mathcal{A}_\ell \setminus \mathcal{C}_\mathcal{O}$ can be represented in a unique way as the sum of two lines that intersect $\mathcal{O}$.*

Now, suppose $s \in \mathcal{C}_\mathcal{O}$, $x \in \mathcal{A}_\ell \setminus \mathcal{C}_\mathcal{O}$ (see Figure 8.3). As already explained,

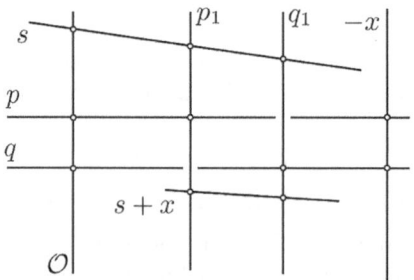

Figure 8.3: Partial operation

take lines $p, q \in \mathcal{C}_\mathcal{O}$ such that $x = p + q$. Construct

$$p_1 = (-s) + (-p), \quad q_1 = (-s) + (-q)$$

as above, since both pairs s, p and s, q intersect $\mathcal{O}$. Now both p_1 and q_1 intersect s. Thus, the three lines belong to a DRC with the fourth line z. We put by definition

$$s + x = x + s = z.$$

The following lemma gives a very important property of the operation.

Lemma 8.11. *Let s, x be lines in $\mathcal{C}_\mathcal{O}$, $\mathcal{A}_\ell$ respectively and $\mathcal{Q}_s$ the quadric from the confocal family such that s and $\mathcal{O}$ reflect to each other on it. Then the lines $s + x$ and $-x$ intersect each other, their intersection point belongs to quadric $\mathcal{Q}_s$, and these two lines satisfy the billiard reflection law on $\mathcal{Q}_s$.*

Proof. Follows from [Don1980, Knö1980] and Lemma 8.4.   □

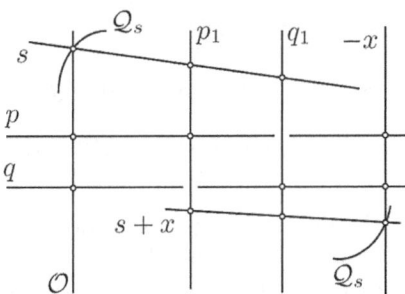

Figure 8.4: Lemma 8.11

Lemma 8.11 is going to play an important role in proving basic properties of the operation:

Lemma 8.12. *For all $p, q, s \in \mathcal{C}_\mathcal{O}$, the associative law holds: $p + (q+s) = (p+q) + s$.*

Proof. Denoting, as on Figure 8.4, by $-x$, p_1, q_1 lines forming DRCs with triplets $(\mathcal{O}, p, q)$, $(\mathcal{O}, p, s)$, $(\mathcal{O}, q, s)$, and applying Lemma 8.11, we see that $(p+q) + s$ is the unique line forming a DRC with triplets (p_1, q_1, s), $(p, p_1, -x)$, $(q, q_1, -x)$. The same holds for line $p + (q + s)$, thus the two lines coincide. $\square$

Lemma 8.13. *Let $p, q, s \in \mathcal{C}_\mathcal{O}$. Then $-(p+q+s) = (-p) + (-q) + (-s)$.*

Proof. It is straightforward to prove that $p + q + s \in \mathcal{C}_\mathcal{O}$ if and only if two of the lines p, q, s are inverse to each other. In this case, the equality we need to prove immediately follows.

Thus, suppose $p + q + s$ does not intersect $\mathcal{O}$ and that a, b are lines forming a DRC with $p + q + s, \mathcal{O}$. By Lemma 8.11, we have that $p + q + s$ is the unique line intersecting $(-p) + (-q)$, $(-p) + (-s)$ and $(-q) + (-s)$. Reflecting p, q, s, a, b on $\mathcal{Q}_\mathcal{O}$ and applying the Double Reflection Theorem, we get that $-(p + q + s)$ intersects lines $p + q$, $p + s$, $q + s$. Thus, it is equal to $(-p) + (-q) + (-s)$, by Lemma 8.11. $\square$

Suppose now two lines $x, y \in \mathcal{A}_\ell$ are given, none of which is intersecting $\mathcal{O}$. We define their sum as follows.

First, we represent x as a sum of two lines intersecting $\mathcal{O}$: $x = s_1 + s_2$, $s_1, s_2 \in \mathcal{C}_\mathcal{O}$. Then, we define

$$x + y := s_1 + (s_2 + y).$$

We need to show that this definition is correct.

Lemma 8.14. *Let $s_1, s_2 \in \mathcal{C}_\mathcal{O}$, and $y \in \mathcal{A}_\ell$. Then*

$$s_1 + (s_2 + y) = s_2 + (s_1 + y).$$

Proof. If $y \in \mathcal{C}_\mathcal{O}$, it is enough to apply Lemma 8.12 and the commutativity property for the addition in $\mathcal{C}_\mathcal{O}$.

If $y \in \mathcal{A}_\ell \setminus \mathcal{C}_\mathcal{O}$, then the lines $-s_2 - y$ and y intersect with billiard reflection, as well as lines $-s_1 - y$ and y do. Thus, there is a unique, fourth line in the intersection of the space generated by $[-s_1 - y, y, -s_2 - y]$ and $\mathcal{A}_\ell$. On one hand, this line is equal to $s_2 - (-s_1 - y)$, on the other it is equal to $s_1 - (-s_2 - y)$.   □

Now, from Lemmata 8.12–8.14, it follows that we have constructed on $\mathcal{A}_\ell$ a commutative group structure that is naturally connected with the billiard law.

Further Properties of the Operation.   We started this section with The Leading Principle, and used it as a natural motivation for the construction of an algebra. Now, we are going to show that this is justified, i.e., that the statement formulated as The Leading Principle really holds in the group structure.

Theorem 8.15. *The sum of the lines in any double reflection configuration is equal to zero.*

Proof. Let a, b, c, d be the lines of a DRC. Assume that pairs a, b and c, d satisfy the reflection law on $\mathcal{Q}_1$, and pairs b, c, d, a on $\mathcal{Q}_2$. Then, by Lemma 8.11, we have:

$$b = -a + s_1, \quad c = -b + s_2, \quad d = -c + \bar{s}_1, \quad a = -d + \bar{s}_2, \tag{8.4}$$

where $s_i, \bar{s}_i$ are obtained from $\mathcal{O}$ by the billiard reflection on $\mathcal{Q}_i$, $i = 1, 2$. Obviously, $\bar{s}_i \in \{s_i, -s_i\}$.

From (8.4), we get

$$a + b + c + d = s_1 + \bar{s}_1 = s_2 + \bar{s}_2. \tag{8.5}$$

Thus, we only need to check the case when $\bar{s}_1 = s_1$ and $\bar{s}_2 = s_2$. Then we have $s_2 = s_1 + s_1 + (-s_2)$ and, from the definition of the addition operation, it follows that lines $\mathcal{O}, s_1, -2s_1, s_2$ constitute a closed billiard trajectory with consecutive reflections on $\mathcal{Q}_1, \mathcal{Q}_1, \mathcal{Q}_2, \mathcal{Q}_2$. On the other hand, $(-s_1) + (-s_2)$ is the unique line in $\mathcal{A}_\ell$ that, besides $\mathcal{O}$, intersects both s_1 and s_2. Thus,

$$-2s_1 = (-s_1) + (-s_2) \;\Rightarrow\; s_1 = s_2 \;\Rightarrow\; \mathcal{Q}_1 = \mathcal{Q}_2.$$

This means that the double reflection configuration a, b, c, d is degenerated, i.e., two of the lines in the configuration coincide. Say $a = c$, then both b and d are obtained from a by the billiard reflection on $\mathcal{Q}_1$. So, $b = -a + s_1$, $d = -a - s_1$ and the theorem follows.   □

Let us consider a billiard trajectory $\mathbf{t} = (\ell_0, \ell_1, \dots, \ell_n)$ with the initial line $\ell_0 = \mathcal{O}$ and the reflections on quadrics $\mathcal{Q}_1, \dots, \mathcal{Q}_n$ from the confocal family. Applying the Double Reflection Theorem to $\mathbf{t}$, we obtain different trajectories with shared initial segment $\mathcal{O}$ and final segment ℓ_n. The reflections on each such a trajectory are also on the same set of quadrics $\mathcal{Q}_1, \dots, \mathcal{Q}_n$, but their order may be changed. Let us denote by $\ell_1^{(1)}, \dots, \ell_n^{(n)}$ all possible lines obtained after the first reflection on all these trajectories. Then, in our algebra, the final segment ℓ_n can be calculated from these ones.

Proposition 8.16. $\ell_n = (-1)^{n+1}(\ell_1^{(1)} + \dots + \ell_n^{(n)})$.

Proof. Follows from Theorem 8.15. $\qquad\qquad\qquad\qquad\qquad\qquad\qquad\square$

The following interesting property also may be proved from Lemmata 8.12–8.14.

Proposition 8.17. *Let line $x \in \mathcal{A}_\ell$ be obtained from $\mathcal{O}$ by consecutive reflections on three quadrics $\mathcal{Q}_1$, $\mathcal{Q}_2$, $\mathcal{Q}_3$. Then the six lines obtained from $\mathcal{O}$ by reflections of these quadrics may be divided into two groups such that:*

- *for each $i \in \{1, 2, 3\}$, the lines obtained from $\mathcal{O}$ by the reflections on $\mathcal{Q}_i$ are in different groups;*
- *all trajectories with three reflections on $\mathcal{Q}_1$, $\mathcal{Q}_2$, $\mathcal{Q}_3$ starting with $\mathcal{O}$ and ending with x contain lines only from one of the groups.*

We will finish this subsection with a very beautiful and non-trivial theorem on confocal families of quadrics.

Theorem 8.18. *Let $\mathcal{F}$ be a family of confocal quadrics in $\mathbf{P}^3$. There exist configurations consisting of 12 planes in $\mathbf{P}^3$ with the following properties:*

- *The planes may be organized in eight triplets, such that each plane in a triplet is tangent to a different quadric from $\mathcal{F}$ and the three touching points are collinear. Every plane in the configuration is a member of two triplets.*
- *The planes may be organized in six quadruplets, such that the planes in each quadruplet belong to a pencil and they are tangent to two different quadrics from $\mathcal{F}$. Every plane in the configuration is a member of two quadruplets.*

Moreover, such a configuration is determined by three planes tangent to three different quadrics from $\mathcal{F}$, with collinear touching points.

Proof. Denote by $\mathcal{O}$ the line containing the three touching points and p, q, s the lines obtained from $\mathcal{O}$ by the billiard reflection on the given quadrics from $\mathcal{F}$. We construct lines $p_1, q_1, -x, x + s$ as explained before Lemma 8.11 (see Figure 8.4). The planes of the configuration are tangent to corresponding quadrics at points of the intersection of the lines.

The configuration of the planes in the dual space $\mathbf{P}^{3*}$ is shown on Figure 8.5. Here, each plane is denoted by two lines that reflect to each other on the corresponding quadric. $\qquad\qquad\qquad\qquad\qquad\qquad\qquad\qquad\qquad\qquad\qquad\square$

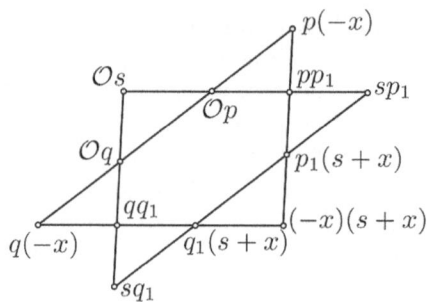

Figure 8.5: The configuration of planes

It would be interesting to describe the variety of such configurations as a moduli-space.

General Case

The genus 2 case we have just considered in detail gives us the necessary experience of using the billiard constructions to build a group structure. Moreover, it provides us with the case $n = 2$ in the general construction we are going to present now.

Billiard Trajectories and Effective Divisors. Proposition 8.16 is going to serve as the motivation for defining the operation in the set $\mathcal{A}_\ell$ for higher dimensions. Thus, let us now describe in detail the construction that preceded this proposition in Section 8.2.

Suppose that quadrics $\mathcal{Q}_1, \ldots, \mathcal{Q}_n$ from the confocal family in $\mathbf{E}^d$ are given. Let $\mathcal{O} = \ell_0, \ell_1, \ldots, \ell_n$ be lines in $\mathcal{A}_\ell$ such that each pair of successive lines ℓ_i, ℓ_{i+1} satisfies the billiard reflection law at the quadric $\mathcal{Q}_{i+1}$ ($0 \leq i \leq n-1$); thus the lines form billiard trajectory $\mathbf{t} = (\ell_0, \ldots, \ell_n)$.

Let us, for $n \geq 1$, define lines $\ell_n^{(n)}, \ell_n^{(n-1)}, \ldots, \ell_n^{(1)}$ by the procedure, as follows:

1. Set $\ell_n^{(n)} = \ell_n$.
2. $\ell_n^{(k)}$, for $n-1 \geq k \geq 1$ is the unique line that constitutes a DRC with ℓ_{k-1}, ℓ_k and $\ell_n^{(k+1)}$.

Let us note that, in this way, each line in the sequence

$$\ell_0, \ \ell_1, \ \ldots, \ \ell_k, \ \ell_n^{(k+1)}, \ \ldots, \ \ell_n^{(n)}$$

is obtained from the previous one by the reflection from

$$\mathcal{Q}_1, \ \ldots, \ \mathcal{Q}_k, \ \mathcal{Q}_n, \ \mathcal{Q}_{k+1}, \ \ldots, \ \mathcal{Q}_{n-1}$$

respectively.

Considering only the initial subsequences $\ell_0, \ldots, \ell_k$ $(1 \leq k \leq n)$, we may in the same way define lines $\ell_k^{(1)}, \ldots, \ell_k^{(k)}$. Notice that, for each k, $\ell_k^{(1)}$ intersects ℓ_0, and these two lines obey the reflection law on $\mathcal{Q}_k$.

Thus, we have constructed a mapping $\mathcal{D}$ from the set $\mathcal{TB}(\mathcal{O})$ of billiard trajectories with the fixed initial line $\ell_0 = \mathcal{O}$ to the ordered sets of lines from $\mathcal{A}_\ell$ which intersect $\mathcal{O}$:

$$\mathcal{D} \ : \ \mathbf{t} = (\ell_0, \ldots, \ell_n) \ \mapsto \ (\ell_1^{(1)}, \ldots, \ell_n^{(1)}),$$

where line $\ell_k^{(1)}$ intersects $\mathcal{O}$ according to the billiard law on the quadric $\mathcal{Q}_k$.

Doing the opposite procedure, we define morphism $\mathcal{B}$, the inverse of $\mathcal{D}$, which assigns to an n-tuple of lines intersecting $\mathcal{O}$ the unique billiard trajectory of length n with $\ell_0 = \mathcal{O}$ as the initial line:

$$\mathcal{B} \ : \ (\ell_1^{(1)}, \ldots, \ell_n^{(1)}) \ \mapsto \ (\ell_0, \ldots, \ell_n).$$

Hence, the mapping $\mathcal{B}$ gives *the billiard representation* of an ordered set of lines intersecting $\mathcal{O}$.

In order to consider just divisors of the curve $\mathcal{C}_\mathcal{O}$, instead of ordered n-tuples of lines, we need to introduce the following relation α between billiard trajectories: we say that two billiard trajectories are α-*equivalent* if one can be obtained from the other by a finite set of *double reflection moves*. The double reflection move transforms a trajectory $p_1 p_2 \ldots p_{k-1} p_k p_{k+1} \ldots p_n$ into trajectory $p_1 p_2 \ldots p_{k-1} p_k' p_{k+1} \ldots p_n$ if the lines $p_{k-1}, p_k, p_k', p_{k+1}$ form a double reflection configuration.

It follows directly from The Double Reflection Theorem that two trajectories $\mathbf{t}_1, \mathbf{t}_2$ with the same initial segment $\mathcal{O} = \ell_0$ are α-equivalent if and only if the n-tuples $\mathcal{D}(\mathbf{t}_1)$ and $\mathcal{D}(\mathbf{t}_2)$ may be obtained from each other by a permutation.

Thus, the mapping $\mathcal{D}$ may be considered as a mapping from

$$\widehat{\mathcal{TB}}(\mathcal{O}) = \mathcal{TB}(\mathcal{O})/\alpha$$

to the set of positive divisors on $\mathcal{C}_\mathcal{O}$ and it represents *the divisor representation* of billiard trajectories.

The following lemma, which follows from [Don1980], is a sort of Riemann–Roch theorem in this approach.

Lemma 8.19. *A minimal billiard trajectory of length s from x to y is unique, up to the relation α. If there are two non α-equivalent trajectories of the same length $k > s$ from x to y, then there is a trajectory from x to y of length $k - 2$.*

Now, we are able to introduce an operation, *summation*, in the set $\widehat{\mathcal{TB}}(\mathcal{O})$. Given two billiard trajectories $\mathbf{t}_1, \mathbf{t}_2 \in \widehat{\mathcal{TB}}(\mathcal{O})$, we define their sum by the equation

$$\mathbf{t}_1 \oplus \mathbf{t}_2 := \mathcal{B}(\mathcal{D}(\mathbf{t}_1) + \mathcal{D}(\mathbf{t}_2)).$$

This operation is associative and commutative, according to The Double Reflection Theorem.

Theorem 8.20. *The set $\widehat{\mathcal{TB}}(\mathcal{O}) = \mathcal{TB}(\mathcal{O})/\alpha$ with the summation operation is a commutative semigroup. This semigroup is isomorphic to the semigroup of all effective divisors on the curve $\mathcal{C}_\mathcal{O}$ that do not contain the points corresponding to the caustics.*

Note that in $\widehat{\mathcal{TB}}(\mathcal{O})$, the trajectory consisting only of the single line $\mathcal{O}$ is a neutral element.

Define now the following equivalence relation in the set of all finite billiard trajectories.

Definition 8.21. Two billiard trajectories are β-*equivalent* if they have common initial and final segments, and they are of the same length.

The set of all classes of β-equivalent billiard trajectories of length n, with the fixed initial segment $\mathcal{O}$, we denote by $\widehat{\mathcal{TB}}(\mathcal{O})(n)$, and

$$\widehat{\mathcal{TB}}(\mathcal{O}) = \bigcup_n \widehat{\mathcal{TB}}(\mathcal{O})(n).$$

Proposition 8.22. *The relation β is compatible with the addition of billiard trajectories.*

To prove this, we will need the following important lemma:

Lemma 8.23. *Let $\mathbf{t} = (\ell_1, \ell_2, \ldots, \ell_{2k}, \ell_{2k+1} = \ell_1)$ be a closed billiard trajectory, and $p_1 \in \mathcal{A}_\ell$ a line that intersects ℓ_1. Construct iteratively lines $p_2, \ldots, p_{2k+1}$ such that quadruples $p_i, \ell_i, p_{i+1}, \ell_{i+1}$ $(1 \le i \le 2k)$ form double reflection configurations. Then $p_{2k+1} = p_1$.*

Proof. We are going to proceed with the induction. For $k = 2$, the lines $\ell_1, \ell_2, \ell_3, \ell_4$ form a double reflection configuration, and the statement follows by The Double Reflection Theorem.

Now, suppose $k > 2$. $(\ell_1, \ldots, \ell_k)$ and $(\ell_1 = \ell_{2k+1}, \ell_{2k}, \ldots, \ell_k)$ are two billiard trajectories of the same length from ℓ_1 to ℓ_k. If they are α-equivalent, then the claim follows by The Double Reflection Theorem. If they are not α-equivalent, then, by Lemma 8.19, there is a billiard trajectory $\ell_1' = \ell_1, \ell_2', \ldots, \ell_{k-2}' = \ell_k$. The statement now follows from the inductive hypothesis applied to trajectories

$$(\ell_1, \ell_2, \ldots, \ell_k = \ell_{k-2}', \ell_{k-3}', \ldots, \ell_1' = \ell_1)$$

and

$$(\ell_1', \ell_2', \ldots, \ell_{k-2}' = \ell_k, \ell_{k+1}, \ldots, \ell_{2k+1} = \ell_1). \qquad \square$$

Proof of Proposition 8.22. We need to prove the following:

$$\mathbf{t}_1 \sim_\beta \mathbf{t}_1', \ \mathbf{t}_2 \sim_\beta \mathbf{t}_2' \ \Rightarrow \ \mathbf{t}_1 + \mathbf{t}_2 \sim_\beta \mathbf{t}_1' + \mathbf{t}_2'.$$

Clearly, it is enough to prove this relation for the case when $\mathbf{t}_2 = \mathbf{t}_2'$ and the length of $\mathbf{t}_2$ is equal to 2. Suppose that $\mathbf{t}_2 = (\mathcal{O}, p)$, where p is obtained from $\mathcal{O}$ by the reflection on quadric $\mathcal{Q}_p$. Then trajectories $\mathbf{t}_1 + \mathbf{t}_2$, $\mathbf{t}_1 + \mathbf{t}_2'$ are obtained by adding one segment to $\mathbf{t}_1$, $\mathbf{t}_1'$ respectively. These segments satisfy the reflection law on $\mathcal{Q}_p$ with the final segments of $\mathbf{t}_1$, $\mathbf{t}_1'$. Since $\mathbf{t}_1 \sim_\beta \mathbf{t}_1'$, their final segments coincide and the statement follows from Lemma 8.23 applied to the trajectory

$$(\ell_1, \ell_2, \ldots, \ell_n = \ell_n', \ell_{n-1}', \ell_{n-2}', \ldots, \ell_1' = \ell_1),$$

where $\mathbf{t}_1 = (\ell_1, \ldots, \ell_n)$, $\mathbf{t}_2 = (\ell_1', \ldots, \ell_n')$. $\qquad \square$

The Group Structure in $\mathcal{A}_\ell$. We wish to use the constructed algebra on the set of billiard trajectories, in order to obtain an algebraic structure on $\mathcal{A}_\ell$, such that the operation is naturally connected with the billiard reflection law.

From [Tyu1975] we get

Theorem 8.24. *For any two given lines x and y from $\mathcal{A}_\ell$, there is a system of at most $d-1$ quadrics from the confocal family, such that the line y is obtained from x by consecutive reflections on these quadrics.*

The divisor representation of the corresponding billiard trajectory of length $s \le d-1$ will be called *the s-brush of y related to x.*

Now, we can define a group structure in $\mathcal{A}_\ell$ associated with a fixed line in this set.

Neutral element

Let us fix a line $\mathcal{O} \in \mathcal{A}_\ell$.

Inverse element

Let x be an arbitrary line in $\mathcal{A}_\ell$, and $\mathcal{D}(x)$ the divisor representation of the minimal billiard trajectory connecting $\mathcal{O}$ with x. Define $-x$ as the final segment of the billiard trajectory $\mathcal{B}(\tau \mathcal{D}(x))$, where τ is the hyperelliptic involution of $\mathcal{C}_\mathcal{O}$.

Addition

For two lines x and y from $\mathcal{A}_\ell$, denote their brushes related to $\mathcal{O}$ as S_1 and S_2. Define

$$x + y := (-1)^{|S_1| + |S_2| + 1} \mathcal{E}\mathcal{B}(\tau^{|S_1|+1}(S_1), \tau^{|S_2|+1}(S_2)),$$

where $\mathcal{E}\mathcal{B}(\tau^{|S_1|+1}(S_1), \tau^{|S_2|+1}(S_2))$ is the final segment of the billiard trajectory $\mathcal{B}(\tau^{|S_1|+1}(S_1), \tau^{|S_2|+1}(S_2))$.

From all above, similarly as in the genus 2 case, we get

Theorem 8.25. *The set $\mathcal{A}_\ell$ with the operation defined above is an Abelian group.*

Let us observe that another kind of divisor representation of billiard trajectories may be introduced, that associates a divisor of degree 0 to a given trajectory. Such a representation was used in [Dar1914] and explicitly described in [DR2004, DR2006b]. The positive part of this representation coincides with the divisor obtained by $\mathcal{B}$, and the negative part is invariant for the hyperelliptic involution. Moreover, as it follows from [Dar1914], two trajectories with the initial segment $\mathcal{O}$ will have the same final segments if and only if their representations are equivalent divisors. Thus, it follows that the group structure on $\mathcal{A}_\ell$ is isomorphic to the quotient of the Jacobian of the curve $\mathcal{C}_\mathcal{O}$ by a finite subgroup which is generated by the points corresponding to the caustic quadrics.

8.3   *s*-weak Poncelet trajectories

Now, as a consequence of Theorem 8.24 we are able to introduce the following hierarchies of notions.

Definition 8.26. For two given lines x and y from $\mathcal{A}_\ell$ we say that they are *s-skew* if s is the smallest number such that there exist a system of $s+1 \le d-1$ quadrics $\mathcal{Q}_k$, $k = 1, \ldots, s+1$ from the confocal family, such that the line y is obtained from x by consecutive reflections on $\mathcal{Q}_k$. If the lines x and y intersect, they are *0-skew*. They are (-1)-*skew* if they coincide.

Definition 8.27. Suppose that a system S of n quadrics $\mathcal{Q}_1, \ldots, \mathcal{Q}_n$ from the confocal family is given. For a system of lines $\mathcal{O}_0, \mathcal{O}_1, \ldots, \mathcal{O}_n$ in $\mathcal{A}_\ell$ such that each pair of successive lines $\mathcal{O}_i, \mathcal{O}_{i+1}$ satisfies the billiard reflection law at $\mathcal{Q}_{i+1}$ $(0 \le i \le n-1)$, we say that it forms an *s-weak Poncelet trajectory of length n associated to the system S* if the lines $\mathcal{O}_0$ and $\mathcal{O}_n$ are *s*-skew.

For *s*-weak Poncelet trajectories we will also sometimes say $(d-s-2)$-*resonant billiard trajectories*. Periodic trajectories or generalized classical Poncelet polygons are (-1)-weak Poncelet trajectories or, in other words, $(d-1)$-resonant billiard trajectories, and they are described analytically in [DR2006b, DR2004], see Theorems 7.14, 7.2, 7.7.

Our next goal is to get a complete analytical description of *s*-weak Poncelet trajectories of length r, generalizing in such a way the results from [DR2006b]. Here, we are going to use fully the tools and the power of billiard algebra.

To fix the idea, let us consider first the system S consisting of r equal quadrics $\mathcal{Q}_1 = \cdots = \mathcal{Q}_r$ from the confocal family. Suppose a system of lines $\mathcal{O}_0, \mathcal{O}_1, \ldots, \mathcal{O}_r$ in $\mathcal{A}_\ell$ forms an *s*-weak Poncelet trajectory of length r associated to the system S. Then

$$\mathcal{O}_r = r\mathcal{O}_1^{(1)},$$

with some line $\mathcal{O}_1^{(1)}$ which intersects $\mathcal{O}_0$. Again, from the condition that $\mathcal{O}_r$ and $\mathcal{O}_0$ are *s*-skew we get

$$\mathcal{O}_r = \mathcal{O}_1'^{(1)} + \cdots + \mathcal{O}_{s+1}'^{(1)}$$

with some lines $P_i = \mathcal{O}_i'^{(1)}$ which intersect $\mathcal{O}_0$. From the last two equations we come to the conclusion

Proposition 8.28. *The existence of an s-weak Poncelet trajectory of length r is equivalent to existence of a meromorphic function f on the hyperelliptic curve $\mathcal{C}_\ell$ such that f has a zero of order r at $P = \mathcal{O}_1^{(1)}$, a unique pole at "infinity" E, and the order of the pole is equal to $n = r + s + 1$.*

Now, we are going to derive an explicit analytical condition of Cayley's type. As in [DR2004] we consider the space $\mathcal{L}(nE)$ of all meromorphic functions on $\mathcal{C}_\ell$ with the unique pole at the infinity point E of order not exceeding n. Let $(f_1, \ldots, f_k)$ be one of the bases of this space, $k = \dim \mathcal{L}(nE)$. Consider the vectors

$$v_1, \ldots, v_r \in \mathbf{C}^k,$$

where $v_i^j = f_j^{(i-1)}(P)$ and vectors

$$u_1, \ldots, u_{s+1} \in \mathbf{C}^k,$$

with $u_i^j = f_j(P_i)$. From the condition (see [DR2004])

$$\mathrm{rank}\,[v_1, \ldots, v_r, u_1, \ldots, u_{s+1}] < n - g + 1$$

we get the condition

$$\mathrm{rank}\,[v_1, \ldots, v_r] < r + s - g + 2 = r + s - d + 3.$$

Now, we can rewrite it in the form usual for Cayley-type conditions.

Theorem 8.29. *The existence of an s-weak Poncelet trajectory of length r is equivalent to*

$$\mathrm{rank} \begin{pmatrix} B_{d+1} & B_{d+2} & \cdots & B_{m+1} \\ B_{d+2} & B_{d+3} & \cdots & B_{m+2} \\ \cdots & \cdots & \cdots & \cdots \\ B_{d+m-s-2} & B_{d+m-s-1} & \cdots & B_{r-1} \end{pmatrix} < m - d + 1,$$

when $r + s + 1 = 2m$, and

$$\mathrm{rank} \begin{pmatrix} B_d & B_{d+1} & \cdots & B_{m+1} \\ B_{d+1} & B_{d+2} & \cdots & B_{m+2} \\ \cdots & \cdots & \cdots & \cdots \\ B_{d+m-s-2} & B_{d+m-s-1} & \cdots & B_{r-1} \end{pmatrix} < m - d + 2,$$

when $r + s + 1 = 2m + 1$.

With $B_0, B_1, B_2, \ldots$, we denoted the coefficients in the Taylor expansion of function $y = \sqrt{\mathcal{P}(x)}$ in a neighbourhood of P, where $y^2 = \mathcal{P}(x)$ is the equation of the generalized Cayley curve, with the polynomial $\mathcal{P}(x)$ given by (8.1).

Proof. Denote by $\mathcal{L}((r+s+1)E)$ the linear space of all meromorphic functions on $\mathcal{C}_\ell$, having a unique pole at the infinity point E, with order not exceeding $r+s+1$. By the Riemann–Roch theorem:

(i) $\dim \mathcal{L}((r+s+1)E) = [\frac{r+s+1}{2}] + 1$ if $r+s+1 \leq 2d-1$,

(ii) $\dim \mathcal{L}((r+s+1)E) = r+s-g+1$ if $r+s+1$ is even and greater than $2d-2$,

(iii) $\dim \mathcal{L}((r+s+1)E) = r+s-g+2$ if $r+s+1$ is even and greater than $2d-2$.

We may choose the following bases in each of the three cases:

(i) $1, x, \ldots, x^m$, where $m = [\frac{r+s+1}{2}] \leq d$;

(ii) $1, x, \ldots, x^m, y, xy, \ldots, x^{m-d}$, for $r+s+1 = 2m \geq 2d-2$;

(iii) $1, x, \ldots, x^m, y, xy, \ldots, x^{m-d+1}$, for $r+s+1 = 2m+1 \geq 2d-2$.

Now, the statement follows from the considerations preceding the theorem. $\qquad\square$

Example 8.30. For $s = -1$, the inequalities in Theorem 8.29 become the conditions for periodic billiard trajectories (see [DR2004, DR2006b]).

Example 8.31. The condition for existence of a $(d-3)$-weak Poncelet trajectory of length r is equivalent to

$$\det \begin{pmatrix} B_{d+1} & B_{d+2} & \cdots & B_{m+1} \\ B_{d+2} & B_{d+3} & \cdots & B_{m+2} \\ \cdots & \cdots & \cdots & \cdots \\ B_{m+1} & B_{m+2} & \cdots & B_{r-1} \end{pmatrix} = 0, \quad r+d-2 = 2m;$$

$$\det \begin{pmatrix} B_d & B_{d+1} & \cdots & B_{m+1} \\ B_{d+1} & B_{d+2} & \cdots & B_{m+2} \\ \cdots & \cdots & \cdots & \cdots \\ B_{m+1} & B_{m+2} & \cdots & B_{r-1} \end{pmatrix} = 0, \quad r+d-2 = 2m+1.$$

Example 8.32. If $r+s < 2d-1$, then, as it follows from the proof of Theorem 8.29, an s-weak Poncelet trajectory of length r may exist only if the hyperelliptic curve $\mathcal{C}_\ell$ is singular.

8.4   On generalized Weyr's theorem and the Griffiths–Harris space Poncelet theorem in higher dimensions

In this section we obtain higher-dimensional generalizations of the results of [Wey1870], [Hur1879], [GH1977]. A nice exposition of those classical results can be found in [BB1996]. The dual version of [GH1977] is given in [CCS1993].

Each quadric $\mathcal{Q}$ in $\mathbf{P}^{2d-1}$ contains at most two unirational families of $(d-1)$-dimensional linear subspaces. Such unirational families are usually called *rulings of the quadric*.

Theorem 8.33. *Let $\mathcal{Q}_1$, $\mathcal{Q}_2$ be two general quadrics in $\mathbf{P}^{2d-1}$ with the smooth intersection V and $\mathcal{R}_1$, $\mathcal{R}_2$ their rulings. If there exists a closed chain*

$$L_1, \; L_2, \; \ldots, \; L_{2n}, \; L_{2n+1} = L_1$$

of distinct $(d-1)$-dimensional linear subspaces, such that $L_{2i-1} \in \mathcal{R}_1$, $L_{2i} \in \mathcal{R}_2$ $(1 \leq i \leq n)$ and $L_j \cap L_{j+1} \in F(V)$ $(1 \leq j \leq 2n)$, then there are such closed chains of subspaces of length $2n$ through any point of $F(V)$. ($F(V)$ is the set of all $(d-2)$-dimensional linear subspaces of V, as defined in Section 8.2.)

Proof. Each of the unirational families $\mathcal{R}_i$ determines an involution τ_i on an Abelian variety $F(V)$. Such an involution interchanges two $(d-2)$-intersections of an element of $\mathcal{R}_i$ with V. Denote by $\mathrm{Tr} : F(V) \to F(V)$ their composition and by $L := L_{2n} \cap L_1 \in F(V)$. Since Tr is a translation on $F(V)$ satisfying $\mathrm{Tr}^n(L) = L$ we see that Tr is of order n and the theorem follows. $\square$

Definition 8.34. We will call the chains considered in Theorem 8.33 *generalized Weyr's chains*.

The theorem can be adjusted for nonsmooth intersections, but we are not going into details. Instead, we consider the case of two quadrics (8.2) and (8.3) as in Section 8.1. By using the projection π^* we get

Proposition 8.35. *A generalized Weyr chain of length $2n$ projects into a Poncelet polygon of length $2n$ circumscribing the quadrics $\mathcal{Q}^p_{\alpha_1}, \ldots, \mathcal{Q}^p_{\alpha_{d-1}}$ and alternately inscribed into two fixed confocal quadrics (projections of $\mathcal{Q}_1, \mathcal{Q}_2$). Conversely, any such a Poncelet polygon of the length $2n$ circumscribing the quadrics $\mathcal{Q}^p_{\alpha_1}, \ldots, \mathcal{Q}^p_{\alpha_{d-1}}$ and alternately inscribed into two fixed confocal quadrics can be lifted to a generalized Weyr chain of length $2n$.*

Proof. It follows from Lemma 4.1 and Corollary 4.2 of [Knö1980] and Lemma 8.4. $\square$

Thus, we obtained in a correspondence between generalized Weyr chains and Poncelet polygons subscribed in $d-1$ given quadrics and alternately inscribed in two quadrics from some confocal family. Such Poncelet polygons have been completely analytically described, among others, in [DR2006b] (see Example 4 from there).

Let us note that a correspondence between the classical Weyr's theorem in $\mathbf{P}^3$ and the classical Poncelet theorem about two conics in a plane was observed by Hurwitz in [Hur1879]. Nevertheless the projection we used here is not a straightforward generalization of the one used by Hurwitz.

Polygonal lines, circumscribed about one conic and alternately inscribed in two conics, appear as an example in [Ves1990]. Our result, applied to the lowest dimension, gives the condition for the closeness of such a polygonal line, see Corollary 1 from [DR2006b].

Let us also mention that if rulings $\mathcal{R}_1$ and $\mathcal{R}_2$ are connected by a generalized Weyr's chain of length $2n$, then the same is true for $\mathcal{R}_2$, $\mathcal{R}_1$ and also for the pair $\mathcal{R}_1'$, $\mathcal{R}_2'$ of the complementary rulings of $\mathcal{Q}_1$ and $\mathcal{Q}_2$.

Now we are able to present a new higher-dimensional generalization of the Griffiths–Harris Space Poncelet theorem from [GH1977].

Theorem 8.36. *Let $\mathcal{Q}_1^*$ and $\mathcal{Q}_2^*$ be the duals of two general quadrics in $\mathbf{P}^{2d-1}$ with the smooth intersection V. Denote by $\mathcal{R}_i, \mathcal{R}_i'$ pairs of unirational families of $(d-1)$-dimensional subspaces of $\mathcal{Q}_i^*$. Suppose there are generalized Weyr's chains between $\mathcal{R}_1$ and $\mathcal{R}_2$ and between $\mathcal{R}_1$ and $\mathcal{R}_2'$. Then there is a finite arrangement inscribed and subscribed in both quadrics $\mathcal{Q}_1$ and $\mathcal{Q}_2$. There are infinitely many such arrangements.*

The arrangements from the previous theorem can be described in more details as arrangements of d-dimensional "faces" which are bitangents of the quadrics $\mathcal{Q}_1$ and $\mathcal{Q}_2$. Their intersections are $(d-1)$-dimensional "edges" which alternately belong to the rulings of $\mathcal{Q}_1^*$ and $\mathcal{Q}_2^*$. The intersections of "edges" are $(d-2)$-dimensional "vertices" which belong to $F(V)$.

The proof of the last theorem is based on the following lemma from [Don1980].

Lemma 8.37. *Let $\mathcal{Q} \subset \mathbf{P}^{2d-1}$ be a quadric of rank not less than $2d-1$ and $x \subset \mathcal{Q}$ a linear subspace of the dimension $d-2$. If $\mathcal{Q}$ is singular, suppose additionally that x does not contain the vertex. Then, for each ruling $\mathcal{R}$ of $\mathcal{Q}$, there is a unique $(d-1)$-dimensional linear subspace $s = s(\mathcal{R}, x)$ such that $s \in \mathcal{R}$ and $x \subset s$.*

8.5 Poncelet–Darboux grid and higher-dimensional generalizations

Historical Remark: on Darboux's Heritage. In [Dar1914, Volume 3, Book VI, Chapter I], Darboux proved that Liouville surfaces are exactly those having an orthogonal system of curves that can be regarded in two or, equivalently, infinitely many different ways, as geodesic conics. These coordinate curves are analogues of systems of confocal conics in the Euclidian plane. Knowing this, essential properties of conics may be generalized to all Liouville surfaces (see [Dar1914], and also [DR2006b] for a review and clarifications). Here is the citation of one of these properties:

"Considerons un polygone variable dont tous les côtés sont des lignes géodésiques tangentes a une même courbe coordonnée. Le dernier sommet de ce polygone décrira aussi une des courbes coordonnées et il en sera de même de points d'intersection de deux côtés quelconques de ce polygone."

This statement is not only the generalization of Poncelet's theorem to Liouville surfaces. Additionally, for a Poncelet polygon circumscribed about a fixed coordinate curve, with each vertex moving along one of the coordinate curves, Darboux stated here that the intersection point of two arbitrary sides of the polygon is also going to describe a coordinate curve.

Let us note that a weaker version of this claim, although with some improvements, has been recently rediscovered in [Sch2007], and a more elementary proof of the result from [Sch2007] has been published in [LT2007]. The main theorem of [Sch2007] corresponds only to a fixed Poncelet polygon associated to a pair of ellipses in the Euclidean plane. In a sense, this gives a new argument in favour of our observation in [DR2004] and [DR2006b], that the work of Darboux connected with billiards and Poncelet's theorem seems to be unknown to today's mathematicians.

This section is devoted to a multi-dimensional generalization of the Darboux theorem, related to billiard trajectories within an ellipsoid in the d-dimensional Euclidean space.

Poncelet–Darboux Grid in Euclidean Plane. Before starting with higher-dimensional generalizations, let us give some improvements together with a simpler proof of the statement corresponding to the plane case.

Theorem 8.38. *Let $\mathcal{E}$ be an ellipse in $\mathbf{E}^2$ and $(a_m)_{m\in\mathbf{Z}}$, $(b_m)_{m\in\mathbf{Z}}$ be two sequences of the segments of billiard trajectories $\mathcal{E}$, sharing the same caustic. Then all the points $a_m \cap b_m$ $(m \in \mathbf{Z})$ belong to one conic $\mathcal{K}$, confocal with $\mathcal{E}$.*

Moreover, under the additional assumption that the caustic is an ellipse, we have: if both trajectories are winding in the same direction about the caustic, then $\mathcal{K}$ is also an ellipse; if the trajectories are winding in opposite directions, then $\mathcal{K}$ is a hyperbola. (See Figure 8.6.)

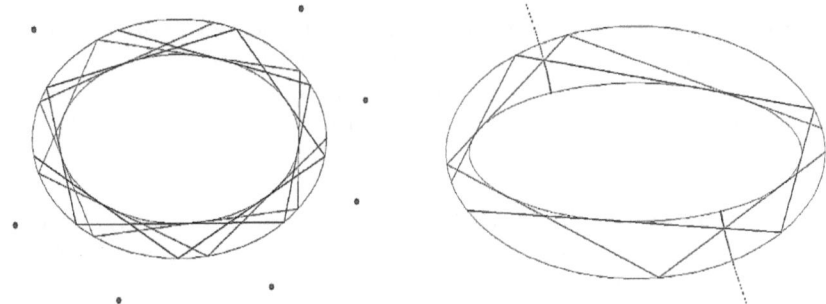

Figure 8.6: Billiard trajectories with an ellipse as a caustic and the intersection points of the corresponding segments

For a hyperbola as a caustic, it holds: if segments a_m, b_m intersect the long axis of $\mathcal{E}$ in the same direction, then $\mathcal{K}$ is a hyperbola, otherwise it is an ellipse. (See Figure 8.7.)

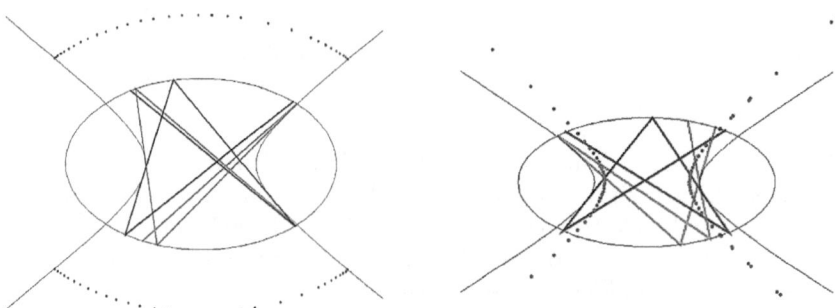

Figure 8.7: Billiard trajectories with a hyperbola as a caustic and the intersection points of the corresponding segments

Proof. The statement follows by application of the Double Reflection Theorem. Namely, the lines a_0 and b_0 intersect at a point that belongs to one ellipse and one hyperbola from the confocal family. They satisfy the reflection law on exactly one of these two curves, depending on the orientation of the billiard motion along the lines. Now, by the Double Reflection Theorem, a_1 and b_1 satisfy the reflection law on the same conic. The same is true, for any pair a_m, b_m, by induction.

For the second part of the theorem, it is sufficient to observe that the winding direction about an ellipse is changed by the reflections on the hyperbolae, and preserved by the reflections on the ellipses from the confocal family. If an oriented line is placed between the foci, then the direction in which it intersects the axis containing the foci is changed by the reflections of the ellipses and preserved by the reflections on the hyperbolae. □

Proposition 8.39. *Let $(a_m)_{m \in \mathbf{Z}}$, $(b_m)_{m \in \mathbf{Z}}$ be two sequences of the segments of billiard trajectories within the ellipse $\mathcal{E}$, sharing the same caustic. If the caustic is an ellipse and the trajectories are winding in the opposite directions about it, then all the points $a_m \cap b_m$ ($m \in \mathbf{Z}$) are placed on two centrally symmetric half-branches of a hyperbola confocal with $\mathcal{E}$.*

Proof. Denote by $\mathcal{E}_c$ the ellipse which is the common caustic of the given billiard trajectories. It is possible to introduce a metric μ on $\mathcal{E}_c$, such that $\mu(AB) = \mu(CD)$ if and only if the tangent lines at A, B and at C, D intersect on the same ellipse of the confocal family. In particular, this means that the points where two consecutive billiard segments touch the caustic $\mathcal{E}'$ are always at a fixed distance from each other.

Take that a_0 and b_0 intersect on the upper left half-branch of the hyperbola $\mathcal{K}$ (see Figure 8.6). Then it may be proved from The Double Reflection Theorem, that each touching points of these two segments with caustic $\mathcal{E}'$ are at equal μ-distances from the upper left intersection point of $\mathcal{K}$ and $\mathcal{E}_c$. Since μ is symmetric with respect to the coordinate centre, they are also equally distanced from the lower right intersection point, but not from the other two intersection points. The same holds for any pair a_m, b_m. Thus, their intersections lie on two centrally symmetric half-branches of the hyperbola. $\qquad\square$

Now, the generalized claim about Poncelet–Darboux grids follows immediately from Theorem 8.38.

Theorem 8.40. *Let $(\ell_m)_{m \in \mathbf{Z}}$ be the sequence of segments of a billiard trajectory within the ellipse $\mathcal{E}$. Then each of the sets $\mathrm{P}_k = \bigcup_{i-j=k} \ell_i \cap \ell_j$, $\mathrm{Q}_k = \bigcup_{i+j=k} \ell_i \cap \ell_j$, $(k \in \mathbf{Z})$ belongs to a single conic confocal to $\mathcal{E}$.*

If the caustic of the trajectory (ℓ_m) is an ellipse, then the sets P_k are placed on ellipses and Q_k on hyperbolae. If the caustic is a hyperbola, then the sets P_k, Q_k are placed on ellipses for k even and on hyperbolae for k odd.

Proof. To show the statement for P_k, take $a_m = \ell_m$, $b_m = \ell_{m+k}$ and apply Theorem 8.38. For Q_k, take $a_m = \ell_m$, $b_m = \ell_{k-m}$. $\qquad\square$

Remark 8.41. Notice that Theorem 8.40 is more general then the one given in [Sch2007, LT2007], since we do not suppose that the billiard trajectory is closed. Also, the statement can be formulated for an arbitrary conic, not only for an ellipse.

The main statement proved in [Sch2007] and [LT2007] is a special case consequence of Theorem 8.40:

Corollary 8.42 ([Sch2007], [LT2007])**.** *Let (ℓ_m) be a closed billiard trajectory within an ellipse $\mathcal{E}$, with the elliptical caustic. Each set P_k lies on an ellipse confocal to $\mathcal{E}$, and Q_k on a confocal hyperbola. (See Figure 8.6.)*

Let us show one interesting property of Poncelet–Darboux grids.

Proposition 8.43. *Let (ℓ_m) be a billiard trajectory within ellipse $\mathcal{E}$, with the elliptical caustic $\mathcal{E}_c$. Then the ellipse containing the set P_k depends only on k, $\mathcal{E}$ and $\mathcal{E}_c$. In other words, this ellipse will remain the same for any choice of billiard trajectory within $\mathcal{E}$ with the caustic $\mathcal{E}_c$.*

Proof. This claim may be proved by the use of the Double Reflection Theorem, similarly as in Theorem 8.38. Nevertheless, we are going to show it in another way, that gives a possibility of explicit calculation of the ellipse containing P_k.

To the billiard within $\mathcal{E}$ with the fixed caustic $\mathcal{E}_c$, we may associate an elliptic curve (see [Leb1942, GH1978a, Mos1980, MV1991]). Each point of the curve corresponds to a conic from the confocal family. On the other hand, the billiard motion may be viewed as the linear motion on the Jacobian of the curve (i.e., on the curve itself, since this is the elliptic case), with translation jumps corresponding to reflections. More precisely, the translation is exactly by this value on the elliptic curve, which is associated to the ellipse that a segment is reflected on.

Thus, if the point M on the elliptic curve is associated to the ellipse $\mathcal{E}$, then the set P_k is placed on the ellipse corresponding to kM.   □

Grids in Arbitrary Dimension.   Although Theorem 8.38 gives certain progress in understanding of Poncelet-Darboux grids, the essential breakthrough in this matter represents, in our opinion, the study of the higher-dimensional situation. This analysis is based on introduction of higher-dimensional analogues of grids and our notion of s-skew lines.

Theorem 8.44. *Let* $(a_m)_{m \in \mathbf{Z}}$, $(b_m)_{m \in \mathbf{Z}}$ *be two sequences of the segments of billiard trajectories within the ellipsoid $\mathcal{E}$ in $\mathbf{E}^d$, sharing the same $d-1$ caustics. Suppose the pair (a_0, b_0) is s-skew, and that by the sequence of reflections on quadrics $\mathcal{Q}^1, \ldots, \mathcal{Q}^{s+1}$ the minimal billiard trajectory connecting a_0 to b_0 is realized.*

Then, each pair (a_m, b_m) is s-skew, and the minimal billiard trajectory connecting these two lines is determined by the sequence of reflections on the same quadrics $\mathcal{Q}^1, \ldots, \mathcal{Q}^{s+1}$.

Proof. This may be proved by the use of the Double Reflection Theorem, similarly as in Theorem 8.38.   □

This theorem also can be stated for an arbitrary quadric, not only for ellipsoids.

Proposition 8.45. *Let $\mathcal{E}$ be an ellipsoid in $\mathbf{E}^d$, $\mathcal{Q}_1, \ldots, \mathcal{Q}_{d-1}$ quadrics confocal to $\mathcal{E}$, and k an integer.*

Suppose that there exists a trajectory (ℓ_m) of the billiard within $\mathcal{E}$, having the caustics $\mathcal{Q}_1, \ldots, \mathcal{Q}_{d-1}$, such that the pair (ℓ_0, ℓ_k) is s-skew and that the minimal billiard trajectory connecting ℓ_0 to ℓ_k is realized by the sequence of reflections on quadrics $\mathcal{Q}^1, \ldots, \mathcal{Q}^{s+1}$ confocal to $\mathcal{E}$.

Then, for any billiard trajectory $(\hat{\ell}_m)$ within $\mathcal{E}$ with the caustics $\mathcal{Q}_1, \ldots, \mathcal{Q}_{d-1}$, the pairs of lines $(\hat{\ell}_m, \hat{\ell}_{m+k})$ are s-skew and the minimal billiard trajectory between $\hat{\ell}_m$ and $\hat{\ell}_{m+k}$ is realized by the sequence of reflections on $\mathcal{Q}^1, \ldots, \mathcal{Q}^{s+1}$.

Figure 8.8: Kandinsky, Grid 1923
(Albertina, Wien;
`www.albertina.at`;
reprinted with
permission)

Chapter 9

Poncelet Theorem and Continued Fractions

Modern algebraic approximation theory with continued fraction theory was established by Chebyshev (Tchebycheff according to the traditional French transcription) and his Sankt Petersburg school in the second half of the XIX century. Chebyshev's motivation for these studies was his interest in practical problems: in the mechanism theory as an important part of mechanical engineering of that time and ballistics. Steam engines were fundamental tools in technological revolution and their kernel part was *the Watt's complete parallelogram*, a planar mechanism to transform linear motion into circular, shown on Figure 9.1.

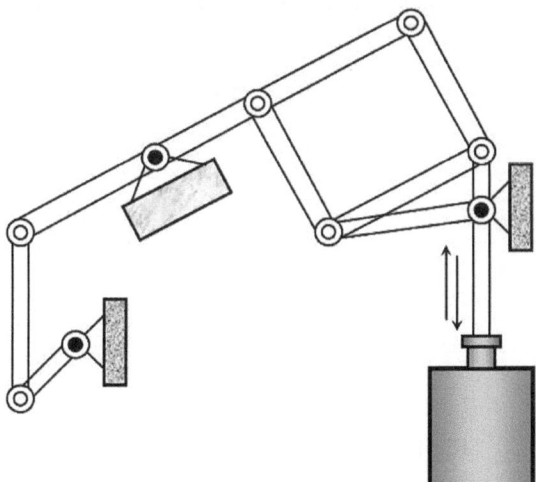

Figure 9.1: Watt's complete parallelogram

The fundamental problem was to estimate error of the mechanism in execution of that transformation.

The starting point of Chebyshev's investigation ([Tch1852]) was work on the theory of mechanisms of the French military engineer, professor of mechanics and academician Jean Victor Poncelet [Pon1822]. In his study of mistakes of mechanisms, Poncelet came to the question of rational and linear approximation of the function

$$\sqrt{x^2 + 1}.$$

In other words he studied approximation of the functions $\sqrt{X_2(x)}$ of the form of the square root of polynomials of the second degree, and he gave two approaches to the posed problems, one based on the analytical arguments and the second one based on geometric consideration.

Although Poncelet was described by Chebyshev as "a well-known scientist in practical mechanics" (see [Tch1852]), nowadays J. V. Poncelet is known first of all as one of the foremost geometers of the XIX century.

However, today it is almost forgotten that there is an amazing connection between Poncelet's Theorem and continued fractions and approximation theory of the functions of the form

$$\sqrt{X_4(x)},$$

where $X_4(x)$ denotes a general polynomial of the fourth degree. This connection of continued fractions and approximations of square root functions of polynomials of fourth degree with the Poncelet configuration and Poncelet theorem was indicated by Halphen [Hal1888].

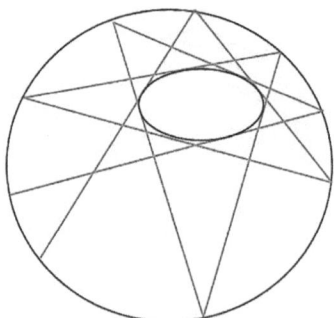

Figure 9.2: Poncelet configuration

The theory of continued fractions of square roots $\sqrt{X_4(x)}$ of polynomials of degree up to 4 started with Abel and Jacobi. Faced with the problem of very complicated algebraic formulae in algorithms, Jacobi [Jac1884b] turned to an approach based on elliptic function theory. Further development of that approach has been done by Halphen [Hal1888]. Halphen studied, instead of the square root of a polynomial, the more general *Halphen element*

$$\frac{\sqrt{X_4} - \sqrt{Y_4}}{x - y},$$

where $Y_4 = X_4(y)$ is the value of polynomial X at given point y.

In this chapter, we are going to present a more general theory of continued fractions of *hyperelliptic Haplhen elements*

$$\frac{\sqrt{X_{2g+2}} - \sqrt{Y_{2g+2}}}{x - y},$$

where X_{2g+2} is a polynomial of degree $2g+2$ and $Y_{2g+2} = X_{2g+2}(y)$. It is obviously related to the theory of functions of the hyperelliptic curve

$$\Gamma : z^2 = X_{2g+2}$$

of genus g. We are also going to refer to this theory as *HH continued fractions*. We started the study of this theory in [Dra2009] We hope that this theory will find its way to concrete applications in modern technology. As a possibility we can mention development of *branched, multivalued algorithms* to be used in future cryptography. However, this requires strong interaction with experts of different fields.

Here, we give two geometric interpretations of the dynamics which lie in the basis of the HH continued fraction developments.

In Section 9.2 we give two geometric realizations of the $2 \leftrightarrow g+1$ dynamics. The first one deals with nets of polynomials and with polygons circumscribed about a conic K. It extends, not only in its flavor, the initial story of the Poncelet polygons. The dynamics is realized as a path of polygons of $g+1$ sides inscribed in a curve B of degree $2g$ and circumscribed about the conic K obtained by successive moves of a certain type – so-called *flips along edges*.

The second geometric realization arising in Section 9.2 starts with the notion of the generalized Cayley curve (see Section 8.1) and leads to a new interpretation of generalized Jacobians of hyperelliptic curves following and continuing Chapter 8.

9.1 Hyperelliptic Halphen-type continued fractions

Basic algebraic lemma

Given a polynomial X of degree $2g + 2$ in x. We suppose that X is not a square of a polynomial. Assuming that the values of y and ε are finite and fixed, we are going to study HH elements in a neighborhood of ε. Then, X can be considered as a polynomial of degree $2g + 2$ in s, where $s = x - \varepsilon$ is chosen as a variable in a neighborhood of ε.

Lemma 9.1 (Basic Algebraic Lemma). *Let X be a polynomial of degree $2g+2$ in x and $Y = X(y)$ its value at a given fixed point y. Then, there exists a unique triplet of polynomials A, B, C with $\deg A = g + 1$, $\deg B = \deg C = g$ in x such that*

$$\frac{\sqrt{X} - \sqrt{Y}}{x - y} - C = \frac{B(x - \varepsilon)^{g+1}}{\sqrt{X} + A}. \qquad (9.1)$$

Proof. Put $s = x - \varepsilon$ and $t = y - \varepsilon$ and denote $X = X'(s) = \sum_{i=0}^{2g+2} p_i s^i$ and

$$A = \sum_{i=0}^{g+1} A_i s^i, \quad B = \sum_{i=0}^{g} B_i s^i, \quad C = \sum_{i=0}^{g} C_i s^i.$$

We are going to determine the coefficients A_i, B_i, C_i in the way that equation (9.1) is satisfied. Taking into account that $\sqrt{X}$ is irrational, the last equation can be separated into two equations:

$$A - \sqrt{Y} - C(s - t) = 0;$$
$$X - A\sqrt{Y} - AC(s - t) - Bs^{g+1}(s - t) = 0. \qquad (9.2)$$

Add the following consequence of (9.2):

$$X - A^2 = Bs^{g+1}(s - t). \qquad (9.3)$$

For $s = 0$, from (9.1) and (9.2), we get

$$C_0 = \frac{\sqrt{Y} - \sqrt{p_0}}{t},$$
$$A_0 = -C_0 t + \sqrt{Y} = \sqrt{p_0}.$$

Then we calculate A_i, $i = 1, \ldots, g$, from equation (9.3). From the first of equations (9.2), by putting $s = t$, we get

$$A(t) = \sqrt{Y}.$$

From the first equation (9.2), for $s = t$, we see that $\deg C = \deg A - 1$ and we compute all the coefficients C_i as functions of the coefficients of the polynomial A. For example, $C_g = A_{g+1}$.

The last step is to compute the polynomial B. Observe that the coefficients $A_0, \ldots A_g$ of the polynomial A are obtained in a manner such that

$$s^{g+1} | X - A^2.$$

The leading coefficient A_{g+1} is such that $X - A^2 = 0$ for $s = t$. Thus, there is a unique polynomial B, $\deg B = g$ such that

$$X - A^2 = Bs^{g+1}(s - t). \qquad \square$$

Hyperelliptic Halphen-type continued fractions

Let us start with the factorization of the polynomial B:

$$B(s) = B_g \prod_{i=1}^{g} (s - t_1^i)$$

and denote $A(t_1^i) = -\sqrt{Y_1^i}$. Then we have

$$\frac{A + \sqrt{X}}{s - t_1^i} = P_A^g(t_1^i, s) + \frac{\sqrt{X} - \sqrt{Y_1^i}}{x - y_1^i},$$

with a certain polynomial P_A^g of degree g in s and with coefficients depending on the coefficients of A and t_1^i.

Let

$$Q_0 = \frac{\sqrt{X} - \sqrt{Y}}{x - y} - C.$$

Then we have

$$Q_0 = \frac{B_g \prod_{j=1, j \neq i}^{g} (s - t_1^j) s^{g+1}}{P_A^g(t_1^i, s) + \frac{\sqrt{X} - \sqrt{Y_1^i}}{x - y_1^i}}.$$

Now, by applying Lemma 9.1 we obtain the polynomials $A^{(1,i)}$, $B^{(1,i)}$, $C^{(1,i)}$ of degree $g + 1$, g, g respectively, such that

$$\frac{\sqrt{X} - \sqrt{Y_1^i}}{x - y_1^i} - C^{(1,i)} = \frac{B^{(1,i)}(x - \varepsilon)^{g+1}}{\sqrt{X} + A^{(1,i)}}.$$

Let

$$\alpha_1^{(i)} := P_A^g(t_1^i, s), \quad \beta_1^{(i)} := B_g \prod_{j=1, j \neq i}^{g} (s - t_1^j) s^{g+1},$$

and introduce $Q_1^{(i)}$ by the equation

$$Q_0 = \frac{\beta_1^{(i)}}{\alpha_1^{(i)} + Q_1^{(i)}}.$$

Observe that $\deg \alpha_1^{(i)} = g$ and $\deg \beta_1^{(i)} = 2g$.

Now, one can go further, step by step: to factorize $B^{(1,i)}$, to choose one of its zeroes t_2^j and to write $B^{i,j} := B^{(1,i)}/(s - t_2^j)$. Further, we let

$$\alpha_2^{(i,j)} := P_{A^{1,i}}g(t_2^j, s), \quad \beta_2^{(i,j)} := B^{i,j} s^{g+1},$$

and calculate $Q_2^{(i,j)}$ from the equation

$$Q_1^{(i)} = \frac{\beta_2^{(i,j)}}{\alpha_2^{(i,j)} + Q_2^{(i,j)}}.$$

Thus we have

$$\frac{\sqrt{X} - \sqrt{Y}}{x - y} = C + \cfrac{\beta_1^{(i)}}{\alpha_1^i + \cfrac{\beta_2^{(i,j)}}{\alpha_2^{(i,j)} + Q_2^{(i,j)}}}.$$

Following the same scheme, in the ith step we introduce polynomials

$$A^{(i,j_1,\ldots,j_i)}, \quad B^{(i,j_1,\ldots,j_i)}, \quad C^{(i,j_1,\ldots,j_i)}$$

of degrees $g + 1$, g, g respectively. They satisfy the equations

$$\begin{aligned}
A^{(i,j_1,\ldots,j_i)} &= C^{(i,j_1,\ldots,j_i)}(s - t_i^{j_1,\ldots,j_i}) + \sqrt{Y_i^{j_1,\ldots,j_i}}, \\
X - A^{(i,j_1,\ldots,j_i)2} &= B^{(i,j_1,\ldots,j_i)}s^{g+1}(s - t_i^{j_1,\ldots,j_i}).
\end{aligned} \tag{9.4}$$

We see that in the case $g > 1$ the formulae of the $(i + 1)$th step depend on the choice of one of the roots of the polynomial $B^{(i)}$ and of the choices from the previous steps. To avoid abuse of notation we are going to omit many times in future formulae the indexes $j_1, \ldots, j_i$, which indicate the choices made in the first i steps, although we assume all the time that the choice has been made.

According to our notation we have

$$s - t_i | B^{(i-1)}$$

and

$$B^{(i-1)} = \frac{\beta_i}{s^{g+1}}(s - t_i)$$

or

$$B^{(i)} = \hat{\beta}_{i+1}(s - t_{i+1}),$$

where $\hat{\beta}_i = \beta_i / s^{g+1}$. From equations (9.4) we have

$$\begin{aligned}
X - A^{(i-1)2} &= \hat{\beta}_i(s - t_{i-1})s^{g+1}(s - t_i), \\
X - A^{(i)2} &= \hat{\beta}_{i+1}(s - t_{i+1})s^{g+1}(s - t_i)
\end{aligned} \tag{9.5}$$

together with

$$\begin{aligned}
A^{(i)}(t_i) &= \sqrt{Y_i}, \\
A^{(i-1)}(t_i) &= -\sqrt{Y_i}.
\end{aligned}$$

We introduce λ_i by the relation

$$A_{g+1}^{(i)} = \sqrt{p_0}\lambda_i.$$

Theorem 9.2. *If λ_i is fixed, then t_i, $t_{i+1}^{(1)}$, $\ldots$, $t_{i+1}^{(g)}$ are roots of the following polynomial equation of degree $g + 1$ in s:*

$$Q_X(\lambda_i, s) = 0.$$

The proof follows from equations (9.5). In the same way, we get

Theorem 9.3. *If t_i is fixed, then λ_i and λ_{i-1} are the roots of the polynomial equation of degree 2 in λ:*

$$Q_X(\lambda_{i-1}, t_i) = 0, \qquad Q_X(\lambda_i, t_i) = 0.$$

One can easily calculate

$$B_g^{(i)} = p_{2g+2} - p_0\lambda_i^2,$$

thus

$$\beta_{i+1} = (p_{2g+2} - p_0\lambda_i^2)\prod_{j=2}^{g}(s - t_i^j)s^{g+1}.$$

We also have

$$A^{(i)} = \sqrt{p_0}\left(1 + q_1 s + \cdots + \lambda_i s^{g+1}\right),$$
$$C^{(i)} = \sqrt{p_0}\left(q_1 + \cdots + \lambda_i(s^g + s^{g-1}t_i + \cdots + t_i^g)\right),$$

and

$$\alpha_i = \sqrt{p_0}\left(2q_1 + \cdots + (\lambda_{i-1} + \lambda_i)(s^g + s^{g-1}t_i + \cdots + t_i^g)\right).$$

According to Theorem 9.3, the sum $\lambda_{i-1} + \lambda_i$ from the last equation, can be expressed through the coefficients of the polynomial $Q_X(\lambda, t_i)$ as a polynomial of the second degree in λ.

Basic examples: genus 1 case

The genus 1 case, or the elliptic case, has been studied by Halphen. Here we reproduce some of his formulae (see [Hal1888] for more details). The elliptic curve is given by a polynomial X of degree 4, in variable s in a neighborhood of ε:

$$X = S(s) = \sum_{i=0}^{4} p_i s^i.$$

The development around the point ε of its square root has the form

$$\sqrt{X} = \sqrt{p_0}\left(1 + q_1 s + q_2 s^2 + q_3 s^3 + \cdots\right),$$

with the following relations between q's and p's:

$$q_1 = \frac{p_1}{2p_0},$$
$$q_2 = \frac{1}{8p_0^2}\left(4p_0p_2 - p_1^2\right),$$
$$q_3 = \frac{1}{4p_0^3}\left(2p_0p_3 - p_0p_1p_2 + \frac{p_1^3}{4}\right),$$
$$q_4 = \frac{1}{2p_0}\left(p_4 - 2q_1q_3p_0 - q_2^2p_0\right).$$

Here we have

$$\frac{X}{p_0} = (1 + q_1 s + q_2 s^2)^2 + 2q_3 s^3 + 2(q_1 q_3 + q_4)s^4.$$

From the Basic Algebraic Lemma, applied to the case $g = 1$, we get the polynomials $A = A_0 + A_1 s + A_2 s^2$, $B = B_0 + B_1 s$, $C = C_0 + C_1 s$ which satisfy

$$A - \sqrt{Y} - C(s - t) = 0,$$
$$X - A\sqrt{Y} - AC(s - t) - Bs^2(s - t) = 0, \qquad (9.6)$$
$$X - A^2 = Bs^2(s - t).$$

From equations (9.6), one gets the formulae for the polynomials A, B, C:

$$A_0 = \sqrt{p_0}, \qquad A_1 = q_1\sqrt{p_0}, \qquad A_2 = \frac{\sqrt{Y} - (1 + q_1 t)\sqrt{p_0}}{t^2},$$

$$C_0 = \frac{\sqrt{Y} - \sqrt{p_0}}{t}, \qquad C_1 = A_2,$$

$$B_0 = \frac{2\sqrt{p_0}}{t^3}\left(\sqrt{Y} - \sqrt{p_0}(1 + q_1 t + q_2 t^2)\right),$$

$$B_1 = \frac{2\sqrt{p_0}}{t^4}\left((1 + q_1 t)\sqrt{Y} - \sqrt{p_0}(1 + 2q_1 t + (q_1^2 + q_2)t^2 + (q_1 q_2 + q_3)t^3)\right).$$

If we let

$$P_A^{(1)}(t, s) := A_1 + A_2(s + t),$$

then we have

$$Q_0 = \frac{B_1 s^2}{P_A^{(1)}(t_1, s) + \frac{\sqrt{X} - \sqrt{Y_1}}{x - y_1}},$$

and, step by step,

$$A^{(i)} = \sqrt{Y_i} - C^{(i)}(s - t_i),$$
$$X - A^{(i)2} = B^{(i)}s^2(s - t_i), \qquad (9.7)$$

where

$$B^{(i-1)}(t_i) = 0,$$
$$A^{(i)}(t_{i+1}) = -\sqrt{Y_{i+1}}.$$

Finally, one gets

$$\beta_i = B_1^{(i-1)}s^2,$$
$$\alpha_i = P_{A^{(i-1)}}^{(1)}(t_i, s) + C^{(i)}.$$

From equation (9.7) we get

$$X - A^{(i-1)2} = ms^2(s - t_{i-1})(s - t_i),$$
$$X - A^{(i)2} = ns^2(s - t_i)(s - t_{i+1}),$$
$$A^{(i)}(t_i) = \sqrt{Y_i},$$
$$A^{(i-1)}(t_i) = -\sqrt{Y_i},$$
$$A_2^{(i)} = \sqrt{p_0}\lambda_i,$$

with some constants m, n, and then we have

$$\lambda_i = \frac{1}{t_i^2}\left(\frac{\sqrt{Y_i}}{\sqrt{p_0}} - (1 + q_1 t_i)\right),$$

$$\lambda_{i-1} = \frac{1}{t_i^2}\left(-\frac{\sqrt{Y_i}}{\sqrt{p_0}} - (1 + q_1 t_i)\right).$$

From the last equations one obtains:

Proposition 9.4. *If λ_i is fixed, then t_i, t_{i+1} are roots of the polynomial $Q_X(\lambda_i, s)$ quadratic in s:*

$$Q_X(\lambda_i, s) := (p_4 - p_0\lambda_i^2)s^2 + (p_3 - p_1\lambda_i)s + 2p_0(q_2 - \lambda_i) = 0.$$

Corollary 9.5. *The product of two consecutive t_i and t_{i+1} is*

$$t_i t_{i+1} = \frac{2p_0(\lambda_i - q_2)}{p_0\lambda_i^2 - p_4},$$

and their sum is equal to

$$t_i + t_{i+1} = \frac{p_1\lambda_i - p_3}{p_4 - p_0\lambda_i^2}.$$

Proposition 9.4 can be reformulated giving a relation between two consecutive λ_{i-1} and λ_i:

Proposition 9.6. *If t_i is fixed, then λ_{i-1}, λ_i are solutions of quadratic equation*

$$\lambda^2(p_0 t_i^2) + \lambda(p_1 t_i + 2p_0) - (p_4 t_i^2 + p_3 t_i + 2p_0 q_2) = 0.$$

For the *normal form* of the elliptic HH continued fraction we consider the case where

$$\alpha_i' = 1 + u_i s, \qquad \beta_i' = v_i s^2.$$

Then we have

$$t_i = -\frac{1}{q_1 + u_i}, \qquad \lambda_i = q_2 - 2v_{i+1}.$$

The recurrent relations are given with

$$u_i + u_{i-1} = -q_1 + \frac{q_2}{2v_i},$$

$$v_i + u_i u_{i-1} = q_2 + \frac{q_4}{2v_i}.$$

The second set of recurrent equations consists of

$$v_i + v_{i+1} = q_2 + q_1 u_i + u_i^2,$$

$$2v_i v_{i+1} = -q_4 + q_3 u_i.$$

Basic examples: genus 2 case

Notation

We start with a polynomial X of degree 6 in x and rewrite it as a polynomial in s in a neighborhood of ε,

$$X = S(s) = \sum_{i=0}^{6} p_i s^i$$

and its square root developed around ε as

$$\sqrt{X} = \sqrt{p_0}(1 + q_1 s + q_2 s^2 + q_3 s^3 + q_4 s^4 + q_5 s^5 + q_6 s^6 + q_7 s^7 + \cdots).$$

Then, the relations between coefficients p_i and q_j are

$$p_1 = 2p_0 q_1,$$
$$p_2 = p_0(2q_2 + q_1^2),$$
$$p_3 = 2p_0(q_3 + q_1 q_2),$$
$$p_4 = p_0(2q_4 + q_2^2 + 2q_1 q_3),$$
$$p_5 = 2p_0(q_5 + q_2 q_3 + q_1 q_4),$$
$$p_6 = p_0(2q_6 + 2q_1 q_5 + 2q_2 q_4 + q_3^2),$$

with relations between q_i such as

$$0 = q_7 + 2q_1 q_6 + 2q_2 q_5 + 2q_3 q_4.$$

Conversely, q_i's can be expressed through p's:

$$q_1 = \frac{p_1}{2p_0},$$

$$q_2 = \frac{1}{2p_0^2}\left(p_2 p_0 - \frac{p_1^2}{4}\right),$$

$$q_3 = \frac{1}{2p_0^3}\left(p_3 p_0^2 - \frac{p_1 p_2 p_0}{2} + \frac{p_1^3}{8}\right),$$

$$q_4 = \frac{1}{2p_0^4}\left\{p_4 p_0^3 - \frac{4p_2 p_0 - p_1^2}{8} - \frac{p_1}{16}(8p_3 p_0^2 - 4p_1 p_2 p_0 + p_1^3)\right\},$$

$$q_5 = \frac{p_5}{2p_0} - q_2 q_3 - q_1 q_4,$$

$$q_6 = \frac{1}{2p_0}(p_6 - 2q_1 q_5 p_0 - 2q_2 q_4 p - 0 - q_3^2 p_0).$$

The initial polynomial X can be expressed through q_i's:

$$\frac{X}{p_0} = (1 + q_1 s + q_2 s^2 + q_3 s^3)^2 + 2q_4 s^4 + 2(q_1 q_4 + q_5)s^5 + 2(q_1 q_5 + q_2 q_4 + q_6)s^6.$$

The case of the Basic Algebraic Lemma

We are going to determine polynomials A, B, C of degrees $\deg A = 3$, $\deg B = \deg C = 2$. Write $A = A(s) = A_0 + A_1 s + A_2 s^2 + A_3 s^3$, $B = B(s) = B_0 + B_1 s + B_2 s^2$, $C = C(s) = C_0 + C_1 s + C_2 s^2$. Then equations (9.2) and (9.3) become

$$A - \sqrt{Y} - C(s - t) = 0,$$
$$X - A\sqrt{Y} - AC(s - t) - Bs^3(s - t) = 0, \qquad (9.8)$$
$$X - A^2 = Bs^3(s - t).$$

For $s = 0$, we obtain

$$C_0 = \frac{\sqrt{Y} - \sqrt{p_0}}{t}, \qquad A_0 = \sqrt{p_0}.$$

Then we calculate A_i, $i = 1, 2$ from the last equation of (9.8) by comparing polynomials X and A term by term up to the second degree:

$$A_1 = \frac{p_1}{2\sqrt{p_0}}, \qquad A_2 = \frac{1}{2\sqrt{p_0}}\frac{4p_2 p_0 - p_1^2}{4p_0},$$

thus

$$A = \sqrt{p_0}(1 + q_1 s + q_2 s^2 + \lambda_1 s^3).$$

From the relation $A(t) = \sqrt{Y}$ we get

$$A_3 = \frac{1}{t^3}\left[\sqrt{Y} - \left(\sqrt{p_0} + \frac{p_1}{2\sqrt{p_0}}t + \frac{4p_0p_2 - p_1^2}{8p_0^{3/2}}t^2\right)\right].$$

The coefficients of C are $C_1 = A_2$ and $C_2 = A_3$. The coefficients of the polynomial B are

$$B_2 = p_6 - A_3^2,$$
$$B_1 = B_2t + p_5 - 2A_2A_3,$$
$$B_0 = B_1t + p_4 - (2A_1A_3 + A_2^2).$$

We factorize it as

$$B = B_2(s - t_1^0)(s - t_1^1),$$

and write

$$A(t_1^0) = -\sqrt{Y_1^0}, \qquad A(t_1^1) = -\sqrt{Y_1^1}.$$

Now, we have

$$\frac{A + \sqrt{X}}{s - t_1^0} = \frac{A + \sqrt{Y_1^0}}{s - t_1^0} + \frac{\sqrt{X} - \sqrt{Y_1^0}}{x - y_1^0}$$
$$= \frac{A(s) - A(t_1^0)}{s - t_1^0} + \frac{\sqrt{X} - \sqrt{Y_1^0}}{x - y_1^0}$$
$$= A_1 + A_2(s + t_1^0) + A_3(s^2 + st_1^0 + t_1^{02}) + \frac{\sqrt{X} - \sqrt{Y_1^0}}{x - y_1^0}.$$

Set

$$P_A^{(2)}(t, s) := A_1 + A_2(s + t) + A_3(s^2 + st + t^2).$$

Then we have finally

$$Q_0 = \frac{B_2(s - t_1^1)s^3}{P_A^{(2)}(t_1^0, s) + \frac{\sqrt{X} - \sqrt{Y_1^0}}{x - y_1^0}}.$$

Step by step we get

$$A^{(i)} = \sqrt{Y_i} - C^{(i)}(s - t_i),$$
$$X - A^{(i)2} = B^{(i)}s^3(s - t_i), \tag{9.9}$$

where

$$B^{(i-1)}(t_i) = 0, \quad t_i := t_i^0,$$
$$A^{(i)}(t_{i+1}) = -\sqrt{Y_{i+1}}.$$

Now, we have

$$\beta_i = B_2^{(i-1)}(s - t_i^1)s^3,$$
$$\alpha_i = P_{A^{(i-1)}}^{(2)}(t_i, s) + C^{(i)}. \tag{9.10}$$

We can represent the hyperelliptic Halphen-type continued fraction in the following manner:

$$\frac{\sqrt{X} - \sqrt{Y}}{x - y} = C + \frac{\beta_1|}{|\alpha_1} + \frac{\beta_2|}{|\alpha_2} + \cdots + \frac{\beta_i|}{|\alpha_i + Q_i},$$

where

$$Q_i = \frac{\sqrt{X} - \sqrt{Y_i}}{x - y_i} - C^{(i)} = \frac{B^{(i)} s^3}{\sqrt{X} + A^{(i)}}$$

and

$$Q_i = \frac{\beta_{i+1}}{\alpha_{i+1} + Q_{i+1}}.$$

Relations between λ_i and t_i

From equation (9.9), we get

$$X - A^{(i-1)2} = B_2^{(i-1)}(s - t_i^1)s^3(s - t_{i-1})(s - t_i),$$
$$X - A^{(i)2} = B_2^{(i)}(s - t_{i+1}^1)s^3(s - t_i)(s - t_{i+1}),$$
$$A^{(i)}(t_i) = \sqrt{Y_i},$$
$$A^{(i-1)}(t_i) = -\sqrt{Y_i},$$
$$A_3^{(i)} = \sqrt{p_0}\lambda_i.$$

(9.11)

From (9.11), we have

$$\lambda_i = \frac{1}{t_i^3}\left(\frac{\sqrt{Y_i}}{\sqrt{p_0}} - (1 + q_1 t_i + q_2 t_i^2)\right),$$

$$\lambda_{i-1} = \frac{1}{t_i^3}\left(-\frac{\sqrt{Y_i}}{\sqrt{p_0}} - (1 + q_1 t_i + q_2 t_i^2)\right),$$

and thus

$$t_i^3\sqrt{Y_{i+1}} + t_{i+1}^3\sqrt{Y_i} = \sqrt{p_0}(t_{i+1} - t_i)[t_{i+1}^2 + t_{i+1}t_i + q_1 t_i t_{i+1}(t_{i+1} + t_i) + q_2 t_i^2 t_{i+1}^2].$$

From equations (9.11) we also get

$$\lambda_{i-1} + \lambda_i = -\frac{2}{t_i^3}(1 + q_1 t_i + q_2 t_i^2),$$

$$\lambda_{i-1}\lambda_i = -\frac{1}{t_i^6}\left[(1 + q_1 t_i + q_2 t_i^2)^2 - \frac{Y_i}{p_0}\right].$$

(9.12)

Finally, we have

Proposition 9.7. *If λ_i is fixed, then t_i, t_{i+1}, t_{i+1}^1 are roots of the polynomial $Q_X(\lambda_i, s)$ of degree 3 in s:*

$$Q_X(\lambda_i, s) := (p_6 - p_0\lambda_i^2)s^3 + (p_5 - 2p_0q_2\lambda_i)s^2 + (p_4 - 2p_0q_1\lambda_i - q_2^2p_0)s$$
$$+ (p_3 - 2p_0\lambda_i - 2p_0q_1q_2) = 0.$$

Corollary 9.8. *The product of two consecutive t_i and t_{i+1} is*

$$t_i t_{i+1} = \frac{p_3 - 2p_0\lambda_i - 2p_0q_1q_2}{t_{i+1}^1(p_6 - p_0\lambda_i^2)}.$$

Proposition 9.7 can be reformulated to give a relation between two consecutive λ_{i-1} and λ_i:

Proposition 9.9. *If t_i is fixed, then λ_{i-1}, λ_i are solutions of quadratic equation*

$$\alpha\lambda^2 + \beta\lambda + \gamma = 0,$$

with

$$\alpha = -p_0 t_i^3,$$
$$\beta = -2p_0q_2 t_i^2 - 2p_0q_1 t_i - 2p_0,$$
$$\gamma = p_6 t_i^3 + p_5 t_i^2 + (p_4 - q_2^2 p_0)t_i + p_3 - 2p_0q_1q_2.$$

Normal form of genus 2 hyperelliptic Halphen-type continued fractions. Recurrent relations

Using equations (9.12) and (9.10), we get formulae for α_i:

$$\alpha_i = \sqrt{p_0}\left(-\frac{2}{t_i} + (2q_2 + \lambda_{i-1}t_i)s - \frac{2}{t_i^3}(1 + q_1 t_i + q_2 t_i^2)s^2\right).$$

Given HH continued fraction with α_i, β_i, it can be transformed to the equivalent one with

$$\alpha_i' = c_i\alpha_i, \qquad \beta_i' = c_{i-1}c_i\beta_i.$$

Here, we chose coefficients

$$c_i = -\frac{t_i}{2\sqrt{p_0}}$$

and get

$$\alpha_i' = 1 + w_i s + u_i s^2,$$
$$\beta_i' = v_i \frac{s - t_i^1}{t_i^1}s^3, \tag{9.13}$$

where

$$u_i = \frac{1 + q_1 t_i + q_2 t_i^2}{t_i^2},$$

$$w_i = -\left(q_2 t_i + \frac{\lambda_{i-1}}{2} t_i^2\right), \tag{9.14}$$

$$v_i = -\frac{\lambda_{i-1}}{2} + q_3.$$

We will refer to the form (9.13) as to *normal form* of the given hyperelliptic Halphen-type continued fraction.

From equations (9.13), we get

$$\lambda_{i-1} = -2v_i + q_3 \tag{9.15}$$

and

$$(q_2 - u_i) t_i^2 + q_1 t_i + 1 = 0,$$

$$\frac{\lambda_{i-1}}{2} t_i^2 + q_2 t_i + w_i = 0. \tag{9.16}$$

From equations (9.16), we have

$$t_i = \frac{(u_i - q_2) w_i + (v_i - \frac{q_3}{2})}{q_2(q_2 - u_i) - q_1(\frac{q_3}{2} - v_i)}.$$

From Proposition 9.9 and equation (9.16), one obtains

$$\lambda^2 \left(-\frac{t_i}{2}\right) - u_i \lambda + [q_6 t_i + q_5(q_1 t_i + 1) + q_3 u_i + q_4 u_i t_i] = 0,$$

having two zeroes λ_{i-1} and λ_i. From the last equation, we get

$$t_i = \frac{u_i(\lambda - q_3) - q_5}{-\frac{\lambda^2}{2} + q_6 + q_5 q_1 + q_4 u_i}.$$

By using the second of equations (9.16) and equating the right sides of the last equation for λ_{i-1} and λ, we get

Lemma 9.10. *The following relation between* u_i *and* v_i, v_{i+1} *holds:*

$$-\frac{1}{2}(2u_i v_{i+1} + q_5)(q_3 - 2v_i)^2 + 2u_i(q_6 + q_5 q_1 + q_4 u_i)(v_{i+1} - v_i)$$

$$+ \frac{1}{2}(2u_i v_i + q_5)(q_3 - 2v_{i+1})^2 = 0.$$

This lemma implies the following

Corollary 9.11. *If* $v_i \neq v_{i+1}$ *then*

$$0 = q_3^2 u_i - 4u_i v_i v_{i+1} - 2q_5(v_i + v_{i+1}) + 2q_5 q_3 - 2u_i(q_6 + q_5 q_1 + q_4 u_i).$$

From equations (9.12), (9.14) and (9.15) we get

Proposition 9.12. *The recurrence equations connecting v_i and v_{i+1} for fixed u_i and t_i are*

$$v_i + v_{i+1} = \frac{u_i}{t_i} + q_3,$$

$$4v_i v_{i+1} = (-2q_6 - 2q_1 q_5 - 2q_4 u_i) + q_3^2 - 2\frac{q_5}{t_i}.$$

Rewrite polynomial $Q_X(\lambda_i, s)$ in the form

$$Q_X(\lambda_i, s) = Q_3 s^3 + Q_2 s^2 + Q_1 s + Q_0,$$

where

$$Q_3 = q_6 + q_1 q_5 + q_4 q_2 + \frac{q_3^2}{2} - \frac{\lambda_i}{2},$$
$$Q_2 = q_5 + q_1 q_4 + q_2(q_3 - \lambda_i),$$
$$Q_1 = q_4 + q_1(q_3 - \lambda_i),$$
$$Q_0 = q_3 - \lambda_i.$$

Summing the relations $Q_X(\lambda_i, t_i) = 0$ and $Q_X(\lambda_i, t_{i+1}) = 0$, we get

$$(u_i + u_{i+1})(q_3 - \lambda_i)$$
$$= (t_i + t_{i+1}) \left(\frac{\lambda_i^2}{2} - q_6 - q_5 q_1 - q_4 q_2 \right) - q_4 \frac{t_i + t_{i+1}}{t_i t_{i+1}} - 2q_5 - 2q_1 q_4.$$

From the last equation, using the Viète formulae for polynomial $Q_X(s)$ and equation (9.15), we get

Proposition 9.13.

$$u_i + u_{i+1} = \frac{1}{2} v_{i+1} \left[\left(\frac{Q_2}{Q_3} - t_{i+1}^1 \right) \left(\frac{(-2v_{i+1} + q_3)^2}{2} - q_6 - q_5 q_1 - q_4 q_2 \right) \right.$$
$$\left. - q_4 \left(\frac{Q_2}{Q_3} - t_{i+1}^1 \right) \frac{Q_3}{Q_0} t_{i+1}^1 - 2q_5 - 2q_1 q_4 \right].$$

Periodicity and symmetry

Definition and the first properties

According to Theorem 9.3, in the case

$$t_h = t_k$$

for some h, k, there are two possibilities:

(I) $\lambda_{h-1} = \lambda_{k-1}, \qquad \lambda_h = \lambda_k,$
(II) $\lambda_{h-1} = \lambda_k, \qquad \lambda_h = \lambda_{k-1}.$

The first possibility leads to *periodicity*:

$$t_{h+s} = t_{k+s}, \qquad \lambda_{h+s} = \lambda_{k+s}$$

for any s and with appropriate choice of roots. If $p = h - k$ and $r \equiv s \;(\bmod\; p)$ then

$$\alpha_r = \alpha_s, \qquad \beta_r = \beta_s.$$

The second possibility leads to *symmetry*:

$$t_{h+s} = t_{k-s}, \qquad \lambda_{h+s} = \lambda_{k-s-1}$$

for any s. More precisely, we introduce

Definition 9.14.

(i) If $h + k = 2n$ we say that a hyperelliptic Halphen-type continued fraction is *even symmetric* with

$$\alpha_{n-i} = \alpha_{n+i}, \qquad \beta_{n-i} = \beta_{n+i-1}$$

for any i and with α_n as the *centre of symmetry*.

(ii) If $h + k = 2n + 1$ we say that a hyperelliptic Halphen-type continued fraction is *odd symmetric* with

$$\alpha_{n-i} = \alpha_{n+i-1}, \qquad \beta_{n-i} = \beta_{n+i}$$

for any i and with β_n as the *centre of symmetry*.

Now we can formulate some initial properties connecting periodicity and symmetry.

Proposition 9.15.

(A) *If a hyperelliptic Halphen-type continued fraction is periodic with period $2r$ and even symmetric with α_n as the centre, then it is also even symmetric with respect to α_{n+r}.*

(B) *If a hyperelliptic Halphen-type continued fraction is periodic with period $2r$ and odd symmetric with respect to β_n, then it is also odd symmetric with respect to β_{n+r}.*

(C) *If a hyperelliptic Halphen-type continued fraction is periodic with period $2r-1$ and even symmetric with respect to α_n, then it is also odd symmetric with respect to β_{n+r}. The converse is also true.*

Proposition 9.16. *If a hyperelliptic Halphen-type continued fraction is double symmetric, then it is periodic. Moreover:*

(A) *If a hyperelliptic Halphen-type continued fraction is even symmetric with respect to α_m and α_n, $n < m$, then the period is $2(n - m)$.*

(B) *If a hyperelliptic Halphen-type continued fraction is odd symmetric with respect to β_m and β_n, $n < m$, then the period is $2(m - n)$.*

(C) *If a hyperelliptic Halphen-type continued fraction is even symmetric with respect to α_n and β_m, then the period is $2(n - m) + 1$ in the case $m \leq n$ and the period is $2(m - n) - 1$ when $m > n$.*

Remark 9.17.

(i) A hyperelliptic Halphen-type continued fraction can be at the same time even symmetric and odd symmetric.

(ii) If $\lambda_i = \lambda_{i-1}$ then the symmetry is even; if $t_i = t_{i+1}$ then the symmetry is odd.

Further results

Theorem 9.18. *A hyperelliptic Halphen-type continued fraction is even-symmetric with the central parameter y if $X(y) = 0$.*

The proof follows from the fact that even-symmetry is equivalent to the condition $\lambda_p = \lambda_{p-1}$, which is equivalent to equality $Y_p = 0$.

For odd-symmetry, let us start with the example of the genus 2 case. From relations

$$Q_X(\lambda, s) = 0, \qquad \frac{d}{ds} Q_X(\lambda, s) = 0$$

we get the system

$$3Q_3 s^2 + 2Q_2 s + Q_1 = 0,$$
$$Q_2 s^2 + 2Q_1 s + 3Q_0 = 0. \qquad (9.17)$$

From the last system we get

$$v_{i+1} = -\frac{s[q_5 + q_1 q_4)s + q_4]}{2q_2 s^2 + 4q_1 s + 6}$$

or, equivalently,

$$\lambda_i = \frac{p_5 s^2 + 2(p_4 - q_2^2 p_0)s + 3(p_3 - 2p_0 q_1 q_2)}{2p_0 q_2 s^2 + 4p_0 q_2 s + 6p_0}.$$

By replacing any of the last two relations in the first equation of (9.17), we get an equation of the sixth degree in s. On the other hand, from (9.17) we get

$$s = \frac{9Q_0 Q_3 - Q_1 Q_2}{2Q_2^2 - 6Q_1 Q_3}.$$

Now, by replacing the last formula in the first equation of (9.17) we get an equation of the eighth degree in λ_i.

General case

Invariant approach

Now we pass to the general case, with polynomial X of degree $2g + 2$. Relation

$$Q_X(\lambda, s) = 0 \tag{9.18}$$

defines *a basic curve* Γ_X. Denote its genus by G and consider its projections p_1 to the λ-plane, and p_2 to the s-plane.

Denote by R_e the ramification points of the second projection and call them *even-symmetric points* of the basic curve.

The set R_{o+r} of the ramification points of the first projection is the union of sets of *the odd-symmetric points* and of *the gluing points*.

The gluing points represent a situation where some of the roots of the polynomial $B^{(i)}$ coincide. For example in the genus 2 case the gluing points correspond to the condition $t_{i+1} = t'_{i+1}$.

From Theorem 9.18, we get

$$\deg R_e = 2g + 2.$$

Applying the Riemann–Hurwitz formula (Theorem 3.64), we have

$$2 - 2G = 4 - \deg R_e,$$
$$2 - 2g = 2(g + 1) - \deg R_{o+r}.$$

Thus

$$\text{genus} (\Gamma_X) = G = g$$

and

$$\deg R_{o+r} = 4g.$$

We get a birational morphism

$$f : \Gamma \to \Gamma_X$$

by the formulae

$$f : (x, s) \mapsto (t, \lambda),$$

where

$$t = x,$$

$$\lambda = \frac{1}{t^{g+1}} \left(\frac{s}{\sqrt{p_0}} - Q_g(t) \right),$$

$$Q_g(t) = 1 + q_1 t + \cdots + q_g t^g.$$

Function f satisfies the commuting relation

$$f \circ \tau_\Gamma = \tau_{\Gamma_X} \circ f,$$

where τ_Γ and τ_{Γ_X} are natural involutions on hyperelliptic curves Γ and Γ_X respectively.

Multi-valued divisor dynamics

The inverse image of a value z of the function λ is a divisor of degree $g + 1$:

$$\lambda^{-1}(z) =: D(z), \quad \deg D(z) = g + 1.$$

Now, the HH-continued fractions development can be described as a multi-valued discrete dynamics of divisors $D_k^j = D(z_k^j)$. Here, the lower index k denotes the kth step of the dynamics and the upper index j varies over the range from 1 to $(g+1)k$ denoting branches of multivaluedness. More precisely, the discrete divisor dynamics which governs HH-continued fraction development can be described as follows.

Suppose the development has started with a point $P_0 = P_0^1$. It leads to the divisor

$$D_0 := D(\lambda(P_0)) = P_0^1 + P_0^2 + \cdots + P_0^{g+1},$$

with $\lambda(P_0^i) = \lambda(P_0^j)$. In the next step, we get $g + 1$ divisors of degree $g + 1$:

$$D_1^j := D\left(\lambda(\tau_\Gamma(P_0^j))\right).$$

And we continue like this. In each step, the divisor

$$D_{k-1}^j = P_{k-1}^{(j,1)} + \cdots + P_{k-1}^{(j,(g+1))}$$

from the previous step, gives $g + 1$ new divisors

$$D_k^{(j-1)(g+1)+l} := D\left(\lambda(\tau_\Gamma(P_{k-1}^{(j,l)}))\right), \qquad l = 1, \ldots, g+1.$$

In the case of genus 1, this dynamics can be traced out from the (2-2)-correspondence $Q_\Gamma(\lambda, t) = 0$. According to [Hal1888], for example, there exist constants a, b, c, d, T such that for every i we have

$$\lambda_i = \frac{ax(u_i + T) + b}{cx(u_i + T) + d},$$

where u is a uniformizing parameter on the elliptic curve. The involution is the symmetry at the origin and since the function x is even, the two parameters corresponding to the fixed value λ_i are u_i and $\bar{u}_i = -u_i - 2T$. Thus

$$u_{i+1} = u_i + 2T,$$

$$\lambda_{i+1} = \frac{ax(u_i + 3T) + b}{cx(u_i + 3T) + d}.$$

In the cases of higher genera the dynamics is much more complicated. Thus we pass to the consideration of generalized Jacobians.

Generalized Jacobians

A natural environment for consideration of divisors of degree $g + 1$ on the curve Γ of genus g is a generalized Jacobian Jac $(\Gamma, \{Q_1, Q_2\})$ of Γ, obtained by gluing a pair of points Q_1, Q_2 of Γ (see [Fay1973]).

This Jacobian can be introduced as a set of classes of relative equivalence among the divisors on Γ of certain degree. Two divisors of the same degree D_1 and D_2 are called *equivalent relative to points* Q_1, Q_2, if there exists a meromorphic function f on Γm such that $(f) = D_1 - D_2$ and $f(Q_1) = f(Q_2)$.

The generalized Abel map is defined as

$$\widetilde{\mathcal{A}}(P) = (\mathcal{A}(P), \mu_1(P), \mu_2(P)), \quad \mu_i(P) = \exp \int_{P_0}^{P} \Omega_{Q_i Q_0}, i = 1, 2,$$

where $\mathcal{A}(P)$ is the standard Abel map. Here $\Omega_{Q_i Q_0}$ denotes the normalized differential of the third kind, with poles at the point Q_i and at an arbitrary fixed point Q_0.

Here we consider the case where $Q_1 = +\infty$ and $Q_2 = -\infty$ on the curve Γ of genus g. The divisors we are going to consider are those of degree $g + 1$ of the form $D_i = D(z_i)$ where usually $z_i = \lambda(P_i)$. The divisors of degree $g + 1$ up to the equivalence relative to the points Q_1 and Q_2 are uniquely determined by their generalized Abel image on the generalized Jacobian.

Thus, in order to measure the distance between relative classes of $D_1 = D(z_1) = D(\lambda(P_1))$ and of $D_2 = D(z_2) = D(\lambda(P_2))$ we introduce the index

$$I(D_1, D_2) = I(z_1, z_2) = I(P_1, P_2) := \frac{\lim_{P \to +\infty} \frac{\lambda(P) - z_1}{\lambda(P) - z_2}}{\lim_{P \to -\infty} \frac{\lambda(P) - z_1}{\lambda(P) - z_2}}.$$

We are interested in the case $P_2 = \tau_\Gamma(P_1)$ and we have

$$I(P_1) := I(P_1, \tau_\Gamma(P_1)) = \lim_{P \to +\infty} \frac{\frac{\lambda(P) - \lambda(P_1)}{\lambda(P) - \lambda(\tau(P_1))}}{\frac{\lambda(\tau(P)) - \lambda(P_1)}{\lambda(\tau(P)) - \lambda(\tau(P_1))}}.$$

After some calculations we get

Lemma 9.19. *The index of the point is given by the formula*

$$I(P_1) = 1 + \frac{2\sqrt{p_{2g+2}}(\lambda(\tau(P_1)) - \lambda(P_1))}{p_{2g+2} - \sqrt{p_{2g+2}}(\lambda(\tau(P_1)) - \lambda(P_1)) - \lambda(P_1)\lambda(\tau(P_1))}.$$

Irregular terms

The parameters t that appear to be infinite or zero, we call *irregular*.

t_h – infinite

Suppose $t_0 = \infty$. We start from the following relation:

$$X - A^2 = Bs^{g+1}.$$

Then, HH continued fraction is based on the relation

$$\sqrt{X} - \sqrt{p_{2g+2}}s^{g+1} = C + \frac{Bs^{g+1}}{\sqrt{X} + A}.$$

Proposition 9.20. *Irregular hyperelliptic Halphen-type continued fraction with $t_h = \infty$ is even symmetric if and only if $p_{2g+2} = 0$.*

$t_h = 0$

Let $t_0 = 0$. In that case the basic relation of HH continued fraction is

$$\frac{\sqrt{X} - \sqrt{p_0}}{x - \varepsilon} - C = \frac{B(x - \varepsilon)^{g+1}}{\sqrt{X} + A}.$$

Then we have also

$$A - \sqrt{p_o} = Cs,$$
$$X - A^2 = Bs^{g+2}.$$

An HH continued fraction is developed through the relations

$$\sqrt{X} = A + \frac{Bs^{g+2}}{\sqrt{X} + A},$$
$$\sqrt{X} = A + \frac{Bs^{g+2}}{P_A^{(g)}} + \frac{\sqrt{X} - \sqrt{Y_1}}{x - y_1}.$$

Proposition 9.21. *The condition $t_h = 0$ is equivalent to $v_{h+1} = \infty$. Such a hyperelliptic Halphen-type continued fraction is odd symmetric with respect to β_{h+1}.*

ε – infinite

The starting relation in the case $\varepsilon = \infty$ is

$$X - A^2 = B(x - y).$$

Changing the variables: $x = 1/s$, $y = 1/t$, we come to

$$X' - A'^2 = -\frac{1}{t}B's^{g+1}(s - t).$$

The hyperelliptic Halphen-type continued fraction takes the form

$$\frac{\sqrt{X} - \sqrt{Y}}{x - y} = C + \frac{B_0|}{|A_1} + \frac{B_0^{(1)}|}{|A_2} + \cdots + \frac{B_0^{(i-1)}|}{|A_i + Q_i},$$

where $\deg B_0^{(i)} = g-1$, $B_0^{(i)} = B^{(i)}/(x-t_i^0)$, $\deg C = g$, $\deg A_i = g$. An appropriate hyperelliptic Halphen-type continued fraction is obtained from the last one after a change of variables.

Lemma 9.22. *The following identity holds:*

$$y^{g+1} \frac{\sqrt{X} - \sqrt{Y}}{x - y} = (x^g + x^{g-1}y + \cdots + xy^{g-1} + y^g)\sqrt{Y} + \frac{y^{g+1}\sqrt{X} - x^{g+1}\sqrt{Y}}{x - y}.$$

Proposition 9.23. *The HH element $(\sqrt{X} - \sqrt{Y})/(x - y)$ around $x = \infty$ has the same coefficient as $(\sqrt{X'} - \sqrt{Y'})/(s - t)$ around $s = 0$.*

Remainders, continuants and approximation

We consider an HH continued fraction of an element f:

$$f = C + \frac{\beta_1|}{|\alpha_1} + \frac{\beta_2|}{|\alpha_2} + \cdots .$$

Together with *the remainder of rank i Q_i*, where

$$Q_i = \frac{B^{(i)}s^{g+1}}{\sqrt{X} + A^{(i)}},$$

we consider *the continuants (G_i) and (H_i)* and *the convergents G_i/H_i* such that

$$\begin{bmatrix} G_m & G_{m-1} \\ H_m & H_{m-1} \end{bmatrix} = T_C T_1 \cdots T_m. \tag{9.19}$$

Here

$$T_i = \begin{bmatrix} \alpha_i & 1 \\ \beta_i & 0 \end{bmatrix}, \quad T_C = \begin{bmatrix} C & 1 \\ 1 & 0 \end{bmatrix}.$$

By taking the determinant of (9.19) we get

$$G_m H_{m-1} - G_{m-1} H_m = (-1)^{m-1}\beta_1\beta_2 \ldots \beta_m$$
$$= \delta_m s^{(g+1)m},$$
$$\deg \delta_m = (g - 1)m.$$

We also have the following relations:

$$f = \frac{(\alpha_m + Q_m)G_{m-1} + \beta_m G_{m-2}}{(\alpha_m + Q_m)H_{m-1} + \beta_m H_{m-2}} = \frac{G_m + Q_m G_{m-1}}{H_m + Q_m H_{m-1}},$$

and

$$Q_m = -\frac{G_m - H_m f}{G_{m-1} - H_{m-1}}.$$

Proposition 9.24. *The degree of the continuants is* $\deg G_m = g(m+1)$, $\deg H_m = gm$.

Let us introduce

$$\widehat{G}_m = G_m + \frac{H_m}{s-t}\sqrt{Y},$$

$$\widehat{H}_m = \frac{H_m}{s-t}.$$

Then we have

$$Q_m = -\frac{\widehat{G}_m - \widehat{H}_m\sqrt{X}}{\widehat{G}_{m-1} - \widehat{H}_{m-1}\sqrt{X}}$$

and also

$$\widehat{G}_m A^{(m)} + \widehat{G}_{m-1} B^{(m)} s^{g+1} = \widehat{H}_m X,$$

$$\widehat{H}_m A^{(m)} + \widehat{H}_{m-1} B^{(m)} s^{g+1} = \widehat{G}_m X.$$

From the last equations we get

$$\delta_m s^{(g+1)m} A^{(m)} = P_1(s),$$

$$\delta_m s^{(g+1)(m+1)} B^{(m)} = P_2(s),$$

with

$$P_1(s) := H_m H_{m-1}\frac{X_Y}{x-y} - (G_m H_{m-1} - G_{m-1} H_m)\sqrt{Y} - G_m G_{m-1}(s-t),$$

$$P_2(s) := G_m^2(s-t) + 2G_m H_m\sqrt{Y} - H_m^2\frac{X-Y}{x-y}.$$

Theorem 9.25.

(A) *The polynomial* $G_m H_{m-1} - H_m G_{m-1}$ *is of degree* $2gm$. *The first* $(g+1)m$ *coefficients are zero.*

(B) *The polynomial* P_1 *is of degree* $2mg+g+1$. *Its first* $(g+1)m$ *coefficients are zero.*

(C) *The polynomial* P_2 *is of degree* $2mg + 2g + 1$ *and its* $(g+1)(m+1)$ *first coefficients are zero.*

Lemma 9.26. *The following relations hold:*

$$\frac{G_{m-1}(t_m)}{H_{m-1}(t_m)} = -A^{(m)}(t_m) = A^{(m-1)}(t_m),$$

$$\widehat{G}_m - \widehat{H}_m\sqrt{X} = (-1)^{m+1} Q_0 Q_1 Q_2 \ldots Q_m.$$

Theorem 9.27. *If $X(\varepsilon) \neq 0$ and $\varepsilon \neq y$, then the element*

$$\widehat{G}_m - \widehat{H}_m \sqrt{X} = G_m - H_m \frac{\sqrt{X} - \sqrt{Y}}{x - y}$$

has a zero of order $(g+1)(m+1)$ at $s = 0$. If $H(0) \neq 0$ then the differences

$$\frac{\sqrt{X} - \sqrt{Y}}{x - y} - \frac{G_m}{H_m}, \qquad \sqrt{X} - \frac{\widehat{G}_m}{\widehat{H}_m}$$

have developments starting with the order of $s^{(g+1)(m+1)}$.

Now, we consider $\sqrt{X}$ and its development as HH continued fraction. In that case, starting from

$$\frac{\sqrt{X} - \sqrt{p_0}}{x - \varepsilon},$$

we have

$$\deg G_0 = g+1, \quad H_0 = 1, \quad H_1 = \alpha_1, \quad G_1 = \alpha_1 G_0 + \beta_1 s^{g+2}$$

and

$$G_m = \alpha_m G_{m-1} + \beta_m G_{m-2},$$
$$H_m = \alpha_m H_{m-1} + \beta_m H_{m-2}.$$

From the last relation, we have

Theorem 9.28.

(A) *The degree of the continuants in this case is $\deg G_m = g(m+1)+1$, $\deg H_m = gm$.*

(B) *If $y = \varepsilon$ then the development of the difference*

$$\sqrt{X} - \frac{\widehat{G}_m}{\widehat{H}_m}$$

starts with the order $s^{(g+1)(m+1)+1}$.

Theorem 9.29.

(A) *The polynomial $G_m H_{m-1} - H_m G_{m-1}$ is of degree $2gm + 1$ in s. The first $(g+1)m + 1$ coefficients are 0.*

(B) *The polynomial $H_m H_{m-1} X - G_m G_{m-1}$ is of degree $2mg + g + 2$. Its first $(g+1)m + 1$ coefficients are zero.*

(C) *The polynomial $G_m^2 - H_m^2 X$ is of degree $2mg + 2g + 2$ and its $(g+1)(m+1) + 1$ coefficients are zero.*

There are infinite ways to calculate $\sqrt{X}$ in the neighborhood of ε, depending on choice of the parameter y. The best approximation one obtains for the choice

$$y = \varepsilon.$$

To conclude the last observation we need to check the case $y = \infty$. In this case we have

$$G_0 = \sqrt{p_0}(1 + q_1 s + \cdots + q_g s^g), \quad H_0 = 1, \quad G_1 = \alpha_1 G_0 + \beta_1, \quad H_1 = \alpha_1,$$

and we write

$$\widehat{G}_m = G_m + \sqrt{a_0} H_m s^{g+1}, \quad \widehat{H}_m = H_m.$$

Then we have

Proposition 9.30.

(A) *The degree of continuants is* $\deg G_m = g(m+1)$, $\deg H_m = gm$.
(B) *If* $X(\varepsilon) \neq 0$ *and if the parameters* $t_1, \ldots, t_{m+1}$ *are finite and different from zero, then*

$$\sqrt{X} - \frac{\widehat{G}_m}{\widehat{H}_m}$$

has the development starting with order $2m + 2$ *in* s.

9.2 Geometric realizations of the $2 \leftrightarrow g + 1$ dynamics

First geometric realization of the $2 \leftrightarrow g + 1$ dynamics

Let us start with equation (9.18) which defines basic curve Γ_X. Consider polynomial $Q_X(\lambda, s)$ as a *quadratic pencil* or a *net* of polynomials a, b, c of degree $g + 1$ in s:

$$Q_X(\lambda, s) = a(s)\lambda^2 + b(s)\lambda + c(s).$$

Linear pencils of polynomials were considered, for example, in [Dar1917, Tra1988, Dra2008].

As in Section 6.2, we start with a fixed conic $\mathcal{K}$ given by the equation $z_1^2 = 4z_0 z_2$, and its rational parametrization $(s^2, 2s, 1)$.

Now, we interpret the net condition

$$a(s)\lambda^2 + b(s)\lambda + c(s) = 0 \tag{9.20}$$

as a correspondence between values of λ and sets of $g + 1$ tangents to the conic $\mathcal{K}$: Denote by $s_1, \ldots, s_{g+1}$ the set of solutions of equation (9.20) for fixed λ and consider the tangents $t_K(s_1), \ldots, t_K(s_{g+1})$.

Moreover, we associate to the polynomial X a plane curve $\mathcal{B}_X$ such that the Darboux coordinates (ρ, ρ_1) of a point of the curve $\mathcal{B}_X$ satisfy equation (9.20) with a fixed λ.

Definition 9.31. We will call the curve $\mathcal{B}_X$ *the boundary curve* associated with the polynomial X and the conic $\mathcal{K}$.

Collecting together the results of classics: Jacobi, Steiner, Liuoville, Hesse, Cremona, Darboux, we may formulate the following statements.

Theorem 9.32.

(a) *The curve $\mathcal{B}_X$ is of degree $2g$ and in general it has $g(g-1)/2$ double points.*

(b) *For a fixed value of λ there correspond $g+1$ solutions $\rho_1, \ldots, \rho_{g+1}$ of equation (9.20) determining a $g+1$-polygon inscribed in $\mathcal{B}_X$, circumscribed about the conic K and satisfying the following system of differential equations:*

$$\frac{\rho_1^i d\rho_1}{\sqrt{b^2(\rho_1) - 4a(\rho_1)c(\rho_1)}} + \cdots + \frac{\rho_{g+1}^i d\rho_{g+1}}{\sqrt{b^2(\rho_{g+1}) - 4a(\rho_{g+1})c(\rho_{g+1})}} = 0, \quad (9.21)$$

where $i = 0, \ldots, g-1$.

(c) *There exist $2g+2$ lines tangent to the conic K which are tangent to every integral curve of the system of equations (9.21). Each of these tangents is tangent to each integral curve in g points.*

We give a more detailed presentation of the cases of genus 1 and 2.

Example 9.33. For $g = 1$, the system (9.21) consists of one equation. This is the Euler equation. The integral curves of the Euler equation are conics, which are, together with the conic $\mathcal{K}$, inscribed in a quadrilateral.

For a given value $s = \rho_1$ there are two solutions of equation (9.20), denote them by λ_1 and λ_2. Let ρ_1, ρ be the solutions of equation (9.20) for λ_1, and ρ_1, ρ_2 the solutions for λ_2. The pairs of lines (ρ, ρ_1) and (ρ_1, ρ_2) form two angles inscribed in a conic $\mathcal{B}$ and circumscribed about the conic $\mathcal{K}$. The involution which corresponds to the shift from λ_1 to λ_2 is realized as passage from the first angle to the second one.

Example 9.34. For $g = 2$, the system (9.21) consists of two equations:

$$\frac{d\rho_1}{\sqrt{b^2(\rho_1) - 4a(\rho_1)c(\rho_1)}} + \frac{d\rho_2}{\sqrt{b^2(\rho_2) - 4a(\rho_2)c(\rho_2)}}$$
$$+ \frac{d\rho_3}{\sqrt{b^2(\rho_3) - 4a(\rho_3)c(\rho_3)}} = 0,$$

$$\frac{\rho_1 d\rho_1}{\sqrt{b^2(\rho_1) - 4a(\rho_1)c(\rho_1)}} + \frac{\rho_2 d\rho_2}{\sqrt{b^2(\rho_2) - 4a(\rho_2)c(\rho_2)}}$$
$$+ \frac{\rho_3 d\rho_3}{\sqrt{b^2(\rho_3) - 4a(\rho_3)c(\rho_3)}} = 0$$

giving the first generalization of the Euler equation. The integral curves are of degree 4 with one double point. Together with the conic $\mathcal{K}$, they are inscribed in a hexagon.

Given one of the integral curves $\mathcal{B}$ and a tangent $t_\mathcal{K}(\rho)$ to the conic $\mathcal{K}$ with the Darboux coordinate ρ.

For the value $s = \rho$ there are two solutions λ_1, λ_2 of equation (9.20). Let ρ_1, ρ_2 be the solutions of (9.20) for λ_1 that are different than ρ, and similarly, let ρ_3, ρ_4 be the solutions for λ_2. The triplets of lines (ρ, ρ_1, ρ_2) and (ρ, ρ_3, ρ_4) form two triangles inscribed in the degree 4 curve $\mathcal{B}$ and circumscribed about the conic $\mathcal{K}$. The involution which corresponds to the shift from λ_1 to λ_2 is realized this time as the passage from the first triangle to the second one.

The line $t_\mathcal{K}(\rho)$ intersects the degree 4 curve $\mathcal{B}$ in four points T_1, T_2, T_3, T_4, $T_i \in \rho_i$. The involution defined by

$$(\rho, \lambda_1) \mapsto (\rho, \lambda_2)$$

corresponds to the decomposition of the set $\{T_1, T_2, T_3, T_4\}$ on two subsets of the same number of elements: $\{T_1, T_2\}$ and $\{T_3, T_4\}$.

The last observation in the previous example gives an insight into how to understand $2 \leftrightarrow g + 1$ dynamics in a general situation.

[Geometric realization of the dynamics 1] *Given a boundary curve $\mathcal{B}$ of degree $2g$ and a tangent to the conic $\mathcal{K}$ with the Darboux coordinate ρ. The line $t_\mathcal{K}(\rho)$ intersects the degree $2g$ curve $\mathcal{B}$ in $2g$ points T_1, ..., T_g, T_{g+1}, ..., T_{2g}. By condition $T_i \in \rho_i$, $2g$ new tangents to the conic $\mathcal{K}$ are determined. The involution defined by*

$$(\rho, \lambda_1) \mapsto (\rho, \lambda_2)$$

corresponds to the decomposition of set $\{T_1, \ldots T_g, T_{g+1}, \ldots, T_{2g}\}$ to two subsets of the same number of elements, say: $\{T_1, \ldots, T_g\}$ and $\{T_{g+1}, \ldots, T_{2g}\}$. This means that ρ, together with ρ_i, $i = 1, \ldots, g$, form the set of solutions of equation (9.20) with $\lambda = \lambda_1$, while ρ with ρ_i, $i = g + 1, \ldots, 2g$ form the set of solutions of (9.20) with $\lambda = \lambda_2$.

*The $(g+1)$-tuples of lines $(\rho, \rho_1, \ldots \rho_g)$ and $(\rho, \rho_{g+1}, \ldots, \rho_{2g})$ form two $(g+1)$-polygons inscribed in the degree $2g$ curve $\mathcal{B}$ and circumscribed about the conic $\mathcal{K}$. These two polygons have a pair of sides belonging to the same line – ρ. The involution which corresponds to the shift from λ_1 to λ_2 is realized as the passage from the first polygon to the second one. We can call this move **the flip along the edge**.*

The dynamics is a path of polygons of $g + 1$ sides inscribed in the curve $\mathcal{B}$ of degree $2g$ and circumscribed about the conic $\mathcal{K}$ obtained by successive flips along edges.

Second geometric realization of the $2 \leftrightarrow g + 1$ dynamics

In order to give another geometric realization of the $2 \leftrightarrow g + 1$ dynamics which is governed by the HH continued fractions, first we are going to realize the given hyperelliptic curve

$$\Gamma : z^2 = X_{2g+2}$$

of genus g, as a generalized Cayley curve (see Section 8.1). By a birational iso-morphism which maps one of the zeros of polynomial X_{2g+2} to infinity, one can realize the curve Γ in the form

$$y^2 = \mathcal{P}_{2g+1}(x),$$

where the polynomial $\mathcal{P}_{2g+1}$ is of odd degree equal to $2g+1$. Assuming that zeros of polynomial $\mathcal{P}_{2g+1}$ are real and different, one can order them as

$$b_1 < b_2 < \cdots < b_{2d-1}.$$

Now, decompose the set of zeros of the polynomial $\mathcal{P}_{2g+1}$ in one of the ways that satisfies

$$\{b_1, \ldots, b_{2d-1}\} = \{a_1, \ldots, a_{g+1}, \alpha_1, \ldots, \alpha_g\},$$

where

$$\alpha_j \in \{b_{2j-1}, b_{2j}\}, \quad \text{for} \quad 1 \le j \le g. \tag{9.22}$$

We introduce the following family of confocal quadrics in the $(g+1)$-dimen-sional Euclidean space $\mathbf{E}^{g+1}$:

$$\mathcal{Q}_\lambda : \quad \frac{x_1^2}{a_1 - \lambda} + \cdots + \frac{x_{g+1}^2}{a_{g+1} - \lambda} = 1 \quad (\lambda \in \mathbf{R}), \tag{9.23}$$

where $a_1, \ldots, a_{g+1}$ are different real constants chosen above.

By the Chasles theorem, we know that a given line in $\mathbf{E}^{g+1}$ is tangent to g quadrics from a given confocal family.

The g constants $\alpha_1, \ldots, \alpha_g$, determine g quadrics from family (9.23). Since the constants satisfy conditions (9.22), there exist lines in $\mathbf{E}^{g+1}$ that are tangent to g distinct non-degenerate quadrics $\mathcal{Q}_{\alpha_1}, \ldots, \mathcal{Q}_{\alpha_g}$ from the confocal family.

Let ℓ be a line not contained in any quadric of the given confocal family and tangent to the given set of g quadrics $\mathcal{Q}_{\alpha_1}, \ldots, \mathcal{Q}_{\alpha_g}$. The *generalized Cayley curve* $\mathcal{C}_\ell$ is the variety of hyperplanes tangent to quadrics of the confocal family at the points of ℓ.

There is a natural involution τ_ℓ on the generalized Cayley's curve $\mathcal{C}_\ell$ which maps to each other the two hyperplanes tangent to the same quadric of the confocal family. It is easy to see that the fixed points of this involution are hyperplanes corresponding to the g quadrics that are touching ℓ and to $g+2$ degenerate quadrics of the confocal family.

Now we come to the essential observation. The generalized Cayley's curve $\mathcal{C}_\ell$ is automatically equipped with a meromorphic function of degree $g+1$, namely with the projection

$$p_\ell : \mathcal{C}_\ell \mapsto \mathbf{P}^1(\ell).$$

The projection p_ℓ maps to a point t from ℓ the $g + 1$ hyperplanes from $\mathcal{C}_\ell$ that contain t.

Now, we can give the second geometric realization of the dynamics governed by the hyperelliptic Halphen continued fractions.

[**Geometric realization of the dynamics 2**] *For a suitably chosen line ℓ, choose a point $t_1 \in \ell$ and a tangent hyperplane $T_{1,1}$ to a quadric $\mathcal{Q}_{1,1}$ at t_1. Find the other intersection of quadric $Q_{1,1}$ and line ℓ, and denote it as t_2. Let $T_{2,1}$ be the tangent hyperplane to $\mathcal{Q}_{1,1}$ at t_2. Denote by $T_{2,j}$ the tangent hyperplanes to quadrics $\mathcal{Q}_{2,j}$ at t_2, $j \in \{2, \ldots, g + 1\}$. Choose one of them, and denote the chosen tangent hyperplane by $T_{2,2}$. Find the other intersection of the quadric $\mathcal{Q}_{2,2}$ with the line ℓ and denote it by t_3. Denote the tangent hyperplane to the quadric $\mathcal{Q}_{2,2}$ at the point t_3 as $T_{3,1}$. Denote all other tangent hyperplanes to quadrics $\mathcal{Q}_{3,j}$ at t_3, by $T_{3,j}$ where $j \in \{2, \ldots, g+1\}$. Choose one of the tangent hyperplanes, say $T_{3,3}$ and find the other intersection point of the quadric $\mathcal{Q}_{3,3}$ with the line ℓ. Denote the intersection point as t_4 and so on.*

By using functional notation, we may say that

$$\tau_\ell(T_{i,1}) = T_{i+1,1}, \quad p_\ell(T_{i+1,1}) = t_{i+1}.$$

Also we have

$$p_\ell^{-1}(t_i) = \{T_{i,1}, T_{i,2}, \ldots, T_{i,g+1}\}.$$

Even the case $g = 1$ gives a new geometric representation of the famous Euler–Chasles correspondence. What we give here is one of its asymmetric realizations. For $g = 1$, the projection p_ℓ is two-to-one, and it induces another involution

$$\mu_\ell : \mathcal{C}_\ell \to \mathcal{C}_\ell$$

which exchanges the elements of the inverse image of p_ℓ:

$$p_\ell(x) = p_\ell(\mu_\ell(x)).$$

The dynamics of Halphen continued fractions is executed by a shift L done by composition of the two involutions:

$$L = \mu_\ell \circ \tau_\ell.$$

Even in this basic case, $g = 1$, the construction we made leads to new geometric properties of lines in the plane and dynamics of points of intersection with a given family of confocal conics. These properties are reflections of the Poncelet porism for confocal conics and the Euler–Chasles correspondences. Thus, if a sequence of the dynamics described above forms a cycle starting from a point of a line, then a cycle of the same length will appear in this dynamics starting from any other point of the line.

As a trivial example, one may consider a horizontal or a vertical line and a standard confocal system of conics in a plane. The confocal system decomposes a horizontal or vertical line on cycles of length 2.

Now, we are going back to a general case. We follow the line of Section 8.2 where the set $\mathcal{A}_\ell$ of lines in $g+1$-dimensional space tangent to the fixed set of g quadrics of a given confocal family is equipped with a structure of Abelian variety. In the same spirit, we may consider tautological line bundle $\mathcal{LA}_\ell$ as a generalized Abelian variety. A tautological bundle consists of pairs (line, point) with incidence relation that a point belongs to a line.

Conclusion: Polynomial growth and integrability

Due to some well-known facts, the Padé approximants of hyperelliptic functions are unique up to the scalar factors. The approximants discussed in the previous section in the case of genus higher than 1 are neither unique nor of the Padé type. At first glance, it seems that, by the construction, they have an exponential growth. However, a more careful analysis of their degrees compared to the degrees of approximation done in the previous section indicates their polynomial growth. After Veselov, one can consider a discrete multi-valued dynamics to be integrable if it has polynomial growth instead of an exponential one. In that sense, we can say that the multi-valued discrete dynamics associated with HH-continued fractions is an integrable dynamics.

In the case of genus 1, it can be seen as multi-valued discrete dynamics associated with the Euler–Chasles (2-2)-correspondence, which has been studied by Veselov (see [Ves1992]) and Veselov and Buchstaber (see [BV1996]). It would be quite interesting to consider higher genus dynamics from the point of view of n-valued groups and their actions, following Buchstaber (see [Buc2006]).

A new application of Darboux coordinates, pencils of conics and two-valued Buchstaber–Novikov groups to another integrable system, the celebrated Kowalevski top, has been described very recently in [Dra2010a]. Moreover, it has been shown there that associativity of the two-valued group on $\mathbf{CP}^1$ is equivalent to the full Poncelet theorem for a triangle.

Chapter 10

Quantum Yang–Baxter Equation and (2-2)-correspondences

10.1 A proof of the Euler theorem. Baxter's R-matrix

In theorems of Poncelet and Darboux we have met an important role of symmetric (2-2)-correspondences, of the form

$$E : ax^2y^2 + b(x^2y + xy^2) + c(x^2 + y^2) + 2dxy + e(x + y) + f = 0. \qquad (10.1)$$

Almost unintentionally, we paraphrased the first sentence of the very last section of Baxter's remarkable book (see [Bax1982], p. 471). Its Section 15.10 starts with:

> "In the Ising, eight-vertex and hard hexagon models we encounter symmetric biquadratic relations, of the form (10.1)".

Such a coincidence is not accidental.

In the sequel of the last section of [Bax1982], Baxter derives an elliptic parametrization of a symmetric biquadratic. By use of an appropriate projective transformation

$$p(z) = \frac{\alpha z + \beta}{\gamma x + \delta},$$

both on x and y, he makes b and e vanish in the relation (10.1). Then by dividing, Baxter reduces the given biquadratic to the canonical form

$$x^2y^2 + c_1(x^2 + y^2) + 2d_1xy + 1 = 0$$

and solves it later as a quadratic equation in y, obtaining

$$y = -\frac{d_1x \pm \sqrt{-c_1 + (d_1^2 - 1 - c_1^2)x^2 - c_1x^4}}{c_1 + x^2}.$$

Then, he observes that the argument of the square root is a polynomial of fourth degree in x. By use of Jacobian elliptic functions, the quartic polynomial is transformed into a perfect square, by change of variables

$$x = k^{1/2}\operatorname{sn} u.$$

The modulus k satisfies $k + k^{-1} = (d_1^2 - c_1^2 - 1)/c_1$. Recall (see Exercises 3.95–3.97 and Theorem 3.100) the well-known identities of the Jacobian functions:

$$\operatorname{cn}^2 u + \operatorname{sn}^2 u = 1,$$
$$\operatorname{dn}^2 u + k^2 \operatorname{sn}^2 u = 1, \tag{10.2}$$

and the addition formula

$$\operatorname{sn}(u - v) = \frac{\operatorname{sn} u \cdot \operatorname{cn} u \cdot \operatorname{dn} u - \operatorname{cn} u \cdot \operatorname{dn} u \cdot \operatorname{sn} v}{1 - k^2 \operatorname{sn}^2 u \cdot \operatorname{sn}^2 v}. \tag{10.3}$$

Using the identities (10.2), the argument of the square root is transformed:

$$-c_1(1 - (k + k^{-1})x^2 + x^4) = -c_1(1 - \operatorname{sn}^2 u)(1 - k^2 \operatorname{sn}^2 u)$$
$$= -c \cdot \operatorname{cn}^2 u \cdot \operatorname{dn}^2 u,$$

and parameter η, such that $c_1 = -1/k\operatorname{sn}^2\eta$, is introduced. Appropriately choosing the sign, she got $d_1 = \operatorname{cn}\eta \operatorname{dn}\eta/(k\operatorname{sn}^2\eta)$, and using the addition formula (10.3), she finally obtained the formula for y:

$$y = k^{1/2}\frac{\operatorname{sn} u \cdot \operatorname{cn}\eta \cdot \operatorname{dn}\eta \pm \operatorname{sn}\eta \cdot \operatorname{cn} u \cdot \operatorname{dn} u}{1 - k^2 \operatorname{sn}^2 u \cdot \operatorname{sn}^2\eta} = k^{1/2}\operatorname{sn}(u \pm \eta).$$

Thus, the following parametrization of the canonical form of a biquadratic is obtained:

$$x = k^{1/2}\operatorname{sn} u, \quad y = k^{1/2}\operatorname{sn} v,$$

where $v = u - \eta$ or $v = u + \eta$.

Of course, to get parametrization of the initial biquadratic, one needs to use the inverse of the projective transformations, to get the final form of parametrization:

$$x = \phi(u), \quad y = \phi(u \pm \eta)$$

for some elliptic function ϕ which is obtained from the Jacobian sn function by application of the projective transformations.

Following this line of thought, we actually sketched an effective proof of the Euler theorem. There is only one more step to be taken – to shift the origin in order to transform elliptic function ϕ into an even one.

By using the above argumentation, Baxter managed to get his celebrated R-matrix, which is also known as XYZ R-matrix and Eight Vertex Model R-matrix because of its fundamental role in both of these very important models of quantum and statistical mechanics respectively. We will say a little bit more about that later.

The Baxter R-matrix is a 4×4 matrix $R_b(t, h)$ of the form

$$R_b(t, h) = \begin{pmatrix} a & 0 & 0 & d \\ 0 & b & c & 0 \\ 0 & c & b & 0 \\ d & 0 & 0 & a \end{pmatrix} \tag{10.4}$$

where

$$a = \operatorname{sn}(t + 2h), \quad b = \operatorname{sn} t, \quad c = \operatorname{sn} 2h, \quad d = k \cdot \operatorname{sn} 2h \cdot \operatorname{sn} t \cdot \operatorname{sn}(t + 2h).$$

The Baxter matrix $R_b(t, h)$ is a solution of the *quantum Yang–Baxter equation*

$$R^{12}(t_1 - t_2, h)R^{13}(t_1, h)R'^{23}(t_2, h) = R^{23}(t_2, h)(R^{13}(t_1, h)R^{12}(t_1 - t_2, h).$$

Here t is a so-called *spectral parameter* and h is the *Planck constant*. Here we assume that $R(t, h)$ is a linear operator from $V \otimes V$ to $V \otimes V$ and

$$R^{ij}(t, h) : V \otimes V \otimes V \rightarrow V \otimes V \otimes V$$

is an operator acting on the ith and jth components as $R(t, h)$ and as identity on the third component. For example $R^{12}(t, h) = R \otimes Id$. In the first nontrivial case, matrix $R(t, h)$ is 4×4 and the space V is two-dimensional. Even in this case, the quantum Yang–Baxter equation is highly nontrivial. It represents a strongly overdetermined system of 64 third-degree equations in 16 unknown functions. Miraculously, solutions exist.

The Quantum Yang–Baxter equation is a paradigm of modern addition relation, and it has been one of the central objects in mathematical physics for the last 25 years.

If the h dependence satisfies the quasi-classical property $R = I + hr + O(h^2)$ the classical r-matrix $r = r(t)$ satisfies the co-called *classical Yang–Baxter equation*. Classification of the solutions of the classical Yang–Baxter equation was done by Belavin and Drinfeld in 1982 [BD1982]. The problem of classification of the quantum R-matrices is still open. However, some classification results have been obtained in the basic 4×4 case by Krichever (see [Kri1981]) and following his ideas in [Dra1993, Dra1992b, Dra1992c]. Before we pass to the exposition of Krichever's ideas, let us briefly recall basic definitions of the Heisenberg XYZ model.

10.2 Heisenberg quantum ferromagnetic model

The Heisenberg ferromagnetic model ([Hei1928]) is defined by its Hamiltonian

$$H = -\sum_{i=1}^{N}(X\sigma_{i+1}^x\sigma_i^x + Y\sigma_{i+1}^y\sigma_i^y + Z\sigma_{i+1}^z\sigma_i^z),$$

where the operator H maps $V^{\otimes N}$ to $V^{\otimes N}$, $V = \mathbf{C}^2$ and σ_i^x denotes the operator which acts as the Pauli matrix on ith V and as the identity on the other components. Following Heisenberg's definition of the model in 1928, Bethe solved the simplest case $X = Y = Z$ in 1931. The next step was taken by Yang in 1967 by solving the XXZ case, obtained by putting $X = Y$. The final step was taken by Baxter, who solved the general XYZ problem in 1971 (see [Bax1982, Bax1971a, Bax1971b, Bax1972a, Bax1972b]).

Both Yang and Baxter exploited the connection with statistical mechanics on the plane lattice. The first one used the six-vertex model and the second one the eight-vertex model. Denote by L and L' the local transition matrices

$$L, L' : W \otimes V \rightarrow W \otimes V,$$

and by R a solution of the Yang–Baxter equation acting on $V \otimes V$. The key role is played by the matrix R. In the Yang case, that was R_y, the XXZ- R matrix of the form

$$R_y(t, h) = \begin{pmatrix} a & 0 & 0 & 0 \\ 0 & b & c & 0 \\ 0 & c & b & 0 \\ 0 & 0 & 0 & a \end{pmatrix}$$

where a, b, c are trigonometrical functions obtained when k tends to 0 from corresponding functions in the Baxter matrix (see equation (10.4)).

Fundamental in Yang's and Baxter's approach was the relation between two tensors Λ_1 and Λ_2 in $W \otimes V \otimes V$ which we are going to call *the Yang equation*:

$$\Lambda_1 = \Lambda_2 \qquad\qquad (10.5)$$

where

$$\begin{aligned}
\Lambda_1 &= \Lambda_{1pq\beta}^{ij\alpha} = L_{p\beta}^{\prime k\gamma} L_{q\gamma}^{l\alpha} R_{kl}^{ij}, \\
\Lambda_2 &= \Lambda_{2pq\beta}^{ij\alpha} = R_{pq}^{kl} L_{k\beta}^{i\gamma} L_{l\gamma}^{\prime j\alpha}.
\end{aligned} \qquad (10.6)$$

We assume summation over repeated indices and we use the convention that Latin indices indicate the space V while the Greek ones are reserved for W. In the simplest case, when W and V are two-dimensional, the Yang equation is again an overdetermined system of 64 equations of degree 3 in 48 unknowns.

Set

$$T = \prod_{n=1}^{N} L_n$$

where

$$L_n : W^{\otimes N} \otimes V \rightarrow W^{\otimes N} \otimes V,$$

acts as L on the nth W and V and as identity otherwise. From the Yang equation (10.5, 10.6) it follows that

$$R(T \otimes T') = (T' \otimes T)R.$$

Therefore, all the operators

$$\operatorname{Tr}_V T$$

commute. The connection between the Heisenberg model and the vertex models mentioned above lies in the fact that the Hamiltonian operator H commutes with all the operators $\operatorname{Tr}_V T$. Thus, they have the same eigen-vectors.

The Algebraic Bethe Ansatz is a method of formal construction of those eigen-vectors. There is a very nice presentation of this method in the work of Takhtadzhyan and Faddeev (see [TF1979]). Starting with matrices R_y and R_b they found vectors X, Y, U, V which satisfy the relation

$$RX \otimes U = Y \otimes V. \tag{10.7}$$

They calculated the vectors X, Y, U, V in terms of theta-functions, and using computational machinery of theta-functions, they produced the eigen-vectors of the Heisenberg model.

We are going to present here a sort of converse approach to the Algebraic Bethe Ansatz. Our presentation is applied uniformly to all 4×4 solutions of rank 1 of the Yang equation. It does not involve computations with theta functions, but uses geometry of the Euler–Chasles correspondence, which is, as we know, an interpretation of the billiard dynamics within ellipses and of the Poncelet geometry. This approach is based on ideas of Krichever (see [Kri1981]), on his classification of rank 1 4×4 solutions of the Yang equation in general situations and on classification of remaining cases, developed in [Dra1993, Dra1992b, Dra1992a]. Thus, we are going to expose now the ideas of Krichever.

10.3 Vacuum vectors and vacuum curves

Basic notions

Having in mind Baxter's considerations leading to the discovery of the Baxter R matrix and Faddeev-Takhtadzhyan study of the vectors of the form given by equation (10.7), Krichever in [Kri1981] suggested a sort of inverse approach, following the best traditions of the theory of "finite-gap" integration (see [DKN2001]).

Krichever's method is based on the *vacuum vector representation* of an arbitrary $2n \times 2n$ matrix L. Such a matrix is understood as a 2×2 matrix with blocks of $n \times n$ matrices. In other words, $L = L_{j\beta}^{i\alpha}$ is a linear operator in the tensor product $C^n \otimes C^2$. *The vacuum vectors X, Y, U, V satisfy, by definition, the relation*

$$LX \otimes U = hY \otimes V$$

or in coordinates

$$L_{j\beta}^{i\alpha} X_i U_\alpha = h Y_j V_\beta,$$

where we assume now that Latin indices run from 1 to n while the Greek ones from 1 to 2. We assume additionally the following convention for affine notation

$$X_n = Y_n = U_2 = V_2 = 1, \quad U_1 = u, \quad V_1 = v,$$

and

$$\tilde{V} = (1, -v).$$

The vacuum vectors are parametrized by *the vacuum curve* Γ_L, which is defined by the affine equation

$$\Gamma_L : P_L(u, v) = \det(L^i_j) = \det(V^\beta L^{i\alpha}_{j\beta} U_\alpha) = 0.$$

The polynomial $P_L(u, v)$, called *the spectral polynomial* of the matrix L, is of degree n in each variable. In general position, the genus of the curve Γ_L is $g = g(\Gamma_L) = (n-1)^2$ and Krichever proved that X understood as a meromorphic function on Γ_L is of degree $N = g + n - 1$. And, following the ideology of "finite-gap" integration, Krichever proved a converse statement.

Theorem 10.1 (Krichever, [Kri1981]). *In the general position, the operator L is determined uniquely up to a constant factor by its spectral polynomial and by the meromorphic vector-functions X and Y on the vacuum curve with pole divisors D_X and D_Y of degree $n(n-1)$ which satisfy*

$$D_X + D_U \sim D_Y + D_V.$$

Proof. The degree of the divisor $D = D_X + D_U$ is equal to n^2. The space $L(D)$ of functions having poles of not higher degree than those prescribed by D has a dimension greater than or equal to

$$\deg D - g + 1 = n^2 - (n-1)^2 + 1 = 2n,$$

by the Riemann–Roch theorem (Theorem 3.68). Moreover, since the degree is greater than the genus of the curve, there is equality in the Riemann–Roch theorem, and the dimension of the space is equal to $2n$. One basis forms the coordinate functions of the vector $X \otimes U$. According to the equivalence condition of the theorem, there exists a function h with the divisor of poles equal to $D_X + D_U$ and the divisor of zeros equal to $D_Y + D_U$. The coordinates of the vector function $hY \otimes U$ now provide another basis of the space $L(D)$ and the operator L is uniquely defined as the transition operator.   □

General rank 1 solutions in the (4×4) case

Now we specialize to the basic 4×4 case. The question is to describe analytical conditions, vacuum curves and vacuum vectors of three 4×4 matrices R, L and L' in order to satisfy the Yang equation, see equations (10.5), (10.6).

Denote by $P = P(u,v)$ and $P_1 = P_1(u,v)$ the spectral polynomial of given 4×4 matrices L and L'. They are polynomials of degree 2 in each variable and they define the vacuum curves $\Gamma = \Gamma_L : P(u,v) = 0$ and $\Gamma_1 = \Gamma_{L'} : P_1(u,v) = 0$. The vacuum curves are of genus not greater than 1.

Let us make one general observation concerning vacuum curves in the 4×4 case. As one can easily see, each of them can naturally be understood as the intersection of two quadrics in $\mathbf{P}^3$. The first one is the Segre quadric, seen as embedding of $\mathbf{P}^1 \times \mathbf{P}^1$ represented by $Y \otimes V$. The second one is the image of the Segre quadric represented as $X \otimes U$ by a linear map, induced by the linear operator under consideration.

In this subsection, following Krichever, we are going to consider only the general case, when those curves are elliptic ones.

Thus, we have

$$L(X(u,v) \otimes U) = h(u,v)(Y(u,v) \otimes V),$$
$$L'(X(u^1,v^1) \otimes U^1) = h_1(u^1,v^1)(Y^1(u,v) \otimes V^1).$$

Each of the tensors Λ_i which represents one of the sides of the Yang equation (see equations (10.5), (10.6)), is by itself a 2×2 matrix of blocks of 4×4 matrices. Denote corresponding spectral polynomials of degree 4 in each variable as Q_1 and Q_2 and the vacuum curves as $\hat{\Gamma}_1$ and $\hat{\Gamma}_2$. Krichever's crucial observation is the following

Theorem 10.2 (Krichever). *If 4×4 matrices L, L', R satisfy the Yang equation, then (2-2)-relations defined by the polynomials P and P_1 commute.*

Proof. Consider (u,v,w) such that

$$P(u,v) = 0, \quad P_1(v,w) = 0.$$

Let

$$\hat{X}(u,w) = R^{-1}(X^1(v,w) \otimes X(u,v)).$$

Then we have

$$\Lambda_1(\hat{X}(u,v) \otimes U) = h(u,v)h_1(v,w)(Y^1(v,w) \otimes Y(u,v) \otimes W.$$

Thus, the vectors $\hat{X}(u,w)$ and $Y^1(v,w) \otimes Y(u,v)$ are vacuum vectors of the matrix Λ_1 and the pair (u,w) belongs to its vacuum curve $\hat{\Gamma}_1$. In other words,

$$P(u,v) = 0, \quad P_1(v,w) = 0 \Rightarrow Q_1(u,w) = 0.$$

If we consider the (2-2)-relations in opposite order, then for the triplet $(u,\hat{v},w)$ such that

$$P_1(u,\hat{v}) = 0, \quad P(\hat{v},w) = 0,$$

we get

$$\Lambda_2(X(\hat{v}, w) \otimes X^1(u, \hat{v}) \otimes U) = h(\hat{v}, w)h_1(u, \hat{v})(\hat{Y}(u, w) \otimes W),$$

with

$$\hat{Y}(u, w) = R(Y(\hat{v}, w) \otimes Y^1(u, \hat{v})).$$

Thus, $X(\hat{v}, w) \otimes X^1(u, \hat{v})$ and $\hat{Y}(u, w)$ are vacuum vectors of Λ_2 and (u, w) is contained in its vacuum curve $\hat{\Gamma}_2$:

$$P_1(u, \hat{v}) = 0, \quad P(\hat{v}, w) = 0 \Rightarrow Q_2(u, w) = 0.$$

The Yang equation $\Lambda_1 = \Lambda_2$ gives $Q_1 = Q_2$ and the relations commute. □

From our experience with Euler–Chasles correspondences, we know that composition of two commuting ones is reducible. Although we have not considered symmetry of spectral polynomials yet, the same can be proven for their composition.

Lemma 10.3 ([Kri1981]). *The polynomial $Q(u, w) = Q_1(u, w) = Q_2(u, w)$ is reducible.*

Proof. Suppose that Q is irreducible. Then, there are vacuum vectors corresponding to every point (u, w) of the vacuum curve of Λ. Thus, we have

$$\Lambda(\hat{X}(u, w) \otimes U) = f(u, w)(\hat{Y}(u, w) \otimes W)$$

giving

$$R(X(\hat{v}, w) \otimes X^1(u, \hat{v})) = g(u, w)(X^1(v, w) \otimes X(u, v)),$$
$$R(Y(\hat{v}, w) \otimes Y^1(u, \hat{v})) = g_1(u, w)(Y^1(v, w) \otimes Y(u, v)). \tag{10.8}$$

For commuting property, there are two possible arrangements of the triplets (u, v, w) and $(u, \hat{v}, w)$. The first one is

$$(u - v_1 - w_1 \& w_2) \quad (u - v_2 - w_3 \& w_4) \quad (u - \hat{v}_1 - w_1 \& w_2) \quad (u - \hat{v}_2 - w_3 \& w_4).$$

In this situation, from equation (10.8), we would get the pair

$$\big(x^1(u, \hat{v}^1), x(u, v_1)\big)$$

of the vacuum curve Γ_R of the matrix R with two corresponding vacuum vectors,

$$X(\hat{v}_1, w_1), \quad X(\hat{v}_1, w_2).$$

This is impossible. The second possibility is

$$(u - v_1 - w_1 \& w_2) \quad (u - v_2 - w_3 \& w_4) \quad (u - \hat{v}_1 - w_1 \& w_3) \quad (u - \hat{v}_2 - w_2 \& w_4).$$

Here, there are two pairs (w_1, w_4) and (w_2, w_3) such that the values of v and $\hat{v}$ which correspond to members of the same pair are different. Thus, there are two components of the curve $\hat{\Gamma}$. One component contains pairs (u, w_1) and (u, w_4) while the other one contains (u, w_1) and (u, w_4). This contradiction concludes the proof. □

Definition 10.4. If the spectral polynomial Q is a perfect square, then the triplet (R, L, L') of solutions of the Yang equation is of rank 2. Otherwise, it is a solution of rank 1.

In this subsection we proceed to consider rank 1 solutions. Let us consider the elliptic component $\hat{\Gamma}'$ of the vacuum curve $\hat{\Gamma}$ of the matrix Λ which contains pairs (u, w_1) and (u, w_4), using terminology of the proof of the last lemma. The curve $\hat{\Gamma}'$ is isomorphic to the vacuum curves Γ and Γ_1 and uses the uniformizing parameter x of the elliptic curve.

Denote by $(u(z), w(z))$, $(u(z), v(z))$ and $(u(z), \hat{v}(z))$ parametrizations of the curves $\hat{\Gamma}'$, Γ and Γ_1 respectively. Taking into account that $(v(z), w(z))$ also parametrizes Γ_1, we get

Proposition 10.5. *There exist shifts η and η_1 on the elliptic curve, such that*

$$v(z) = u(z - \eta), \quad \hat{v}(z) = u(z - \eta_1).$$

From equations (10.8) now we have

$$
\begin{aligned}
R(X(z - \eta_1) \otimes X^1(z)) &= g(z)(X^1(z - \eta) \otimes X(z)), \\
Y(z) &= X(z - \eta_2), \\
Y^1(z) &= X^1(z - \eta_2), \\
L(X(z) \otimes U(z)) &= h(z)(X(z + \eta_2) \otimes U(z - \eta)), \\
L^1(X^1(z) \otimes U(z)) &= h^1(z)(X^1(z + \eta_2) \otimes U(z - \eta_1)),
\end{aligned}
\tag{10.9}
$$

where η_2 is a shift, which like the shift η_1, differs from η by a half-period of the elliptic curve.

If we denote by G_X, G_{X^1}, G_U arbitrary invertible 2×2 matrices, then they define a *weak gauge transformation* which transforms a triplet (R, L, L_1) of solutions of the Yang equation to a triplet $(\tilde{R}, \tilde{L}, \tilde{L}_1)$ by the formulae

$$
\begin{aligned}
\tilde{L} &= (G_X \otimes G_U)L(G_X^{-1} \otimes G_U^{-1}), \\
\tilde{L}_1 &= (G_{X^1} \otimes G_U)L_1(G_{X^1}^{-1} \otimes G_U^{-1}), \\
\tilde{R} &= (G_{X^1} \otimes G_X)R(G_X^{-1} \otimes G_{X^1}^{-1}),
\end{aligned}
$$

which are again solutions of the Yang equation.

Denote by G_i the 2×2 matrices which correspond to shifts for half-periods: $U(z + 1/2) = G_1 U(z)$ and $U(z + 1/2\tau) = f G_2 U(z)$, where

$$G_1 = \begin{pmatrix} -1 & 0 \\ 0 & 1 \end{pmatrix},$$

$$G_2 = \begin{pmatrix} 0 & 1 \\ 1 & 0 \end{pmatrix}.$$

Then the shift of η_1 for half-periods transforms solutions according to the formula

$$T_i \colon (L, L_1 R) \mapsto (L, (I \otimes G_i)L_1, R(G_i \otimes I)). \tag{10.10}$$

In the same way, the shift of η_2 for half-periods transforms solutions according to the formula

$$\hat{T}_i \colon (L, L_1 R) \mapsto (L(G_i \otimes I), L_1(G_i \otimes I)). \tag{10.11}$$

Summarizing, we get

Theorem 10.6 (Krichever). *Given an arbitrary elliptic curve Γ with three points η, η_1, η_2 which differ up to a half-period, and three meromorphic functions of degree 2, x, x^1, u. Then formulae (10.9) define solutions of the Yang equation. All 4×4 rank 1 general solutions of the Yang equation are of that form. Moreover, up to weak gauge transformations and transformations (10.10) and (10.11) all such solutions are equivalent to Baxter's matrix R_b.*

For the last part of the theorem, exact formulae for vacuum vectors for the Baxter matrix from [TF1979] have been used.

Krichever also formulated a corollary:

Corollary 10.7. *In case $\eta = \eta_1 = \eta_2$ all 4×4 rank 1 general solutions of the Yang equation are weak-gauge equivalent to the Baxter solution R_b.*

Rank 1 solutions in nongeneral (4×4) cases

But, as it was observed in [Dra1993, Dra1992c], there exist 4×4 rank 1 solutions of the Yang–Baxter equation which are not equivalent to the Baxter solution. The modal example is the so-called Cherednik matrix R_{ch} (see [Che1980]) with the formula

$$R_{ch}(t, h) = \begin{pmatrix} 1 & 0 & 0 & 0 \\ 0 & b & c & 0 \\ 0 & c & b & 0 \\ d & 0 & 0 & 1 \end{pmatrix}$$

where

$$b = \frac{\sinh t}{\sinh(t+h)}, \qquad c = \frac{\sinh t}{\sinh(t+h)}, \qquad d = -4\sinh t \cdot \sinh h.$$

It was calculated in [Dra1993] that the spectral polynomial is of the form

$$P_{ch}(u, v) = A u^2 v^2 + B(u^2 + v^2) + C uv \tag{10.12}$$

and analytic properties have been studied. It was proven that the vacuum curve in this case is rational with an ordinary double point. For such a sort of solutions, description as in Theorem 10.6 has been established in [Dra1993] where it was summarized as

Theorem 10.8. *All* (4×4) *rank* 1 *solutions of the Yang equation with rational vacuum curve with ordinary double point are gauge equivalent to the Cherednik solution.*

The Cherednik and the Baxter solutions of the Yang–Baxter equation are essentially different and are not gauge equivalent. Moreover, the second one is $\mathbf{Z}_2$-symmetric, while the first one is not. It is interesting to note that, nevertheless, the first one can be obtained from the last one in a nontrivial gauge limit.

Lemma 10.9 ([Dra1993]). *Denote by* $T(k)$ *the following family of matrices depending on the modulus* k *of the elliptic curve:*

$$T(k) = \left(\begin{array}{cc} (-4/k)^{1/4} & 0 \\ 0 & (-k/4)^{1/4} \end{array} \right)$$

and denote by $R_b(t, h; k)$ *the Baxter matrix. Then*

$$\lim_{k \to 0} (T(k) \otimes T(k)) R_b(t, h; k) (T(k)^{-1} \otimes T(k)^{-1}) = R_{ch}(it, h).$$

A similar analysis of vacuum data as for the Cherednik solution was done for the Yang solution R_y. In [Dra1992c] it was shown that the vacuum curve consists of two rational components. For such a kind of solution, a similar statement was proven there.

Theorem 10.10. *All* (4×4) *rank* 1 *solutions of the Yang equation with vacuum curve reducible on two rational components are gauge equivalent to the Yang solution.*

It was observed in [Dra1993] that such an approach does not lead to a solution of the Yang–Baxter equation with rational vacuum curve with cusp singularity.

Relationship with Poncelet–Darboux theorems

Let us go back to considerations from [Kri1981], around Lemma 10.3 and Theorem 10.2. We are going to underline a couple of observations which have not been stressed there.

Denote by $P_2(u, w) = 0$ the relation which corresponds to $\hat{\Gamma}'$.

Proposition 10.11. *If* $P(u, v) = 0$, $P_1(u, v) = 0$ *and* $P_2(u, v) = 0$ *are* (2-2)-*relations corresponding to rank* 1 *solutions of the Yang equation, then these correspondences are symmetric.*

Thus, these correspondences are of the Euler–Chasles type. According to geometric interpretations of such correspondences, we may associate a pair of conics $(\mathcal{K}, \mathcal{C})$ to P, pair $(\mathcal{K}, \mathcal{C}_1)$ to P_1, and finally pair $(\mathcal{K}, \mathcal{C}_2)$ to P_2. Since correspondences commute, they form a pencil χ of conics. Since we assume in this subsection that the underlying curve is elliptic, the pencil χ is of general type, having four distinct points in the critical divisor. In other words, conics $\mathcal{K}$, $\mathcal{C}$, $\mathcal{C}_1$, $\mathcal{C}_2$ intersect at four distinct points.

Having in mind the Darboux coordinates and related geometric interpretation, we come to the following

Theorem 10.12. *Suppose a general 4×4 rank 1 solution of the Yang equation is given. Then there exist a pencil of conics χ and four conics $\mathcal{K}, \mathcal{C}, \mathcal{C}_1, \mathcal{C}_2 \in \chi$ intersecting at four distinct points such that triplets (u, v, w) and $(u, \hat{v}, w)$ satisfying $P(u, v) = 0$, $P_1(v, w) = 0$ and $P_1(u, \hat{v}) = 0$, $P(\hat{v}, w) = 0$ form sides of two Poncelet triangles that are circumscribed about $\mathcal{K}$ and have vertices on $\mathcal{C}$, $\mathcal{C}_1$ and $\mathcal{C}_2$. Moreover, quadruplet of lines $(u, v, \hat{v}, w)$ forms a double reflection configuration at $\mathcal{C}$ and $\mathcal{C}_1$.*

Such type of pencils is sometimes denoted also as $(1, 1, 1, 1)$. In cases of 4×4 rank one solutions with rational vacuum curves, the situation is practically the same. One can easily calculate directly that spectral polynomials for the Cherednik R matrix (see equation (10.12)) and for the Yang R matrix are symmetric.

Proposition 10.13.

(a) *In the case of Cherednik type solutions, the corresponding pencil of conics has one double critical point and two ordinary ones – type $(2, 1, 1)$. Conics are tangent in one point and intersect in two other distinct points.*

(b) *In the case of the Yang type solutions, the corresponding pencil of conics is bitangential: it has two double critical points. Conics are tangent in two points – type $(2, 2)$.*

(c) *In both rational cases, the statement about two Poncelet triangles and double reflection configuration holds as in Theorem 10.12.*

It would be interesting to check if geometrically possible degeneration of a pencil to a superosculating one, the one of type (4), can make nontrivial contribution to solutions of the Yang–Baxter equation.

In all cases, elliptic or rational, the vacuum vector representation of rank 1 (4×4) solutions of the Yang–Baxter equation has the following form:

$$LX_l \otimes U_l = hX_{l+1} \otimes U_{l-1},$$

where all the functions are meromorphic on a vacuum curve Γ of degree 2 and we use notation

$$X_{l+n} = X_l \circ \Psi^n.$$

Here Ψ is an automorphism of the vacuum curve Γ and geometric interpretation of the last formula is the following:

the tangent to conic $\mathcal{K}$ with Darboux coordinate x_l reflects n times at the conic $\mathcal{C}_1$ and gives a new tangent to $\mathcal{K}$ with Darboux coordinate x_{l+n}.

The last formula with the last interpretation plays a crucial role in our presentation of the Algebraic Bethe Ansatz.

10.4 Algebraic Bethe Ansatz and Vacuum Vectors

Covector representation

Our starting point, following ([Dra1994]) is the last formula giving general vacuum vector representation of all rank 1 solutions of the Yang–Baxter equation in the 4×4 case.

As in ([Dra1994]), together with vacuum vector representation, we consider *the vacuum covector representation*:

$$A^j B^\beta L^{i\alpha}_{j\beta} = l C^i D^\alpha.$$

We will use a more general notion of spectral polynomial. Let P^{ij}_L denote a polynomial in two variables obtained as determinant of the matrix generated by L, contracting the ith bottom and jth top index. The spectral polynomial we used up to now, in this notation becomes P^{22}_L.

It can easily be shown that vacuum covectors, under the same analytic conditions as vacuum vectors, uniquely define matrix L. Thus, there is a relation between vacuum vectors and vacuum covectors.

Lemma 10.14 ([Dra1994]). *If a 4×4 matrix L satisfies the condition*

$$L(X_l \otimes U_l) = h(X_{l+1} \otimes U_{l-1}),$$

then

$$(\tilde{X}_{l+1} \otimes \tilde{U}_{l+1})L = g(\tilde{X}_{l+2} \otimes \tilde{U}_l).$$

We use notation $X = [x \ 1]^t$ and $\tilde{X} = [1 \ -x]$.

Proof. Denote vacuum covectors of the matrix L by A, B, C, D. From

$$A^j B^\beta L^{i\alpha}_{j\beta} = l C^i D^\alpha,$$

we get $P^{12}_L(a, d) = 0$. By the assumption we have $P^{12}_L(x \circ \psi, u) = 0$. Thus, by reparametrization we can put

$$A = \tilde{X} \circ \Psi, \quad D = \tilde{U}.$$

In the same way, comparing pairs (B, C) and $(U \circ \Psi, X)$, we see that there is a shift Φ such that

$$B = \tilde{U} \circ \Psi \circ \Phi, \quad C = \tilde{X} \circ \Phi.$$

We have to calculate Φ. Consider relations

$$P^{12}_L(u \circ \psi, u) = 0, \quad P^{12}_L(b, u) = 0.$$

There are two points on the curve Γ with the same second coordinate equal to $u(z)$. The corresponding first coordinates are $(u \circ \psi)(z)$ and $(u \circ \psi \tau_u)(z)$ where τ_u is the involution associated with u.

Thus, we have two possibilities for B:

$$B = \tilde{U} \circ \Psi, \quad B = \tilde{U} \circ \Psi \circ \tau_u.$$

In the first case the transformation would be identity, while in the second it would correspond to the shift

$$U_l \longmapsto U_{l+2}, \quad X_l \longmapsto X_{l+2}.$$

Comparing divisors of matrix elements in the equation

$$A^j L_{j\beta}^{i\alpha} = k\tilde{B}X$$

we conclude that the appropriate formula for B is the second one. The last matrix equality follows from the fact that the matrices on both sides are of dimension 2, of rank 1, with the same kernel and image. □

Let us just mention that the proof of the last lemma extends automatically to arbitrary 4×4 matrices.

Local vacuum vectors

The matrices L, solutions of the Yang equation, are local transition matrices. As it was suggested in [TF1979] we change them according to the formula

$$L_n^l(\lambda) = M_{n+l}^{-1}(\lambda)L_n(\lambda)M_{n+l-1} = \begin{pmatrix} \alpha_n^l(\lambda) & \beta_n^l(\lambda) \\ \gamma_n^l(\lambda) & \delta_n^l(\lambda) \end{pmatrix},$$

where the matrix M_l has vacuum vectors as columns:

$$M_l = \begin{pmatrix} x_l & x_{l+1} \\ 1 & 1 \end{pmatrix}.$$

By this transformation, the monodromy matrix $T(\lambda) = \prod_{n=1}^N L_n(\lambda)$ transforms into

$$T_N^l(\lambda) = M_{N+l}^{-1}T(\lambda)M_l.$$

Denote the elements of the last matrix by

$$A_N^l(\lambda), \quad B_N^l(\lambda), \quad C_N^l(\lambda), \quad D_N^l(\lambda).$$

The aim of the Algebraic Bethe Ansatz is to find local vacuum vectors w^l independent of λ such that

$$\gamma_n^l(\lambda)w_n^l = 0,$$
$$\alpha_n^l(\lambda)w_n^l = g(\lambda)w_n^{l-1},$$
$$\delta_n^l(\lambda)w_n^l = g'(\lambda)w_n^{l+1}.$$

Having the last relations satisfied, vector $\Omega_N^l = \omega_1^l \otimes \cdots \otimes \omega_N^l$ would satisfy

$$A_N^l(\lambda)\Omega_N^l = g^N(\lambda)\Omega_N^{l-1},$$
$$D_N^l(\lambda)\Omega_N^l = g'^N(\lambda)\Omega_N^{l+1},$$
$$C_N^l(\lambda)\Omega_N^l = 0,$$

and provide a family of generating vectors.

Theorem 10.15 ([Dra1994]). *The following relations are valid:*

$$\gamma_n^l(\lambda)U_l = 0, \quad \alpha_n^l(\lambda)U_l = g(\lambda)U_{l-1}, \quad \delta_n^l(\lambda)\omega_n^l = g'(\lambda)U_{l+1}.$$

Proof. One can easily check that $\gamma^l(\lambda) = \tilde{X}_{l+1}L(\lambda)X_l$. Thus

$$\gamma^l(\lambda)U_l = \tilde{X}_{l+1}L(\lambda)X_lU_l = \tilde{X}_{l+1}X_{l+1}U_{l-1} = 0.$$

In the same manner,

$$\alpha^l(\lambda)U_l = \tilde{X}_{l+2}L(\lambda)X_lU_l = g(\lambda)U_{l-1}.$$

From the previous Lemma 10.14 we have

$$\tilde{X}_{l+1}\tilde{U}_{l+1}LU_l = 0.$$

Thus,

$$\delta^l(\lambda)U_l = \tilde{X}_{l+1}LX_{l+1}U_l = g'(\lambda)U_{l+1}.$$

And the proof is completed. $\square$

It is also necessary to calculate images of shifted vacuum vectors. The formulae are given in the following

Lemma 10.16 ([Dra1994]). *The image of the shifted vacuum vector is a combination of two shifted vacuum vectors:*

$$LX_{l+1} \otimes U_l = hX_{l+1} \otimes U_l + h'X_l \otimes U_{l+1}.$$

The proof follows from Lemma 10.14 using the same arguments as at the end of the proof of the last theorem.

From the last two lemmata one can easily derive commuting relations between matrix elements of the transfer matrix. We need them for application of the Algebraic Bethe Ansatz method and we state them in the following

Proposition 10.17. *The matrix elements of the transfer matrix commute according to the formulae*

$$B_{l+1}^k(\lambda)B_l^{k+1}(\mu) = B_{l+1}^k(\mu)B_l^{k+1}(\lambda),$$
$$B_{l-2}^k(\lambda)A_{l-1}^{k+1}(\mu) = h'A_l^k(\mu)B_{l-1}^{k+1}(\lambda) + h''B_{l-2}^k(\mu)A_{l-1}^{k+1}(\lambda),$$
$$B_l^{k+2}(\lambda)D_{l-1}^{k+1}(\mu) = k'D_l^k(\mu)B_{l-1}^{k+1}(\lambda) + k''B_l^{k+2}(\mu)D_{l-1}^{k+1}(\lambda).$$

Using the last statements, one can derive relations among λ_i in order that the sum of vectors

$$\Psi_l(\lambda_1, \ldots, \lambda_n) = B_{l+1}^{l-1}(\lambda_1) \cdots B_{l+n}^{l-n}(\lambda_n)\Omega_N^{l-n}$$

be an eigen-vector of the operator

$$\operatorname{Tr} T(\lambda) = A_l^l(\lambda) + D_l^l(\lambda).$$

For more details see [TF1979].

10.5   Rank 2 solutions in (4×4) case

The main example of solutions of rank 2 is the Felderhof R matrix, see [Fel1973, Kri1981]. We will use the following parametrization of this matrix (see [BS1985]):

$$F_f(\phi|q, p|k) = \begin{pmatrix} b_1 & 0 & 0 & d \\ 0 & b_2 & c & 0 \\ 0 & c & b_3 & 0 \\ d & 0 & 0 & b_0 \end{pmatrix}$$

with

$$b_0 = \rho(1 - pqe(\phi)),$$
$$b_1 = \rho(e(\phi) - pq),$$
$$b_2 = \rho(q - pe(\phi)),$$
$$b_3 = \rho(p - qe(\phi)),$$
$$c = \frac{i\rho}{2 \cdot \operatorname{sn}\left(\frac{\phi}{2}\right)} \sqrt{(1 - p^2)(1 - q^2)}(1 - e(\phi)),$$
$$d = \frac{k\rho}{2} \sqrt{(1 - p^2)(1 - q^2)}(1 - e(\phi)) \cdot \operatorname{sn}\left(\frac{\phi}{2}\right),$$
$$e(\phi) = \operatorname{cn}\phi + i \cdot \operatorname{sn}\phi,$$

where ρ is a trivial common constant, p, q are arbitrary constants and cn and sn are Jacobian elliptic functions of modulus k.

The key property of the Felderhof R-matrix is *the free-fermion condition*

$$b_0 b_1 + b_2 b_3 = c^2 + d^2.$$

The free-fermion six-vertex R-matrix F_{XXZ} is given by the limit:

$$F_{XXZ}(\phi|p, q) = \lim_{k \to 0} F_f(\phi|p, q|k).$$

To get a free-fermion analogous to the Cherednik R-matrix, one applies the trick of Lemma 10.9. More detailed, denoting by $T(k)$ the family of matrices

$$T(k) = \begin{pmatrix} (-1/k)^{1/4} & 0 \\ 0 & (-k)^{1/4} \end{pmatrix},$$

we have

Theorem 10.18 ([Dra1998a]). *A rank 2 solution of the Yang–Baxter equation* $F_1(\phi|p,q)$ *is obtained in a limit*

$$F_1(\phi|p,q) = \lim_{k \to 0} (T(k)^{-1} \otimes T(k)^{-1}) F_f(\phi|p,q|k)(T(k) \otimes T(k)),$$

with explicit formulae

$$F_1(\phi|p,q) = \begin{pmatrix} \hat{b}_1 & 0 & 0 & 0 \\ 0 & \hat{b}_2 & \hat{c} & 0 \\ 0 & \hat{c} & \hat{b}_3 & 0 \\ \hat{d} & 0 & 0 & \hat{b}_0 \end{pmatrix}$$

where

$$\hat{b}_0 = \rho(1 - pq\hat{e}(\phi)),$$
$$\hat{b}_1 = \rho(\hat{e}(\phi) - pq),$$
$$\hat{b}_2 = \rho(q - p\hat{e}(\phi)),$$
$$\hat{b}_3 = \rho(p - q\hat{e}(\phi)),$$
$$c = \frac{i\rho}{2 \cdot \sin(\frac{\phi}{2})} \sqrt{(1 - p^2)(1 - q^2)}(1 - \hat{e}(\phi)),$$
$$d = \frac{k\rho}{2} \sqrt{(1 - p^2)(1 - q^2)}(1 - e(\phi)) \cdot \sin\left(\frac{\phi}{2}\right),$$
$$e(\phi) = \cos\phi + i \cdot \sin\phi.$$

Proposition 10.19.

(a) [Kri1981] *The vacuum curve of the Felderhof R matrix is elliptical.*
(b) [Dra1998a] *The vacuum curve of the six-vertex free-fermion R matrix consists of two rational components.*
(c) [Dra1998a] *The vacuum curve of the free-fermion R matrix F_1 is rational with ordinary double point.*

Proof. It follows by simple calculations. The vacuum curves for (a), (b) and (c) are given by

$$P_a(u,v) = u^2 v^2 + Au^2 + Bv^2 + 1,$$
$$P_b(u,v) = Au^2 + Bv^2,$$
$$P_c(u,v) = u^2 v^2 + Au^2 + Bv^2. \qquad \square$$

Consider two polynomials of type P_a:

$$P_{a1} = u^2 v^2 + A_1 u^2 + B_1 v^2 + 1, \quad P_{a2} = u^2 v^2 + A_2 u^2 + B_2 v^2 + 1.$$

Krichever proved that they induce (2-2)-correspondences which commute if and only if

$$A_1 + B_1 = A_2 + B_2.$$

The same is true in the rational case, (see [Dra1998a]).

10.6   Conclusion

Previous considerations were devoted to the basic, most simple 4×4 case of solutions of the Yang–Baxter equation. The real challenge is to develop something similar for higher dimensions. Some attempts were made, for example see [Dra1997]. However, only a very specific situation, connected with the so-called Potts model, was considered there.

In general, the essential problem is that Krichever's vacuum vector approach works only for even-dimensional matrices.

In the first odd-dimensional case, when the dimension of the space V is equal to 3 and R matrices are 9×9, one may try the following.

We introduce the notion of *vacuum locus* as an analogue of the vacuum curve.

Now, the matrices $L = \mathcal{L}_{j\beta}^{i\alpha}$ are considered as a linear operator in the tensor product $\mathbf{C}^3 \otimes \mathbf{C}^3$. The same is for matrices R. As before, we want to parametrize the vacuum vectors, i.e., vectors of the form $X \otimes U$, which L maps to vectors of the same form $Y \otimes V$, where $X, Y, U, V \in \mathbf{C}^3$. Assume the notation

$$U^t = (u_1, u_2, 1), \; V^t = (v_1, v_2, 1), \; \tilde{V}_1 = (1, 0, -v_1), \; \tilde{V}_2 = (0, 1, -v_2).$$

The vacuum locus is the set which parametrizes the vacuum vectors.

From [Dra2006], we have the following

Lemma-Definition 10.20. The affine part of the vacuum locus is the set of

$$(u_1, u_2, v_1, v_2) \in \mathbf{C}^4$$

such that

$$P(u_1, u_2, v_1, v_2) := \det L(\lambda) = 0$$

identically in λ, where $L_j^i(\lambda) = (\tilde{V}_1 + \lambda \tilde{V}_2)^\beta L_{j\beta}^{i\alpha} U_\alpha$.

The lemma follows from the fact that, if two regular matrix binomials of the first degree are equivalent, then they are strictly equivalent (see [Gan1959]). The condition $\det L(\lambda) = 0$ identically in λ gives four equations in $\mathbf{C}^4$ since $\det L(\lambda)$ is a polynomial of the third degree in λ. So, for the general matrix L, the set

$P(u_1, u_2, v_1, v_2)$ is a finite subset of $\mathbf{C}^4$. The working hypothesis among the specialists was that, in a case of the solutions of the quantum Yang–Baxter equation which depend on spectral parameter, there should be an algebraic curve which parametrizes some of the vacuum vectors. However, even in the case of the solutions of the Yang–Baxter equation it is possible that the vacuum locus is a finite set. A good example to start with is the famous 9×9 R-matrix. The definition of the last R matrix can be found in [IK1981].

The structure of this set is still not clear. In order to apply some of the Krichever ideas, such a set should have a subset which satisfies two conditions:

- it is closed for the composition of relations properly defined;
- it is big enough to give a possibility to reconstruct matrices R, L, L' and their products.

This could lead to a construction of the solutions of the Yang–Baxter equation in which the spectral parameter belongs to some discrete group.

However, we saw that there is a quite remarkable similarity in the 4×4 case of study of solutions of the Yang–Baxter equation and of the Poncelet theorem in the plane together with billiards within conics. In previous chapters we developed the last subject to arbitrary dimension and genera. A hope is that careful study of higher-dimensional billiards within pencils of quadrics and generalizations of Poncelet–Darboux theorems could provide us with geometric intuition with possible applications to the Yang–Baxter equations and their higher-dimensional solutions.

In these and other further investigations, important role will be played by unification of synthetic approach to addition theorems developed in the present book with the approach founded on the theory of σ-functions, see [BE1996, BEL1997b, BL2005a, EEM$^+$2008, EEP2003]. For some other aspects of σ-functions theory and applications of the addition theorems, see for example [BK1996, BEL1997a, BEL1997c, BLE2000, BL2002, BL2005b, BK2006, BL2008].

Bibliography

[Akh1970]   N.I. Akhiezer, *Elements of elliptic functions theory*, Nauka, Moscow, 1970
              (Russian).

[App1880]   Paul Appell, *Sur les séries hypergéométriques de deux variables, et sur
              des équations différentielles linéaires simultanées aux dérivées partielles*,
              Comptes Rendus **90** (1880), 296–298, 731–734.

[AKdF1926]   Paul Appell and Joseph Kampé de Fériet, *Fonctions hypergéométriques et
              hyperspheriques*, Polynomes d'Hermite, Gauthier Villars, Paris, 1926.

[Arn1978]   Vladimir Arnold, *Mathematical Methods of Classical Mechanics*, Springer
              Verlag, New York, 1978.

[Aud1994]   Michèle Audin, *Courbes algébriques et systèmes intégrables: géodesiques des
              quadriques*, Expo. Math. **12** (1994), 193–226.

[BB1996]   W. Barth and Th. Bauer, *Poncelet Theorems*, Expo. Math. **14** (1996), 125–
              144.

[BM1993]   W. Barth and J. Michel, *Modular curves and Poncelet polygons*, Math.
              Ann. **295** (1993), 25–49.

[Bax1971a]   R.J. Baxter, *Eight-vertex model in lattice statistics*, Phys. Rev. Let. **26**
              (1971), 832–833.

[Bax1971b]   _____, *One-dimensional anisotropic Heisenberg chain*, Phys. Rev. Let. **26**
              (1971), 834.

[Bax1972a]   _____, *Partition functions of the eight-vertex lattice model*, Ann. Phys. **70**
              (1972), 193–228.

[Bax1972b]   _____, *One-dimensional anisotropic Heisenberg chain*, Ann. Phys. **70**
              (1972), 323–337.

[Bax1982]   _____, *Exactly Solved Models in Statistical Mechanics*, Academic Press,
              1982.

[BS1985]   V.V. Bazhanov and Yu.G. Stroganov, *Hidden symmetry of free-fermion
              model.I*, Teoret. Mat. Fiz. **62** (1985), 377–387.

[BD1982]   A.A. Belavin and V.G. Drinfel'd, *Solutions of the classical Yang–Baxter
              equation for simple Lie algebras*, Funktsional'nyi Analiz i Ego Prilozheniya
              16 (1982), no. 3, 1–29 (Russian); English transl., Functional Analysis and
              Its Applications **16** (1982), no. 3, 159–180.

[Ber1987]   Marcel Berger, *Geometry*, Springer Verlag, Berlin, 1987.

[Ber1852]   J. Bertrand, *Mémoire sur les intégrales communes à plusieurs problèmes de Mécanique*, Jour. de Math. **17** (1852), 121–174.

[Bol1990]   S.V. Bolotin, *Integrable Birkhoff billiards*, Vestnik Moskov. Univ. Ser. I Mat. Mekh. (1990), no. 2, 33–36.

[BF1994]   A.V. Bolsinov and A.T. Fomenko, *The geodesic flow of an ellipsoid is orbitally equivalent to the Euler integrable case in the dynamics of a rigid body*, Dokl. Akad. Nauk SSSR **339** (1994), no. 3, 293–296.

[BF2004]   ———, *Integrable Hamiltonian Systems: Geometry, Topology, Classification*, Chapman and Hall/CRC, Boca Raton, Florida, 2004.

[BMF1990]   A.V. Bolsinov, S.V. Matveev, and A.T. Fomenko, *Topological classification of integrable Hamiltonian systems with two degrees of freedom. List of systems with small complexity*, Russian Math. Surveys **45** (1990), no. 2, 59–94.

[BO2006]   A.V. Bolsinov and A.A. Oshemkov, *Singularities of integrable Hamiltonian systems*, Topological Methods in the Theory of Integrable Systems, Cambridge Scientific Publ., 2006, pp. 1–67.

[BKOR1987]   H.J.M. Bos, C. Kers, F. Oort, and D.W. Raven, *Poncelet's closure theorem*, Expo. Math. **5** (1987), 289–364.

[Buc2006]   V. Buchstaber, *n-valued groups: theory and applications*, Moscow Mathematical Journal **6** (2006), no. 1, 57–84.

[BE1996]   V.M. Bukhshtaber and V.Z. Enolskii, *Explicit algebraic description of hyper-elliptic Jacobians based on Klein's σ-function*, Funktsional. Anal. i Prilozhen. **30** (1996), no. 1, 57–60 (Russian); English transl., Funct. Anal. Appl. **30** (1996), no. 1, 44–47.

[BEL1997a]   V.M. Buchstaber, V.Z. Enolskii, and D.V. Leikin, *Hyperelliptic Kleinian functions and applications*, Solitons, geometry, and topology: on the crossroad, Amer. Math. Soc. Transl. Ser. 2, vol. 179, Amer. Math. Soc., Providence, RI, 1997, pp. 1–33.

[BEL1997b]   ———, *Kleinian functions, hyperelliptic Jacobians and applications*. 2, Reviews in Mathematics and Math. Physics **10** (1997), 3–120.

[BEL1997c]   ———, *Integrable systems with pairwise interactions and functional equations*. 2, Reviews in Mathematics and Math. Physics **10** (1997), 121–166.

[BLE2000]   V.M. Bukhshtaber, D.V. Leikin, and V.Z. Enolskii, *Uniformization of Jacobi manifolds of trigonal curves, and nonlinear differential equations*, Funktsional. Anal. i Prilozhen. **34** (2000), no. 3, 1–16, 96 (Russian, with Russian summary); English transl., Funct. Anal. Appl. **34** (2000), no. 3, 159–171.

[BK1996]   V. Buchstaber and I. Krichever, *Multidimensional vector addition theorems and the Riemann theta functions*, Internat. Math. Res. Notices (1996), no. 10, 505–513.

[BK2006]   V.M. Bukhshtaber and I.M. Krichever, *Integrable equations, addition theorems, and the Riemann-Schottky problem*, Uspekhi Mat. Nauk **61** (2006), no. 1(367), 25–84 (Russian, with Russian summary); English transl., Russian Math. Surveys **61** (2006), no. 1, 19–78.

[BL2002]   V.M. Bukhshtaber and D.V. Leikin, *Polynomial Lie algebras*, Funktsional.
             Anal. i Prilozhen. **36** (2002), no. 4, 18–34 (Russian, with Russian summary);
             English transl., Funct. Anal. Appl. **36** (2002), no. 4, 267–280.

[BL2005a]   Victor Buchstaber and Dmitry Leykin, *Hyperelliptic addition law*, J. Non-
             linear Math. Phys. **12** (2005), no. suppl. 1, 106–123.

[BL2005b]   V.M. Bukhshtaber and D.V. Leikin, *Addition laws on Jacobians of plane
             algebraic curves*, Tr. Mat. Inst. Steklova **251** (2005), no. Nelinein. Din., 54–
             126 (Russian, with Russian summary); English transl., Proc. Steklov Inst.
             Math. (2005), no. 4 (251), 49–120.

[BL2008]   _____, *Solution of the problem of the differentiation of abelian functions
             with respect to parameters for families of (n, s)-curves*, Funktsional. Anal.
             i Prilozhen. **42** (2008), no. 4, 24–36, 111 (Russian, with Russian summary);
             English transl., Funct. Anal. Appl. **42** (2008), no. 4, 268–278.

[BV1996]   V. Buchstaber and A. Veselov, *Integrable correspondences and algebraic
             representations of multivalued groups*, Inter. math. Res. Notices (1996),
             no. 8, 381–400.

[Cay1853]   Arthur Cayley, *Note on the porism of the in-and-circumscribed polygon*,
             Philosophical magazine **6** (1853), 99–102.

[Cay1854]   _____, *Developments on the porism of the in-and-circumscribed polygon*,
             Philosophical magazine **7** (1854), 339–345.

[Cay1855]   _____, *On the porism of the in-and-circumscribed triangle, and on an
             irrational transformation of two ternary quadratic forms each into itself*,
             Philosophical magazine **9** (1855), 513–517.

[Cay1857]   _____, *On the porism of the in-and-circumscribed triangle*, Quarterly
             Mathematical Journal **1** (1857), 344–354.

[Cay1858]   _____, *On the a posteriori demonstration of the porism of the in-and-
             circumscribed triangle*, Quarterly Mathematical Journal **2** (1858), 31–38.

[Cay1861]   _____, *On the porism of the in-and-circumscribed polygon*, Philosophical
             Transactions of the Royal Society of London **51** (1861), 225–239.

[CCS1993]   S.-J. Chang, Bruno Crespi, and K.-J. Shi, *Elliptical billiard systems and the
             full Poncelet's theorem in n dimensions*, J. Math. Phys. **34** (1993), no. 6,
             2242–2256.

[Che1980]   I. Cherednik, *About a method of constructing factorized S-nmatrix in ele-
             mentary functions*, TMPh **43** (1980), 117–119.

[Dar1870]   Gaston Darboux, *Sur les polygones inscrits et circonscrits à l'ellipsoïde*,
             Bulletin de la Société philomathique **7** (1870), 92–94.

[Dar1901]   _____, *Sur un Problème de Méchanique*, Archives Neerlandaises (2) **6**
             (1901), 371–376.

[Dar1914]   _____, *Leçons sur la théorie générale des surfaces et les applications
             géométriques du calcul infinitésimal*, Vol. 2 and 3, Gauthier-Villars, Paris,
             1914.

[Dar1917]   _____, *Principes de géométrie analytique*, Gauthier-Villars, Paris, 1917.

[DFRR2001] A. Delshams, Y. Fedorov, and R. Ramírez-Ros, *Homoclinic billiard orbits inside symmetrically perturbed ellipsoids*, Nonlinearity **14** (2001), 1141–1195.

[Don1980]   Ron Donagi, *Group Law on the Intersection of Two Quadrics*, Ann. Sc. Norm. Sup. Pisa (1980), 217–239.

[Dra1992a] V.I. Dragovich, *Solutions to the Yang equation with rational spectral curves*, Algebra i Analiz **4** (1992), no. 5, 104–116 (Russian); English transl., St. Petersb. Math. J. **4** (1993), no. 5, 921–931.

[Dra1992b] ———, *Baxter reduction of rational Yang solutions*, Vestn. Mosk. Univ., Ser. I (1992), no. 5, 84–86 (Russian); English transl., Mosc. Univ. Math. Bull. **47** (1992), no. 5, 59–60.

[Dra1992c] ———, *Solutions to the Yang equation with rational spectral curves*, Algebra i Analiz, Sankt-Petersburg **4** (1992), no. 5, 104–116 (Russian); English transl., St. Petersb. Math. J. **4** (1993), no. 5, 921–931.

[Dra1993]   ———, *Solutions to the Yang equation with rational irreducible spectrual curves*, Izv. Ross. Akad. Nauk, Ser. Mat **57** (1993), no. 1, 59–75 (Russian); English transl., Russ. Acad. Sci., Izv., Math. **42** (1994), no. 1, 51–65.

[Dra1994]   Vladimir Dragović, *The Algebraic Bethe Ansatz and Vacuum Vectors*, Publ. de l'Institute Math. **55 (69)** (1994), 105–110.

[Dra1996]   ———, *On integrable potential perturbations of the Jacobi problem for the geodesics on the ellipsoid*, J. Phys. A: Math. Gen. **29** (1996), no. 13, L317–L321.

[Dra1997]   V.I. Dragovich, *Solutions to the Yang equation and algebraic curves of genus greater than 1*, Funkts. Anal. Appl. **31** (1997), no. 2, 70–73 (Russian).

[Dra1998a] Vladimir Dragović, *A new rational solution of the Yang–Baxter equation*, Publ. de l'Institute Math. **63 (77)** (1998), 147–151.

[Dra1998b] V.I. Dragovich, *Integrable perturbations of the Birkgof billiard inside an ellipse*, Prikladnaya matematika i mehanika **62** (1998), no. 1, 166–169 (Russian).

[Dra2002]   Vladimir Dragović, *The Appell hypergeometric functions and classical separable mechanical systems*, J. Phys. A: Math. Gen. **35** (2002), no. 9, 2213–2221.

[Dra2003]   ———, *Integrable systems and algebraic curves. Part 1: Ellipsoidal billiards and hyperelliptic curves*, Lecture notes, SISSA, Trieste, 2003, preprint no. 80/2003/FM.

[Dra2006]   ———, *Algebro-geometric integration in classical and statistical mechanics*, Three topics from contemporary mathematics (Bogoljub Stanković, ed.), Selected topics, Mathematical Institute SANU, Beograd, 2006, pp. 121–154.

[Dra2008]   ———, *Marden theorem and Poncelet–Darboux curves* (2008), preprint, available at `arXiv:0812.4829v1[math.CA]`.

[Dra2009]   _____, *Multi-valued hyperelliptic continued fractions of generalized Halphen type*, Int. Math. Res. Notices (2009), 1891–1932.

[Dra2010a]   _____, *Geometrization and generalization of the Kowalevski top*, Communications in Mathematical Physics **298** (2010), no. 1, 37–64, available at `arXiv:0912.3027`. DOI: 10.1007/s00220-010-1066-z.

[Dra2010b]   _____, *Poncelet-Darboux curves, their complete decomposition and Marden theorem*, Int. Math. Res. Notes (2010), to appear.

[DJR2003]   Vladimir Dragović, Božidar Jovanović, and Milena Radnović, *On elliptical billiards in the Lobachevsky space and associated geodesic hierarchies*, J. Geom. Phys. **47** (2003), no. 2-3, 221–234.

[DR1998a]   Vladimir Dragović and Milena Radnović, *Conditions of Cayley's type for ellipsoidal billiard*, J. Math. Phys. **39** (1998), no. 1, 355–362.

[DR1998b]   _____, *Conditions of Cayley's type for ellipsoidal billiard*, J. Math. Phys. **39** (1998), no. 11, 5866–5869.

[DR2004]   _____, *Cayley-type conditions for billiards within k quadrics in $\mathbf{R}^d$*, J. of Phys. A: Math. Gen. **37** (2004), 1269–1276.

[DR2005]   _____, *Corrigendum: Cayley-type conditions for billiards within k quadrics in $\mathbf{R}^d$*, J. of Phys. A: Math. Gen. **38** (2005), 7927.

[DR2006a]   _____, *A survey of the analytical description of periodic elliptical billiard trajectories*, Journal of Mathematical Sciences **135** (2006), no. 4, 3244–3255.

[DR2006b]   _____, *Geometry of integrable billiards and pencils of quadrics*, Journal Math. Pures Appl. **85** (2006), 758–790.

[DR2008]   _____, *Hyperelliptic Jacobians as Billiard Algebra of Pencils of Quadrics: Beyond Poncelet Porisms*, Adv. Math. **219** (2008), no. 5, 1577–1607.

[DR2009]   _____, *Bifurcations of Liouville tori in elliptical billiards*, Regular and Chaotic Dynamics **14** (2009), no. 4-5, 479–494.

[DR2010]   _____, *Integrable Billiards and Quadrics*, Russian Math. Surveys **65** (2010), no. 2, 136–197.

[Dub1981]   B.A. Dubrovin, *Theta-functions and nonlinear equations*, Uspekhi Math. Nauk. **36** (1981), 11–80 (Russian).

[DKN2001]   B.A. Dubrovin, I.M. Krichever, and S.P. Novikov, *Integrable systems I*, Dynamical systems IV, Springer-Verlag, 2001, pp. 173–280.

[1911]   *Enciclopædia Britannica*, 11th ed., 1911.

[EEM⁺2008]   J.C. Eilbeck, V.Z. Enolski, S. Matsutani, Y. Ônishi, and E. Previato, *Addition formulae over the Jacobian pre-image of hyperelliptic Wirtinger varieties*, J. Reine Angew. Math. **619** (2008), 37–48.

[EEP2003]   J.C. Eilbeck, V.Z. Enolskii, and E. Previato, *On a generalized Frobenius-Stickelberger addition formula*, Lett. Math. Phys. **63** (2003), no. 1, 5–17.

[Eul1766]   Leonard Euler, *Evolutorio generalior formularum comparationi curvarum inservientum*, Novi commentarii acad. sc. Petrop. **12** (1766/7), 42–86.

[Fay1973]   J.D. Fay, *Theta functions on Riemann surfaces*, Lecture Notes in Mathematics, vol. 352, Springer-Verlag, Berlin, Heidelberg, New York, 1973.

[Fed2001]   Yuri Fedorov, *An ellipsoidal billiard with quadratic potential*, Funct. Anal.
              Appl. **35** (2001), no. 3, 199–208.

[Fel1973]   B. Felderhof, *Diagonalization of the transfer matrix of the free-fermion
              model*, Physica **66** (1973), 279–298.

[Fla1974]   H. Flaschka, *The Toda Lattice*. I, Phys. Rev. B **9** (1974), 1924–1925.

[Gan1959]   F.R. Gantmacher, *The theory of matrices*, Chelsea, New York, 1959.

[GZ1976]   Moshe Goldberg and Gideon Zwas, *On inscribed circumscribed conics*, El-
              emente der Mathematik **31** (1976), no. 2, 36–38.

[GH1977]   Philip Griffiths and Joe Harris, *A Poncelet theorem in space*, Comment.
              Math. Helvetici **52** (1977), no. 2, 145–160.

[GH1978a]   _____, *On Cayley's explicit solution to Poncelet's porism*, Enseign. Math.
              24 (1978), no. 1-2, 31–40.

[GH1978b]   _____, *Principles of Algebraic Geometry*, John Wiley, New York, 1978.

[Gun1966]   R.C. Gunning, *Lectures on Riemann Surfaces*, Princeton University Press,
              Princeton, NJ, 1966.

[GT2002]   Eugene Gutkin and Serge Tabachnikov, *Billiards in Finsler and Minkowski
              geometries*, Journal of Geometry and Physics **40** (2002), no. 3–4, 277–301.

[Hal1888]   G.-H. Halphen, *Traité des fonctiones elliptiques et de leures applications*.
              deuxieme partie, Gauthier-Villars et fils, Paris, 1888.

[Har1978]   Robin Hartshorne, *Stable vector bundles of rank 2 on* $\mathbb{P}_3$, Math. Ann. **238**
              (1978), 229–280.

[Hei1928]   W. Heisenberg, *Zur Theorie des Ferromagnetismus*, Z. Phys. **49** (1928),
              619–636.

[HN1982]   A. Hirschowitz and M.S. Narasimhan, *Fibres de 'tHooft spéciaux et appli-
              cations*, (Nice, 1981), 1982, pp. 142–163.

[Hur1879]   A. Hurwitz, *Ueber unendlich-vieldeutige geometrische Aufgaben, insbeson-
              dere aber die Schliesungsprobleme*, Math. Ann. **15** (1879), 8–15.

[IK1981]   A. Izergin and V. Korepin, *The inverse scattering method approach to the
              quantuum Shabat–Mikhailov Model*, Comm. Math. Phys. **79** (1981), 303–
              316.

[Jac1829]   Carl Jacobi, *Fundamenta nova theoriae functiorum ellipticarum*, 1829.

[Jac1884a]   _____, *Vorlesungen über Dynamic. Gesammelte Werke, Supplementband*,
              Berlin, 1884.

[Jac1884b]   _____, *Note sur une nouvelle application de l'Analyse des fonctions ellip-
              tiques à l'Algèbre*, Werke, Vol. I, 1884, pp. 329.

[Jak1993]   B. Jakob, *Moduli of Poncelet polygons*, J. reine angew. Math. **436** (1993),
              33–44.

[Kal2008]   D. Kalman, *An Elementary Proof of Marden's Theorem*, The American
              Mathematical Monthly **115** (2008), 330–337.

[Knö1980]   Horst Knörrer, *Geodesics on the ellipsoid*, Inventiones Math. **59** (1980),
              119–143.

[Kow1889]   S. Kowalevski, *Sur le problème de la rotation d'un corp solide autour d'un point fixe*, Acta Math. **12** (1889), 177–232.

[KT1991]   Valery Kozlov and Dmitry Treshchëv, *Billiards*, Amer. Math. Soc., Providence RI, 1991.

[Koz1995]   V.V. Kozlov, *Some integrable generalizations of the Jacobi problem on geodesics on an ellipsoid*, Prikl. Mat. Mekh. **59** (1995), no. 1, 3–9 (Russian); English transl., J. Appl. Math. Mech. **59** (1995), no. 1, 1–7.

[Koz2003]   _____, *Rationality conditions for the ratio of elliptic integrals and the great Poncelet theorem*, Vestnik Moskov. Univ. Ser. I Mat. Mekh. **71** (2003), no. 4, 6–13 (Russian); English transl., Moscow Univ. Math. Bull. **58** (2004), no. 4, 1–7.

[Kri1981]   I.M. Krichever, *Baxter's equation and algebraic geometry*, Func. Anal. Appl. **15** (1981), 92–103 (Russian).

[Leb1942]   Henri Lebesgue, *Les coniques*, Gauthier-Villars, Paris, 1942.

[LT2007]   Mark Levi and Serge Tabachnikov, *The Poncelet grid and the billiard in an ellipse*, Amer. Math. Monthly. (December 2007).

[Mar1966]   Morris Marden, *Geometry of Polynomials*, 2nd ed., Math. Surveys, vol. 2, AMS, 1966.

[MB1951]   D. Mordukhai-Boltovskoi, *The theorem of Poncelet in the Lobachevskii plane and elliptic integrals*, Doklady Akad. Nauk SSSR (N.S) **77** (1951), 961–964 (Russian).

[MV1991]   Jürgen Moser and Alexander Veselov, *Discrete versions of some classical integrable systems and factorization of matrix polynomials*, Comm. Math. Phys. **139** (1991), no. 2, 217–243.

[Mos1975a] Jürgen Moser, *Finitely many mass points on the line under the influence of an exponential potential – an integrable system*, Lecture Notes in Physics, vol. 28, Springer, 1975, pp. 467–497.

[Mos1975b] _____, *Three integrable Hamiltonian systems connected with isospectral deformations*, Adv. Math. **16** (1975), 197–220.

[Mos1980]   _____, *Geometry of quadrics and spectral theory*, The Chern Symposium, Springer, New York-Berlin, 1980, pp. 147–188.

[Mum1983]   David Mumford, *Tata Lectures on Theta*, Birkhäuser, Boston, 1983; Russian transl. in Mir, Moscow, 1988.

[NR1969]   M. Narasimhan and Ramanan, *Moduli of vector bundles on compact Riemann surfaces*, Ann. of Math. (2) **89** (1969), 14–51.

[NT1990]   M.S. Narasimhan and G. Trautmann, *Compactification of $M_{P_3}(0,2)$ and Poncelet pairs of conics*, Pacific J. of Math. **145** (1990), 255–365.

[Pon1822]   Jean Victor Poncelet, *Traité des propriétés projectives des figures*, Mett, Paris, 1822.

[Pre1999]   Emma Previato, *The Poncelet's theorem and generalizations*, Proc. Amer. Math. Soc. **127** (1999), 2547–2556.

[Pre]   _____, *Some integrable billiards*, SPT2002: Symmetry and Perturbation
                    Theory (S. Abenda, G. Gaeta, and S. Walcher, eds.), World Scientific,
                    Singapore, 2002, pp. 181–195.

[Rad2003]   Milena Radnović, *A note on billiard systems in Finsler plane with elliptic
                    indicatrices*, Publications de l'Institut Mathématique **74** (2003), no. 88, 97–
                    102.

[RRK2008]   Milena Radnović and Vered Rom-Kedar, *Foliations of isonergy surfaces
                    and singularities of curves*, Regular and Chaotic Dynamics **13** (2008), no. 6,
                    645–668.

[Rei1972]   M. Reid, *The complete intersection of two or more quadratics*, Ph.D. Thesis,
                    Cambridge, June 1972, unpublished.

[Sie1864]   J. Siebeck, *Ueber eine neue analytische Behandlungweise der Brennpunkte*,
                    J. Reine Angew. Math. **64** (1864), 175–182.

[Sch2007]   Richard Schwartz, *The Poncelet grid*, Advances in Geometry **7** (2007), 157–
                    175.

[SRK2005]   E. Shlizerman and V. Rom-Kedar, *Hierarchy of bifurcations in the trun-
                    cated and forced nonlinear Schrödinger model*, Chaos **15** (2005), no. 1.

[Shi1986]   T. Shiota, *Characterization of Jacobian varieties in terms of soliton equa-
                    tions*, Invent. Math. **83** (1986), 333–382.

[Tab2002]   Serge Tabachnikov, *Ellipsoids, complete integrability and hyperbolic geom-
                    etry*, Moscow Mathematical Journal **2** (2002), no. 1, 185–198.

[TF1979]   L.A. Takhtadzhyan and L.D. Faddeev, *The quantum method of the inverse
                    problem and the Heisenberg XYZ-model*, Usp. Mat. Nauk **34** (1979), 13–64
                    (Russian); English transl., Russian Math. Surveys **34** (1979), 11–68.

[Tch1852]   P.L. Tchebycheff, *Report of the Extaordinary Professor of St Petersburg
                    University Tchebycheff about the Trip Abroad*, Complete Collected Works,
                    Vol. 5, AN SSSR, Moscow-Leningrad, 1946, 1852, pp. 246–255.

[Tod1947]   Todd, *Classical and projective geometry*, Sir Isaac Pitman & Sons, London,
                    1947.

[Tra1988]   G. Trautmann, *Poncelet curves and theta characteristics*, Expositiones
                    Mathematicae **6** (1988), 29–64.

[Tra1997]   _____, *Decomposition of Poncelet curves and instanton bundles*, A. St.
                    Univ. Ovidius Constanta **5** (1997), 105–110.

[Tru1853]   N. Trudi, *Rappresentacione geometrica immediata dell' equazione fonda-
                    menta della teoria delle funzioni ellitiche con diverse applicazioni*, Memoria
                    della R. Accademia delle Scienze di Napoli, Napoli, 1854, 1853, pp. 63–99.

[Tru1863]   _____, *Studii intorno ad una singolare eliminazione, con applicazione
                    alla ricerca delle relazione tra gli elementi di due coniche, l'una iscritta,
                    l'altra circoscritta ad un poligono, ed ai corrispondenti teoremi di Poncelet*,
                    Atti della R. Accademia delle Scienze fisiche e matematiche di Napoli **1**
                    (1863).

[TRK2003]   D. Turaev and V. Rom-Kedar, *Soft billiards with corners*, Journal of Sta-
                    tistical Physics **112** (2003), no. 3/4, 765–813.

[Tyu1975]   A.N. Tyurin, *On intersection of quadrics*, Russian Math. Surveys **30** (1975), 51–105.

[Ves1990]   Alexander Veselov, *Confocal surfaces and integrable billiards on the sphere and in the Lobachevsky space*, J. Geom. Phys. **7** (1990), no. 1, 81–107.

[Ves1992]   _____, *Growth and integrability in the dynamics of mappings*, Comm. Math. Phys. **145** (1992), 181–193.

[VK1995]   N. Ja. Vilenkin and A.U. Klimyk, *Representations of Lie groups and special functions*, Recent Advances, Kluwer Academic Publishers, Dordrecht, 1995.

[WD2002]   H. Waalkens and H.R. Dullin, *Quantum Monodromy in Prolate Ellipsoidal Billiards*, Annals of Physics **295** (2002), no. 1, 81–112.

[Wey1870]   E. Weyr, *Über einige Sätze von Steiner und ihren Zusammenhang mit der zwei und zweigliedrigen Verwandtschaft der Grundebilde ersten Grades*, Crelles Journal für die Reine und Angewandte Mathematik **71** (1870), 18–28.

[Whi1927]   E.T. Whittaker, *A treatise on the analytical dynamics of particles and rigid bodies*, 3rd ed., The University Press, Cambridge, 1927.

[WW1990]   E.T. Whittaker and G.N. Watson, *A Course in Modern Analysis*, 4th ed., Cambridge University Press, Cambridge, England, 1990.

Index

Abel
  differentials
    of the first kind, 59
    of the second kind, 59
    of the third kind, 59
  integrals, 58
  mapping, 47, 58, 59
  theorem, 49, 57, 59, 113, 115
  variety, 58
Abelian group, 47, 48
addition theorem, 1, 51, 108
  for Weierstrass function, 49
affine
  conic, 69
  quadric, 69
  space, 69
Algebraic Bethe Ansatz, 263, 270, 272
algebraic curve, 27, 31, 35, 37, 43, 89, 116
analytic continuation principle, 28
Appell functions, 199

Baker–Akhiezer function, 62
base point, 74, 151
Baxter matrix, 268
Bertrand–Darboux equation, 202
Bézout theorem, 43, 44, 69, 100
bicentric polygon, 9, 17
bicentric quadrilateral, 9, 19
bifolium, 34, 44
billiard, 3, 12, 14, 117, 165, 173, 181, 186
  law, 14
  ordered game, 170, 171
  reflection, 11
  trajectory, 11
  within ellipse, 139
Birkhoff theorem, 17
bitangent, 152

bitangent pencil, 75, 90
blow-up, 38
boundary curve, 253
Brianchon theorem, 87

caustic, 117, 118, 144, 201
Cayley's condition, 3, 115, 116, 179
Cayley's cubic, 113, 115
Chapple formula, 8
Chapple–Euler formula, 18, 117
characteristic polynomial, 82
Chasles theorem, 71, 84, 96, 201
Cherednik matrix, 268
circle, 70, 74, 75, 107, 117
circumscribed polygon, 109, 110
circumscribed triangle, 109
class of algebraic curve, 89
classical Yang–Baxter equation, 6, 261
collineation, 68
collision point, 150
commutative group, 48
complete conic, 87, 88
complex torus, 44
confocal
  conics, 88, 117, 187
  curves, 90
  family, 89
  pencil of conics, 89
  quadrics, 95, 129
conic, 31, 69, 82, 138
  envelope, 88
  locus, 88
  matrix, 69
conjugate spaces, 76
continued fraction theory, 227
coordinate curves, 128
cross-ratio, 67, 71, 83, 102

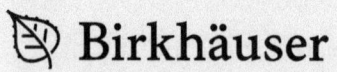 **Birkhäuser** | **www.birkhauser-science.com**

Frontiers in Mathematics

This series is designed to be a repository for up-to-date research results which have been prepared for a wider audience. Graduates and postgraduates as well as scientists will benefit from the latest developments at the research frontiers in mathematics and at the "frontiers" between mathematics and other fields like computer science, physics, biology, economics, finance, etc.

Advisory Board

■ **Elworthy, K.D., Le Jan, Y., Li, X.-M.,**
The Geometry of Filtering (2010).
ISBN 978-3-0346-0175-7

The geometry which is the topic of this book is that determined by a map of one space N onto another, M, mapping a diffusion process, or operator, on N to one on M.
Filtering theory is the science of obtaining or estimating information about a system from partial and possibly flawed observations of it. The system itself may be random, and the flaws in the observations can be caused by additional noise. In this volume the randomness and noises will be of Gaussian white noise type so that the system can be modelled by a diffusion process; that is it evolves continuously in time in a Markovian way, the future evolution depending only on the present situation.
We consider the geometry of this situation with special emphasis on situations of geometric, stochastic analytic, or filtering interest. The most well studied case is of one Brownian motion being mapped to another with a consequent skew product decomposition (or equivalently the case of Riemannian submersions). This sort of decomposition is used to study in particular, classical filtering, (semi-)connections determined by stochastic flows, and generalised Weitzenbock formulae.

■ **Østvær, P.A.,** Homotopy Theory of C*-Algebras (2010). ISBN 978-3-0346-0564-9

Homotopy theory and C*-algebras are central topics in contemporary mathematics. This book introduces a modern homotopy theory for C*-algebras.

One basic idea of the setup is to merge C*-algebras and spaces studied in algebraic topology into one category comprising C*-spaces. These objects are suitable fodder for standard homotopy theoretic moves, leading to unstable and stable model structures. With the foundations in place one is led to natural definitions of invariants for C*-spaces such as homology and cohomology theories, K-theory and zeta-functions. The text is largely self-contained. It serves a wide audience of graduate students and researchers interested in C*-algebras, homotopy theory and applications.

■ **Borsuk, M.,** Transmission Problems for Elliptic Second-Order Equations in Non-Smooth Domains (2010).
ISBN 978-3-0346-0476-5

The goal of this book is to investigate the behavior of weak solutions of the elliptic transmission problem in a neighborhood of boundary singularities: angular and conic points or edges. This problem is discussed for both linear and quasilinear equations. A principal new feature of this book is the consideration of our estimates of weak solutions of the transmission problem for linear elliptic equations with minimal smooth coeciffients in n-dimensional conic domains. Only few works are devoted to the transmission problem for quasilinear elliptic equations. Therefore, we investigate the weak solutions for general divergence quasilinear elliptic second-order equations in n-dimensional conic domains or in domains with edges.
All results are given with complete proofs. The book will be of interest to graduate students and specialists in elliptic boundary value problems and applications.